CLIMATE CHANGE GEOENGINEERING

The international community is not taking the action necessary to avert dangerous increases in greenhouse gases. Facing a potentially bleak future, humanity is confronting the question of whether the best of bad alternatives may be to counter global warming through human-engineered climate interventions. In this book, nineteen prominent authorities on climate change consider the legal, policy, and philosophical issues presented by geoengineering. The book asks: When, if ever, are decisions to embark on potentially risky climate modification projects justified? If such decisions can be justified, in a world without a central governing authority, who should authorize such projects and by what moral and legal right? If states or private actors undertake climate modification ventures absent the blessing of the international community, what recourse do the rest of us have?

Wil C. G. Burns is the Associate Director of the Energy Policy and Climate Program at Johns Hopkins University in Washington, DC. He serves as the Editor-in-Chief of the *Journal of International Wildlife Law & Policy* and as Co-Chair of the International Environmental Law Committee of the American Branch of the International Law Association. He is also the former Co-Chair of the International Environmental Law Interest Group of the American Society of International Law and Chair of the International Wildlife Law Interest Group of the Society. His academic appointments have included Williams College, Colby College, Santa Clara University School of Law, and the Monterey Institute of International Studies of Middlebury College. Prior to becoming an academic, he served as Assistant Secretary of State for Public Affairs for the State of Wisconsin and worked in the nongovernmental sector for twenty years, including as Executive Director of the Pacific Center for International Studies.

Andrew L. Strauss is the Associate Dean for Faculty Research and Development and a Professor of Law at Widener University School of Law. Professor Strauss is a co-author of the fourth edition of *International Law and World Order*, and his articles have appeared in international journals such as *Foreign Affairs*, the *Harvard Journal of International Law*, and the *Stanford Journal of International Law*. He has been a Visiting Professor at the University of Notre Dame Law School and taught on the law faculties of the National University of Singapore and Rutgers Camden Law School. In addition, he has been a lecturer at the European Peace University in Austria, served as the Director of the Geneva/Lausanne International Law Institute and the Nairobi International Law Institute, and been an Honorary Fellow at New York University School of Law's Center for International Studies.

Climate Change Geoengineering

PHILOSOPHICAL PERSPECTIVES, LEGAL ISSUES, AND GOVERNANCE FRAMEWORKS

Edited by

WIL C. G. BURNS
Johns Hopkins University

ANDREW L. STRAUSS
Widener University School of Law

CAMBRIDGE
UNIVERSITY PRESS

32 Avenue of the Americas, New York NY 10013-2473, USA

Cambridge University Press is part of the University of Cambridge.

It furthers the University's mission by disseminating knowledge in the pursuit of education, learning, and research at the highest international levels of excellence.

www.cambridge.org
Information on this title: www.cambridge.org/9781107023932

First published 2013
Reprinted 2013

A catalog record for this publication is available from the British Library.

Library of Congress Cataloging in Publication data
Climate change geoengineering : philosophical perspectives, legal issues, and governance frameworks / edited by Wil C. G. Burns, The Johns Hopkins University; Andrew L. Strauss, Widener University School of Law, Delaware.
pages cm
Includes index.
ISBN 978-1-107-02393-2 (hardback)
1. Environmental geotechnology. 2. Climatic changes. 3. Environmental engineering – Law and legislation. I. Burns, William C. G. II. Strauss, Andrew L.
TD171.9.C55 2013
628.5´32–dc23 2012049895

ISBN 978-1-107-02393-2 Hardback

Contents

Contributors

Randall S. Abate, Associate Professor of Law and Project Director, Environment, Development and Justice Program, Florida A&M University College of Law, Orlando

Wil C. G. Burns, Associate Director, Energy Policy and Climate Program, Johns Hopkins University, Baltimore

Gareth Davies, Professor of European Law, Department of Transnational Legal Studies, VU University, Amsterdam

Stephen M. Gardiner, Professor of Philosophy and Ben Rabinowitz Endowed Professor of the Human Dimensions of the Environment, University of Washington, Seattle

Kerstin Güssow, Walther Schücking Institute for International Law, Christian-Albrechts University of Kiel, Kiel

Clive Hamilton, Charles Sturt Professor of Public Ethics at the Centre for Applied Philosophy and Public Ethics, Canberra

Tracy Hester, Visiting Assistant Professor at the University of Houston Law Center and Director of its Environment, Energy and Natural Resource Center, Houston

Joshua B. Horton, DNV KEMA Energy & Sustainability, Burlington, Massachusetts

Robert E. Kopp, AAAS Science and Technology Policy Fellow, American Association for the Advancement of Science, Washington, DC

Lee Lane, Visiting Fellow, Hudson Institute, Washington, DC

Albert C. Lin, Professor of Law, School of Law, University of California, Davis

Jay Michaelson, PhD Candidate, Hebrew University, Jerusalem

David R. Morrow, Department of Philosophy, University of Alabama, Birmingham

Michael Oppenheimer, Woodrow Wilson School of Public and International Affairs and Department of Geosciences, Princeton University, Princeton

Andreas Oschlies, IFM-GEOMAR, Leibniz Institute of Marine Sciences, Kiel

Alexander Proelss, Department of Law, University of Trier, Trier

Katrin Rehdanz, Department of Economics, Christian-Albrechts University of Kiel, Kiel

Wilfried Rickels, Department of Economics, Christian-Albrechts University of Kiel; Corresponding Author, Kiel Institute for the World Economy, Kiel

Andrew L. Strauss, Associate Dean for Faculty Research and Development and Professor of Law, Widener University School of Law, Wilmington

Introduction

The Emerging Salience of Geoengineering

Wil C. G. Burns and Andrew L. Strauss

What has become increasingly clear over the last few years is that the international community is not even close to tackling the global warming problem in a way that will avert profound climatic consequences. Paragraph 1 of the 2009 Copenhagen Accord formally incorporates "the scientific view that the increase in global temperature should be below two degrees Celsius."[1] In fact, that scientific view is changing as more and more climate researchers come to realize that a two degrees Celsius increase over preindustrial levels threatens serious disruptions of the earth's biosphere.

The current increase in global temperatures of .8 degrees Celsius is already having a significant deleterious effect. Glaciers are melting,[2] sea levels are rising,[3] a third of Arctic sea ice is disappearing in the summer,[4] the oceans are 30 percent more acidic,[5] and the average moisture content of the earth's air has increased by 5 percent, leading to more extreme weather.[6] Prominent NASA scientist Jim Hansen echoed the views of many climatologists when he declared, "warming [of two degrees Celsius] is a guarantee of global disasters."[7]

1 Copenhagen Accord, art. 1, Dec. 18, 2009, *available at* http://unfccc.int/files/meetings/cop15/application/pdf/cop15cphauv.pdf (last visited Aug. 7, 2012).

2 CLIMATE CHANGE 2007: *The Physical Science Basis: Contribution of Working Group I to the Fourth Assessment Report of the Intergovernmental Panel on Climate Change* 109 (Z. Manning et al. eds., 2007).

3 *Id.* at 111; CLIMATE CHANGE & SEA LEVEL RISE: CONSEQUENCES OF CLIMATE CHANGE ON THE OCEANS, Climate Institute, http://www.climate.org/topics/sea-level/index.html (last visited Aug. 7, 2012).

4 ARCTIC REPORT CARD: UPDATE FOR 2011, TRACKING RECENT ENVIRONMENTAL CHANGES, SEA ICE (D. Perovich et al. eds., 2011), *available at* http://www.arctic.noaa.gov/reportcard/sea_ice.html (last visited Aug. 7, 2012).

5 ROYAL SOCIETY, OCEAN ACIDIFICATION DUE TO INCREASING ATMOSPHERIC CARBON DIOXIDE: POLICY DOCUMENT 12/05 25–30 (J. Raven et al. eds., 2005), *available at* http://royalsociety.org/uploadedFiles/Royal_Society_Content/policy/publications/2005/9634.pdf (last visited Aug. 7, 2012).

6 CLIMATE CHANGE 2007, *supra* note 2, at 105.

7 Interview by World Watch Institute with James Hansen, 21 WORLD WATCH MAG. 6, (July/Aug. 2008), *available at* http://www.worldwatch.org/node/5775.

But even the prospects of keeping global warming within the two degrees threshold seem extremely unlikely from today's vantage point. The best scientific estimates are that we can collectively release roughly 565 more gigatons of carbon into the atmosphere by midcentury and stay within the two degrees Celsius threshold; however at current growth rates of approximately 3 percent per year (which show no signs of abating) we are on track to considerably exceed that threshold.[8] Despite considerable scientific consensus about the dangers we are facing, and a well-funded climate change movement that has galvanized citizens from around the world, on balance the political will to make the necessary effort to reduce carbon emissions does not exist. What is more, it does not seem likely to come about within the time frame necessary to stave off very serious consequences.

With this political reality in the foreground, we asked eleven of the world's most prominent students of climate change law and policy to contribute to this book on "the deliberate large-scale manipulation of the planetary environment to counteract anthropogenic climate change,"[9] commonly called *geoengineering* or *climate modification*. Although the prospect of global actors embarking upon major climate modification projects in the hope of countering climate change terrifies some and excites others, few doubt that it could well be in our collective future. As long as the threat of climate change continues to grow and geoengineering technologies are within reach, the tantalizing hope of a geoengineering "fix" will only grow more attractive to many.

This consideration of large-scale geoengineering projects raises many serious legal, policy, and philosophical issues that are explored in the pages that follow. We did not intend this volume to be an advocacy book to either promote or discredit geoengineering as a response to climate change. Rather, in the hopes of helping inform the debate that is emerging, we invited contributors with a wide range of perspectives. At the most general level the questions break down into two broad categories: how do we decide and who decides. Is a decision to embark upon a large-scale and potentially risky project to modify the global climate ever justified? If so, in a world that lacks a global legislature capable of making collective climate modification decisions, who should determine whether to authorize potentially risky projects? To the extent states or private actors undertake such ventures without the blessings of the international community generally, what rights do those who oppose such actions have? Although the methodologies used by our contributors are diverse, and there is considerable overlap in their approaches, generally speaking, the first

[8] P. Friedlingstein, R.A. Houghton, G. Marland, J Hackler, T.A. Boden, T.J. Conway, J.G. Canadell, M.R. Raupach, P. Ciais & C. Le Quéré, *Update on CO2 Emissions*, 3 NATURE GEOSCIENCE 811 (Dec. 2010).

[9] The Royal Society, *Geoengineering the Climate: Science, Governance and Uncertainty* (Sept. 2009), at 11, http://royalsociety.org/Geoengineering-the-climate/ (last visited on Mar. 28, 2011).

three contributors to this volume ground their chapters in ethics and philosophy whereas the remaining contributors ground theirs in law and governance. We have, therefore, chosen to organize the volume along those lines.

In Chapter 1, "Geoengineering and Moral Schizophrenia: What Is the Question?," Stephen M. Gardiner contends that two questions are central to the ethics of geoengineering. The justificatory question asks: "Under what future conditions might geoengineering become justified?" The nature of the future conditions he considers include, for example, the nature and extent of the climate change threat to be confronted, and other background global circumstances, including the existing governance mechanisms, individual protections, and compensation provisions. The contextual question asks: "What is the ethical context of the push toward geoengineering, and what are its implications?" Gardiner argues that early discussions of geoengineering often marginalized both questions because participants in those discussions tended to view their consideration as luxuries that we could not afford given the emergency nature of the climate change problem. Gardiner concludes that such emergency arguments are ethically shortsighted, and morally schizophrenic. In reaching this conclusion, Gardiner employs two abstract examples. Although both are extreme and idealized, according to Gardiner even the imperfect analogies provide reasons for concern about our current predicament. Ethically serious discussion of geoengineering should confront ethical problems, rather than hide behind overly simplistic appeals to moral emergency. As Michael Stocker puts it in his seminal discussion of moral schizophrenia, "to refuse to do so bespeaks a malady of the spirit."

In Chapter 2, "The Ethical Foundations of Climate Engineering," Clive Hamilton argues that the idea that the planet's optimal temperature should be set through a process of calculation reflects a particular conception of the world and the nature of humans that emerged first with the Scientific Revolution and later Enlightenment philosophy. This conception, according to Hamilton, holds that the human being is a self-legislating subjective entity, distinct from the rest of the world and guided by its cognitive abilities. It is, says Hamilton, the basis of the technological thinking now being applied in plans to engineer the climate. Hamilton suggests that solar radiation management is the culmination of the transition to the mechanical conception of nature and the parallel emergence of philosophies built on the idea of the autonomous rational subject exercising control over an inert environment. These conceptions, and the consequentialist ethics they gave rise to, are now challenged by earth-system science itself. The earth under the Anthropocene is not mere putty to be shaped at will by humans.

In Chapter 3, "The Psychological Costs of Geoengineering: Why It May Be Hard to Accept even if It Works," Gareth Davies observes that debates about climate change and geoengineering often revolve around "quantitative and concrete considerations," such as economic and environmental impacts. Such considerations, however, he argues "are often quite divorced from their real psychological importance for most people, the fear, uncertainty, and hope that they may inspire." Davies suggests that an assessment of the psychological "losses" associated with climate geoengineering may explain far more than economic, climatic, or material factors about the basis of the opposition to geoengineering. The primary three losses, Davies argues, are: relative status, security, and hope. Davies suggests that many members of the environmental movement would suffer a diminution of relative status if their moral and political standing was undercut by a solution that did not require fundamentally transforming society. In terms of security, geoengineering could undermine security by offering only partial solutions "between mitigation and climate management" and "entail a probabilistic approach to policy" that many would find disconcerting. Finally, if geoengineering were to remove climate change as a threat, Davies contends that the hope of deep ecologists that climate change would justify their fundamental tenets would be dashed.

In Chapter 4, "Geoengineering and Climate Management: From Marginality to Inevitability," Jay Michaelson makes the case that geoengineering, or *climate management* as he calls it, "is the only approach to climate change that can act as a compromise between liberals and libertarians, greens and browns." It appeals to conservatives, he argues, because it protects economic interests, is in line with market ideology, uses technology rather than restraints on behavior, and avoids government regulation. He argues that to liberals, its appeal may not be intuitive, but that their acceptance of it is necessary if they wish to actually make progress on climate change, given real world political realities. Michaelson acknowledges that liberals have legitimate concerns about embarking on climate management initiatives. Those concerns range from equitable considerations, including the giving of "free passes" to polluters, to the potential risks and costs of projects, including cataclysmic warming in the case of cessation of solar radiation management and the dangers that rogue actors could pose. He argues, however, that these concerns are answerable in every case.

In Chapter 5, "Climate Change and the Anthropocene Era," Lee Lane advocates assessing the judiciousness of climate geoengineering through the lens of a Weberian "ethic of responsibility." He focuses "on knowing the likely consequences of our policy choices and accepting responsibility for them rather than on more abstract ethical precepts." Lane argues that greenhouse gas control measures would yield minimal net financial gains and impose extremely high costs; moreover, such controls could upset existing trade regimes, depress agricultural production, and roil bilateral relationships between major states, including the United States and China. Lane also argues that there are many imposing political barriers to effective implementation of international greenhouse gas controls. The case for geoengineering lies in the fact that the potential benefits are "very large compared to the estimated costs of developing and deploying it." Although Lane acknowledges risk associated with deployment, including potential shutdown of monsoons in Asia, he argues that the benefits would still substantially outweigh such costs, especially if such costs are weighed against the impacts of climate change under a business-as-usual scenario. Finally, Lane outlines a way forward for developing a regime to govern climate geoengineering, suggesting that regime structure will be dependent "on both the distribution of relative power as well as the need to hold down the transaction costs of managing the system."

In Chapter 6, "Political Legitimacy in Decisions about Experiments in Solar Radiation Management," David Morrow, Robert Kopp, and Michael Oppenheimer maintain that making good policy decisions about solar radiation management (SRM) requires a better understanding than we currently have of the effectiveness and side effects of various SRM technologies. The authors argue, however, that gaining such understanding would require multiyear global trials. Observing that such trials would be ethically problematic because they would expose persons, animals, and ecosystems to serious risks, the authors go on to explore under what conditions such trials would be ethically acceptable. They conclude that such acceptability depends upon approval of the trials by an appropriate international body (i.e., one with the political legitimacy to authorize the trial). The authors endorse Buchanan and Keohane's "Complex Standard" for global political legitimacy: a global political institution is legitimate if it enjoys widespread support from democratic states; meets certain substantive conditions, such as avoidance of serious injustices and the production of better outcomes than feasible alternative institutions; and has certain epistemic virtues, such as transparency and accountability. Morrow, Kopp and Oppenheimer survey several global institutions as possible analogs for an SRM governance institution, including those for governing nuclear weapons and for managing the Antarctic environment.

In Chapter 7, "Geoengineering and the Myth of Unilateralism: Pressures and Prospects for International Cooperation," Joshua Horton addresses one of the primary concerns of geoengineering opponents (as well as some proponents): the specter of unilateral deployment. Horton argues that unilateral deployment is unlikely for several reasons. To begin with, a state that chooses to unilaterally deploy a geoengineering option would face the possibility of deployment of the same or other geoengineering options by other states, potentially impairing the effectiveness of this approach. This would, Horton argues, necessitate coordination of deployment with other actors. Moreover, in the case of SRM, the so-called termination problem (the potential for a huge spike in warming should solar deflection once embarked upon be terminated; see Burns, Chapter 9, infra) would encourage states reluctant to make an indefinite commitment on their own to coordinate their efforts internationally. Finally, Horton contends that the availability of countermeasures "would serve as perhaps the most potent check on unilateral deployment of geoengineering technologies such as stratospheric aerosol injections." Horton also maintains that multilateralism in geoengineering research and potential deployment can be fostered by a portfolio of tactics known as "international management theory."

In Chapter 8, "International Legal Regimes and Principles Relevant to Geoengineering," Albert Lin assesses the potential role of international law in governing potential research and development and deployment of geoengineering options. Although concluding that no international agreement directly regulates geoengineering, Lin argues that a number of relevant treaties and principles of international law may play a role in geoengineering governance. Lin initially discusses a series of treaties that may extend to geoengineering options in a general sense, including the United Nations Framework Convention on Climate Change, the Convention on the Prohibition of Military or Any Other Hostile Use of Environmental Modification Techniques, and the Convention on Biological Diversity. Lin then turns to "media-specific" treaties that may apply only to particular types of geoengineering projects, such as the London Convention/London Protocol, and the Law of the Sea Convention (ocean iron fertilization (OIF)); the Convention on Long-Range Transboundary Air Pollution; the Montreal Protocol (SRM options injecting particles into the atmosphere), and the Outer Space Treaty (space-based options). Finally, Lin suggests that there are several international norms that might be apposite, including norms calling for transboundary environmental impact assessment, and the prohibition on inflicting transboundary harm, as well as norms with less certain application, including the precautionary principle and the principle of intergenerational equity.

In Chapter 9, "Climate Geoengineering: Solar Radiation Management and its Implications for Intergenerational Equity," this book's coeditor, William Burns, examines the extent to which the emerging global norms requiring that our present-day actions take into account intergenerational equity legally constrain SRM geoengineering options. Burns contends that ceasing the use of SRM technologies would pose the threat of a "termination effect," a huge multi-decadal pulse of warming that could overwhelm many ecosystems and human institutions. Moreover, some SRM approaches could delay replenishment of the stratospheric ozone layer by as much as seventy years. Such long-term deleterious consequences, the author argues, would violate the principle of intergenerational equity by potentially denying future generations an environment of commensurate quality to that we currently enjoy because of either technological failure or societal choice. Moreover, the threat of a termination effect might compel future generations to continue the use of SRM technologies, even if they deemed these technologies to be morally unacceptable because of the collateral effects. This would violate the intergenerational principle of conservation of options. The chapter concludes that viable options exist to reduce greenhouse gas emissions, which would preclude the need to threaten the interests of future generations.

In Chapter 10 "Ocean Iron Fertilization: Science, Law, and Uncertainty," Randall Abate adds his voice to the discussion of OIF. However, in contrast to Chapter 11, the author expresses considerable skepticism about the potential effectiveness of OIF's ability enhance the oceanic sink for carbon dioxide through the addition of iron to stimulate phytoplankton growth, as well as our capability of meeting the substantial monitoring and verification challenges. In addition to examining the role that existing international regimes could play in the regulation of OIF, Abate addresses potentially applicable domestic laws in the United States (i.e., the Marine Protection, Research and Sanctuaries Act and the National Environmental Policy Act). The chapter concludes with detailed recommendations for establishing "an effective international law framework to regulate OIF." Abate outlines two broad options in this context. The first is the establishment of an independent regime to address geoengineering. Such a regime could be patterned on the UN Environmental Modification Convention. Alternatively, geoengineering research in particular could be regulated under a new international treaty regime, or a less-formal international research consortia. A second option would be to harmonize existing treaties, with the International Maritime Organization serving as the implementing body given its oversight of several relevant regimes, including the London Convention and the London Protocol. At the domestic level in the United States, Abate also suggests coordination of federal responses, including the possibility of establishing a working group.

In Chapter 11, "Ocean Iron Fertilization: Time to Lift the Research Taboo," Kirsten Güssow, Andreas Oschlies, Alexander Proelss, Katrin Rehdanz, and Wilfried Rickels make the case for pursuing research of OIF. Although concluding that OIF may have the potential to sequester comparable amounts of carbon dioxide as forest sequestration techniques, the authors acknowledge substantial uncertainties that necessitate further research. The remainder of the chapter is devoted to legal issues related to potential deployment of an OIF approach. The authors set forth a framework that could integrate OIF into the Clean Development Mechanism of the Kyoto Protocol and include a discussion of methods to account for permanence and leakage. The chapter examines the applicability of international treaty regimes to OIF, including the United Nations Convention for the Law of the Sea, the Convention on Biological Diversity, and the London Convention and London Protocol. The authors conclude that the application of the precautionary principle, often invoked by those who oppose climate geoengineering because of their potential negative impacts could cut in favor of OIF deployment given the threat posed by unchecked climate change.

In Chapter 12, "Remaking the World to Save It: Applying U.S. Environmental Laws to Climate Engineering Projects," Tracy Hester examines the potential applicability of U.S. environmental laws to climate geoengineering research. Pertinent statutes cited by the author include the National Weather Modification Policy Act of 1972, the Clean Air Act, the Clean Water Act, the Endangered Species Act, the Marine Protection, the Research and Sanctuaries Act, and the National Environmental Policy Act. The chapter also examines the potential for judicial review of geoengineering via common law nuisance claims. Hester concludes by noting that the federal government may need to begin drafting strategies and establishing standards for approval or rejection of projects, and that specific agencies may wish to explore options to stop projects that pose excessive dangers or evoke strong public reactions.

PART I

Ethics and Philosophy

1

Geoengineering and Moral Schizophrenia

What Is the Question?

Stephen M. Gardiner

Not to be moved by what one values – what one believes good, nice, right, beautiful, and so on – bespeaks a malady of the spirit.

Michael Stocker

Humanity stands on a precipice. Mainstream science tells us that climate change is real, accelerating, and might credibly result in global catastrophe. For decades it has warned that greenhouse gas emissions should be reduced (mitigation) and that we should prepare for those impacts that are no longer avoidable (adaptation). Yet global emissions of the main culprit, carbon dioxide, continue to grow at a startling rate, and very little action has been taken to prepare us.

In the face of this escalating threat, a previously marginalized proposal has reemerged and become mainstream. Geoengineering – roughly "the intentional manipulation of planetary systems at a global scale"[1] – is now being seriously discussed. Especially prominent are approaches that might provide a quick fix to hold off imminent climate catastrophe. Currently, the leading proposal is that humanity try to offset the heating effects of increases in greenhouse gases by injecting sulfates into the stratosphere, as a way of reducing incoming solar radiation (i.e., "planetary sunblock"). Although most believe this form of "solar radiation management" to be "risky" and probably also "unsustainable"[2] over the long term, respected researchers and institutions are urging national governments to create research programs and to begin envisioning mechanisms of governance. Given

[1] Thomas Schelling, *The Economic Diplomacy of Geoengineering*, 33 CLIMATIC CHANGE 303 (1996); David Keith, *Geoengineering: History and Prospect*, 25 ANNUAL REVIEW OF ENERGY AND THE ENVIRONMENT 245 (2000).

[2] JOHN SHEPHERD ET AL., GEOENGINEERING THE CLIMATE: SCIENCE, GOVERNANCE AND UNCERTAINTY, xi (2009); Stephen Schneider, *Geoengineering: Could We or Should We Make It Work?*, 366 PHILOSOPHICAL TRANSACTIONS OF THE ROYAL SOCIETY A 3843 (2008).

the looming threat of catastrophe, we are told, geoengineering simply must be taken seriously.[3]

At first glance, such arguments, and the emergency framing more generally, appear straightforward, irresistible, and overtly ethical. Clearly, global environmental catastrophe would be very bad for many things we value. If so, do we not have a strong moral obligation to do "whatever it takes" to prevent it, including encouraging the would-be geoengineers? In the face of such a threat, what ethical objections could possibly be strong enough to rule out geoengineering?

This chapter considers whether, in context, these are the most important questions to be asking. Its central claim is that they are not. Although the issue of whether to pursue geoengineering *as such* is relevant, focusing on it obscures much of what is at stake morally speaking, and in ways that threaten to trivialize our understanding of our predicament. One way to illustrate this is by showing how the currently dominant framing of the geoengineering debate in terms of "whatever it takes"-style emergency arguments is often ethically shortsighted and morally schizophrenic.[4] It is ethically shortsighted (in the sense of "missing the bigger picture") insofar as it arbitrarily marginalizes central moral issues such as how we got into this predicament and why we are not seriously pursuing better ways out. It is also frequently morally schizophrenic (in the sense of being "a state characterized by the coexistence of contradictory or incompatible elements"[5]) because it tends to bring on a form of creative myopia: it requires us to emphasize and endorse strong ethical concerns that we are *otherwise unwilling to act on*, and which would, if earnestly and coherently embraced, lead us to approach both climate policy in general and geoengineering in particular in very different ways. In short, the worry is that, even if ethically serious people have reason to support (some forms of) geoengineering research and perhaps even deployment in the abstract, *their approach would look very different from anything currently under consideration, let alone actually likely to transpire.*

This diagnosis has three important implications. First, it threatens to undermine the superficial appeal of the emergency arguments and to render them seriously misleading in practice. Second, it has explanatory value: it seems likely many people's ethical unease about the current push toward geoengineering rests in part on concerns about ethical shortsightedness and moral schizophrenia. Third,

[3] Ken Caldeira & David W. Keith, *The Need for Climate Engineering Research*, ISSUES IN SCIENCE AND TECHNOLOGY 57 (Fall 2010); David Victor et al., *The Geoengineering Option: A Last Resort against Global Warming?* FOREIGN AFFAIRS 64–76 (March/April 2009); J.J. BLACKSTOCK ET AL., CLIMATE ENGINEERING RESPONSES TO CLIMATE EMERGENCIES 45 (2009); Paul Crutzen, *Albedo Enhancement by Stratospheric Sulphur Injections: A Contribution to Resolve a Policy Dilemma?*, 77 CLIMATIC CHANGE 211 (2006).

[4] Note that I do not claim that all emergency arguments must have these flaws.

[5] ENCARTA WORD DICTIONARY.

importantly, it suggests that not all ethical resistance to geoengineering relies on potentially controversial theses about its moral status. For example, those troubled by the geoengineering turn in climate policy need not believe geoengineering as such to be inherently bad, or a violation of the appropriate relationship of humanity to the rest of the natural world, or even ultimately morally impermissible.[6] Instead, some ethical resistance can involve far narrower fears about the context within which geoengineering is currently being pursued, and how this is likely to evolve in the foreseeable future.

Before proceeding, some clarifications may be helpful. First, the target of most of the chapter is our *collective* reasoning and behavior, where the most salient collectives are humanity as such, the dominant nations, and especially the current generation of the world's affluent, who wield most of the political power and to whom most arguments for engaging in geoengineering are in practice ultimately addressed. One consequence of this is that the chapter is not directly concerned with the issue of whether and how ethical responsibility is transmitted from collectives to individuals. Even though I am inclined to think that individuals are normally accountable to some extent for what they do together, the issue is complex, and I will neither argue for that view here nor sketch its implications for geoengineering.[7]

Second, when discussing collectives, my focus is on improving the quality of public argument rather than on prosecuting claims of ethical responsibility. In particular, I am not concerned with questions of who should be *accused* of moral schizophrenia, and how they should be held accountable. Instead, my interest is in how we (collectively) should best think and talk about the challenge that confronts us in a setting where the integrity of public discussion is itself at risk.

Third, in focusing on undifferentiated collectives such as humanity as such and the current generation, I am not claiming that such collectives are currently unified in appropriate structures of agency (e.g., by competent institutions). Instead, I assume only that there is a sense in which they should be so unified, and that public argument often proceeds on this assumption.

Nevertheless, I also do not assume that thinking about collectives is the only or most central way in which geoengineering (or climate policy more generally) should be understood from the ethical point of view. Indeed (as we shall see in Section 1), I have argued elsewhere that one of the key features of climate change and similar problems is the way in which they complicate and potentially undermine effective

[6] I am not dismissing such worries; my point is merely that these are not the only considerations.

[7] For some discussion, see Dale Jamieson, *Ethics, Public Policy and Global Warming*, 17 SCI., TECH. & HUMAN VALUES 139 (1992); Dale Jamieson, *Climate Change, Responsibility and Justice*, 16 SCI., ENG'G & TECH. ETHICS 431 (2010); Stephen M. Gardiner, *Is No One Responsible for Global Environmental Tragedy? Climate Change as a Challenge to Our Ethical Concepts*, *in* THE ETHICS OF GLOBAL CLIMATE CHANGE (Denis Arnold ed., 2011).

collective agency. My remarks should thus be seen as picking out only one salient dimension of our ethical problem (and so in the context of that overall picture rather than in competition with it).

The chapter proceeds as follows. Section 1 considers the general ethical context in which the push toward geoengineering emerges, and clarifies the problem of ethical shortsightedness. Section 2 identifies the problem of moral schizophrenia, introduces a provocative hypothetical case, and suggests that it is analogous to geoengineering. Section 3 briefly sketches some implications of the analysis. Section 4 clarifies the analogy by responding to three basic objections. Section 5 summarizes the main claims of the chapter.

1. ETHICAL SHORTSIGHTEDNESS AND THE CONTEXT OF CLIMATE ENGINEERING

1.1 *Two Questions*

The question people usually ask about geoengineering is: Are you for it or against it? Unfortunately, as stated, this question is not very well formulated. On the one hand, no one favors geoengineering under just any circumstances, for just any reason. For example, a geoengineering program aimed at preventing rain from ever falling on a sitting president of the United States would be morally absurd. More realistically, very few scientists (that I know of) believe that sulfate injection is justified right now. Instead, the vast majority are united in thinking that such an intervention would be too risky given our present state of ignorance about the consequences.[8]

On the other hand, it seems likely that most people would accept geoengineering under *some* circumstances, if the only alternative were truly dire enough, and the consequences of geoengineering were sufficiently benign and understood with very high confidence.[9] However, much then depends on what counts as "dire," "benign," and "high confidence." Given these concerns, a better question would be: Under what conditions do you think geoengineering might become justified? – where the conditions to be considered would include, for example, the threat to be confronted, the background circumstances, the governance mechanisms, the individual protections to be provided, the compensation provisions to be made, and so on. Call this the *Justificatory Question*.

[8] These words were first written in 2010. By late 2011, a few scientists had begun to advocate publicly for quick deployment of a regional engineering scheme over the Arctic. Still, for now, the general consensus remains. How long that will continue is an open question.

[9] However, see Ben Hale, Fixing the Wrong Wrong: Geoengineering and the End of the World (unpublished).

In a way, the emergency arguments for geoengineering (mentioned earlier in the chapter) aim to dodge the justificatory question. For one thing, they tend to be stated in a very general form, with the key term – usually "catastrophe" – left undefined. In addition, they do not explicitly address any of the other potential conditions just mentioned (such as compensation and individual protections). Instead, they appear implicitly to assume either that such conditions are met, or else that their relevance is overwhelmed by whatever "catastrophe" lurks in the background, so that anything else is a side issue at best. In the latter case, the thought is that whatever the catastrophe is, it is sufficiently bad to justify geoengineering, even if (for example) the geoengineering is not at all benign, its wider implications are not well understood, and other protections are not in place.

This dodging of the justificatory question gives rise to a number of problems.[10] First, as stated, the emergency arguments are *opaque*. It is left to the audience to fill in the key content of "catastrophe" and to make their own implicit judgments about tolerable negative impacts, confidence levels, and so on, as well as the importance of other issues such as governance and compensation. Second, this suggests that the widespread initial sympathy for the argument may be *shallow*. For example, some scientists may be drawn to it because they fear the "catastrophe" of a mass extinction, and would favor a robust system of international compensation to offset any negative effects of geoengineering, whereas some economists may approve of the emergency arguments because they fear the much more modest "catastrophe" of a short-term global recession, and would vigorously oppose climate policies that involve any form of compensation or liability for damages.

Taking the justificatory question seriously both exposes such difficulties and helps bring out important underlying issues at stake in geoengineering. The justificatory question is, then, both intellectually interesting and at least potentially of extreme practical and political importance. Nevertheless, I argue that it is neither the only nor probably the most pressing ethical question to be asking about geoengineering. In particular, I propose that we also examine the moral context of the push toward geoengineering, and the shape of the policies likely to emerge given that context. Call this the *Contextual Question*.

One reason for the importance of the contextual question is that the pursuit of geoengineering is not some abstract venture being embarked on in an idealized world by agents with no history. Instead, it is being done for particular reasons, by particular agents, and in response to specific problems. This has implications for how we understand what geoengineering is and is likely to become. It also affects what our ethical obligations are. Indeed, I suggest that the contextual question is not only relevant to the justificatory question, but also is neither subsumed nor dwarfed

[10] Gardiner, *supra* note 7.

by it. On the contrary, in the real world any ethically respectable answer to the justificatory question will have to take seriously the contextual question, or else risk superficiality and dangerous shortsightedness.

1.2 A *Perfect Moral Storm*

So, what is the ethical context for climate engineering? In my view, climate change is "a perfect moral storm."[11] It is genuinely global, it is seriously intergenerational, and it poses deep theoretical problems. Each of these characteristics presents dangerous obstacles to ethical action. Taken together, they constitute a profound, and perhaps hitherto unprecedented, challenge.

At the heart of the matter is the fact that those most responsible for past and current emissions, the relatively affluent, and especially those in the developed nations, benefit (or at least believe that they benefit) from high and unsustainable carbon emissions, but most of the costs of such emissions, and especially the most severe ones, are projected to fall on future generations and nonhuman nature, and especially the future poor. In short, the current generation of the affluent face strong temptations to pass the buck for their behavior on to others who are extremely vulnerable to them. This buck-passing is especially problematic when the benefits taken by the affluent are relatively modest or superficial (e.g., the joy of wearing shorts and T-shirts indoors in winter) in comparison to the severe risks (e.g., famine) that they impose on others.

The ethical challenge of the perfect moral storm is to overcome the temptation of such buck-passing. So far, the current generation has been slow to respond. As a matter of substance, since the first Intergovernmental Panel on Climate Change (IPCC) report in 1990, carbon dioxide emissions have risen by more than 30 percent globally. This is partly because of explosive economic growth in developing nations, and especially China. However, most developed nations have also seen increases. In addition, developed countries usually produce substantially more emissions per capita, their historical contributions to the problem are considerably larger, and much of the growth in the developing countries is fueled by goods that they consume.[12] As a matter of procedure, things are hardly more encouraging. Despite more than twenty years of serious diplomatic activity, including a global treaty, the world still awaits a convincing effort to confront the challenge. The diplomatic circuit has progressed from Rio to Kyoto to Bali to Copenhagen to Cancun

[11] STEPHEN M. GARDINER, A PERFECT MORAL STORM: THE ETHICAL TRAGEDY OF CLIMATE CHANGE (2011).

[12] Tom Boden, Gregg Marland & Robert J. Andreas, *Global CO2 Emissions from Fossil-Fuel Burning, Cement Manufacture, and Gas Flaring: 1751–2007*, CARBON DIOXIDE INFORMATION CENTER (2010); Steven J. Davis & Ken Caldeira, *Consumption-Based Accounting of CO2 Emissions*, 107 PROC. NAT'L ACAD. SCI. 5687 (2010).

and Durban. Yet the key questions remain unresolved and perpetually deferred until later.[13]

Some may object that the reasons for the slow response have more to do with the fact that many remain unconvinced that climate change is a real, or major, problem. However, elsewhere I have argued that this may also involve buck-passing, because the perfect moral storm threatens corruption of the understanding, both epistemic and moral.[14] In essence, for those able to exploit their spatial and temporal position, passing the buck to future generations and the world's poor is tempting, but having to acknowledge that this is what one is doing is ethically uncomfortable. Far better, then, to try to seize the moral high ground, denying that there is a problem at all or that it will be severe, arguing that continuing to do nothing actually benefits future people, asserting that these people should clean up the mess because they will be richer, and so on. Even if such arguments do not bear close scrutiny, they at least provide moral cover under which buck-passing can continue without too much embarrassment. As Robert Samuelson puts it in another intergenerational setting: *"There's a quiet clamor for hypocrisy and deception; and pragmatic politicians respond with ... schemes that seem to promise something for nothing. Please, spare us the truth."*[15]

If we take the perfect moral storm analysis seriously, much of the recent political history of climate change is worrying. Nevertheless, we should not expect buck-passing always to take the form of escalating emissions and political misdirection. For example, perhaps at some point, denial and claims that the future can and should take care of itself begin to strain all credibility. More important in the present context, the current generation might begin to worry that their misbehavior may have gone far enough to pose real threats to themselves (i.e., in the short- to-medium term). For such reasons, we can expect a buck-passing strategy to evolve over time. This worry is probably not enough to provoke a truly ethical policy (e.g., one that takes future generations seriously), but it may make a difference. For example, if catastrophe may be coming quickly, then even a buck-passing generation has reason to do something. In particular, at such a point, any number of quick, short-term fixes (i.e., ones good for one or two generations) are sure to be attractive, especially if they appear to have low start-up costs and to impose most of their risks on others (e.g., in the future or in other parts of the world).

In a perfect moral storm, then, it is easy to see why we might be drawn toward the age of geoengineering, and why this might suggest a deep ethical problem. In context, there is every reason to expect a buck-passing generation to be tempted by

[13] GARDINER, *supra* note 11, chs. 3–4.

[14] *Id.* at chs. 9 and 11, appendix 2.

[15] Robert J. Samuelson, *Lots of Gain and No Pain!*, NEWSWEEK, Feb. 21, 2005, at 41. (emphasis in original)

geoengineering interventions that do not constitute real solutions to the genuine global, intergenerational, and ecological problem of the perfect moral storm, but rather constitute "shadow solutions" that address their own distinct concerns, while being disguised as the real thing. In particular, we should be wary of *parochial geoengineering*, in which the current generation secures short-term benefits for itself only by passing on much more serious long-term risks to the future; and *predatory geoengineering*, in which a country chooses a particular form of geoengineering mainly to set back the interests of a geopolitical rival. These temptations are too easily hidden behind appeals to ethical emergency, yet they are threats that an ethics of geoengineering – or indeed any reasonable appraisal – must take seriously.

At first glance, the possibility of buck-passing geoengineering may seem far-fetched. However, consider the following. It is often said that two of the major advantages of the sulfate injection strategy are that it is reversible and that we have "proof of concept" from past volcanic eruptions. Nevertheless, each of these claims is readily contestable. First, once sulfate injection is masking a significant temperature effect, it may not be reversible in any meaningful sense, because withdrawing the intervention might then result in a rapid bounce-back at least as dangerous as the climate change it is aimed to prevent.[16] Second, it is not clear that we have proof of the relevant concept: sulfates injected into the stratosphere by volcanoes typically wash out in a year or two, whereas effective geoengineering would need to be in place for many decades and perhaps centuries. Both contestations seem relevant to the perfect moral storm. These objections are highly salient if we care about long-term impacts and long-term reversibility. If we do not, then the standard claims look more relevant. If we forget the perfect moral storm, then we may miss the importance of this observation.

2. MORAL SCHIZOPHRENIA

We can get a sharper sense of how the emergency framing of geoengineering puts our focus in the wrong place and on the wrong questions by drawing attention to a problem closely connected to ethical shortsightedness, namely moral schizophrenia.

2.1 *Moral Schizophrenia*

The term "schizophrenia" is formed from the Greek σχιζειν ("to split") and φρην ("mind"); hence, to be schizophrenic is "to have a split mind." As a medical

[16] Damon Matthews & Kenneth Caldeira, *Transient Climate – Carbon Simulations of Planetary Geoengineering*, 104 PROC. NAT'L ACAD. SCI. 9949 (2007).

condition, schizophrenia is defined as "a mental disease occurring in various forms, all characterized by a breakdown in the relation between thoughts, feelings, and actions,"[17] and is "frequently accompanied by delusions and retreat from social life."[18] When used more broadly, it is reasonable to say that schizophrenia is characterized by contradictory propositions.

In a classic paper, Michael Stocker describes the general phenomenon of moral schizophrenia as follows:

> One mark of a good life is a harmony between one's motives and one's reasons, values, justifications. Not to be moved by what one values – what one believes good, nice, right, beautiful, and so on – bespeaks a malady of the spirit. Not to value what moves one also bespeaks a malady of the spirit. Such a malady, or such maladies, can properly be called moral schizophrenia – for they are a split between one's motives and one's reasons.[19]

Stocker, then, is concerned with the split between (on the one hand) an agent's underlying reasons, values, or justifications for action,[20] and (on the other hand) what in fact moves her (her "motives"). Moreover, Stocker treats the problem as one of disharmony, and harmony as a mark of a good human life.

Presumably, the basic phenomenon of splitting can arise in distinct ways in different situations, and might be accounted for by a variety of views in moral psychology. Nevertheless, for our purposes, the details of exactly how the relevant split occurs, and the psychological ontology underlying it, will not be too important,[21] as it will be sufficient to gesture at the problem in a general way. Moreover, we will focus on just one kind of moral schizophrenia, which I will call *creative myopia*. This arises when an agent invokes a set of strong moral reasons to justify a given course of action, but this course of action is supported by these reasons *only because* the agent has ruled out a number of alternative courses of action more strongly supported by the same reasons, and where this is due to motives she has that are less important, and are condemned by those reasons. We shall examine the abstract shape of this case more carefully in a moment. However, the key contextual claim will be that some emergency arguments for geoengineering require the agent to endorse

[17] OXFORD ENGLISH DICTIONARY.

[18] OXFORD AMERICAN DICTIONARY.

[19] Michael Stocker, *The Schizophrenia of Modern Ethical Theories*, 14 J. PHIL. 453 (1976).

[20] Stocker uses "reasons" in his subsequent discussion, treating the three terms as interchangeable for his purposes. I shall do the same.

[21] For instance, we need not assume any deep ontological dualism between what Stocker calls "reasons" and what he calls "motives." Perhaps they are basically the same kind of thing, and disharmony results from how they are treated by the agent. For example, perhaps both express genuinely normative claims (e.g., because they are both "reasons" in a more generic sense), but the conflict between them is caused by the agent's failure to treat them consistently.

strong ethical concerns that she is *otherwise unwilling to act on*, in the sense that if she were truly committed to those concerns, then she would not be *so* interested in geoengineering, and/or she would be interested in geoengineering policies of a markedly different kind (e.g., those reflecting a much broader set of ethical and other concerns).

In order to motivate the thought that this kind of schizophrenia poses an ethical problem, I will now consider two analogous cases – one highly abstract and the other more concrete – and then compare them to the situation facing us with geoengineering. Both cases are intended to function as paradigms of creative myopia, and so are rather extreme, in order to provoke the relevant moral intuitions more clearly. Moreover, although they are designed so as to be broadly relevant to the current geoengineering debate, the situation with geoengineering need not be anywhere near so stark for the analogies to be relevant. Perhaps our collective behavior is not as morally problematic as that of our protagonists; still, that does not mean that it is not problematic at all, and (in the abstract) in a similar general way. The analogies do not have to be perfect to have force, and I am not claiming that geoengineering is a paradigm case.

2.2 *Agent 1*

Here is the highly abstract case. Suppose Agent 1 is engaged in activities that he morally ought not to be engaged in. He has a large number of options available to him to address the situation. When these are ranked according to strong moral values he acknowledges, he faces a set of possible responses, from A to Z, where A is the best and Z the worst. Suppose then that, despite recognizing A as the best option, he nevertheless refuses to take it, but offers no very serious – let alone adequate – reason for doing so (moral or otherwise).[22] In addition, suppose that, although he acknowledges that B and C are also good options, he dismisses these too, and in the same way. Indeed, Agent 1 will not consider any of the good or decent options put in front of him, and neither will he consider the best of the more flawed alternatives. Instead, he rejects every option suggested from A to X, and (again) without serious grounds for doing so. Nevertheless, despite this, Agent 1 is not quite comfortable doing nothing; instead, he is willing to consider options Y and Z (and only these). Y and Z are pretty bad options (although not necessarily inherently morally bad, or absolutely prohibited).[23] In particular, although it is possible that Y and Z may help a lot, they also bring with them very serious risks,

[22] The same issue can arise when serious but inadequate reasons are presented. However, for simplicity I lead with the clearer case here.

[23] Those who think that geoengineering is problematic in itself might add that as part of the description of the badness of the option, but it is neither intended nor required by the schematic example.

including a realistic threat that they may make matters much worse.[24] However, Agent 1 claims that, as Y and Z are (arguably) better than nothing, his pursuit of them is entitled to some moral respect, and that (therefore) he deserves some praise for being open to, and then ultimately choosing, one of these options. In particular, he cannot understand why some are so keen to criticize him for focusing on Y and Z. Indeed, he protests: "Can't people see that not doing Y and Z might result in a catastrophe that we should desperately try to avoid?," and "Why (then) can't they stop being fussy about the "ethics" of the situation, and support a solution that might actually help?"

It seems clear that Agent 1 is behaving badly. A good deal of this is just because he is refusing options A–X without serious grounds for doing so. However, his protests also manifest a further problem. In particular, his appeals to emergency are ethically misleading, and distract our attention from what is really at stake, morally speaking. Questions about whether Y or Z are indeed justified, and under what conditions, are perhaps interesting ones (intellectually and practically); nevertheless, they miss much of what is actually going on, and in ways that suggest ethical shortsightedness and moral schizophrenia.

To begin with, consider matters from an external point of view. First, Agent 1's emergency argument for Y or Z seems to require an *arbitrary narrowing of the ethical point of view*. For example, his questions implicitly set aside a whole host of central issues, such as that his ethical problem is self-inflicted, that he is responsible for radically circumscribing the live options for responding, that in doing so he has picked out much worse options than are otherwise available, and so on.

Second, this narrowing strongly implies that Agent 1 is guilty of a *severe abdication of moral responsibility*, and that this is itself a central feature of the case.

Third, given this, Agent 1's appeal to the importance of Y and Z – and especially the way he admonishes his critics for failing to appreciate their importance – seems at the very least *ethically out of place*. Indeed, in an extreme case, Agent 1's myopia may be so profound as to *cast doubt on whether his is a genuine appeal to moral reasons* at all. From the external perspective, we might begin to wonder whether Agent 1 is merely using the language of morality as a cloak to cover other motives, and as a ruse to confuse others.

Of course, although myopic appeals to morality are often consciously disingenuous, this need not be so. In some situations, Agent 1 will be sincere, but only at the cost of internal disharmony, and pronounced moral schizophrenia. From the internal perspective, there seem to be at least three markers of the relevant disharmony.

[24] In the example below, we might say that the AIDS medicine may have bad side effects, including causing cancer. Similarly, there are worries that sulfate injection may bring on other serious environmental problems (e.g., a collapse of the food supply in Asia, ozone depletion).

First, Agent 1's general attitude *lacks internal coherence*. On the one hand, he is invoking strong moral reasons, and demanding that we all take responsibility ("We must do whatever we can to avoid catastrophe!"). On the other hand, his commitment to these reasons is called into question by the twin facts (a) that taken by themselves the moral reasons suggest very different courses of action (i.e., A–X), and (b) that these courses of action are ruled out only by Agent 1's other motives, which are morally indefensible in light of the same moral reasons.

Second, the internal incoherence threatens to *undermine the force* of the intended moral requirement. How is Agent 1 supposed to register the moral reason to do the radically circumscribed options (i.e., Y or Z) as authoritative or weighty, when he has already dismissed that reason when it supports the better options (i.e., A–X)? At best, Agent 1 can be *acknowledging the force of moral reasons only in a severely attenuated sense*. Although he is asserting that they matter to him, they do so only in a very limited domain where most of what is at stake in terms of those values has already been determined by him, and in ways that do not respect those same values. Given this, it is not clear how the moral reasons gain their purchase, or what their apparently "residual" role is. How does Agent 1 himself understand the pull of the moral values in the case of Y and Z, when he has refused them on dubious or insufficient grounds in the other cases? Some explanation is needed, not least by Agent 1 to himself.

Third, the first two markers of disharmony suggest that from the internal perspective Agent 1 is at risk of *losing his grip* on moral reasons, so that his attitude can be maintained only at some further cost. The prime candidates here are *self-deception* and *delusion*. (Note that medical schizophrenia is often accompanied by delusion.) However, we should also consider another possibility. Perhaps Agent 1 admits that he is being incoherent, but nevertheless persuades himself that he is not capable of doing better, perhaps because (he tells himself) the contrary motives are just too powerful for the moral reasons to have the purchase that he recognizes they should. This claim also implies a "malady of the spirit," a *profound alienation* of the agent's moral reasons from what actually moves him.

2.3 *Wayne's Folly*

The moral problems with Agent 1's moral schizophrenia can be made more vivid if we turn to a more concrete paradigm case:

> Wayne is married to a wonderful woman. On the face of it, they have a great relationship, and his wife assumes that this is so. Unfortunately for her, she is mistaken. Wayne has a secret. He likes to play the field and sleep around. He especially enjoys having sex with women in demographic groups at high risk of contracting serious sexually transmitted diseases, such as HIV and AIDS. He does this on a regular

basis, but also continues to have sex with his wife. In neither case does he take any precautions. His wife is ignorant of what is going on.

What Wayne is doing is morally wrong, for a number of reasons. His activities impose risks of severe harm on his wife and many of his other sexual partners as well. They also violate the relationship he has with his wife. In addition, in his more lucid moments, Wayne firmly believes that his behavior is frivolous as well as immoral. His life is very good already, including his relationship with his wife. Moreover, he admits to himself that the time and resources he consumes while engaging in his infidelity would be much better employed elsewhere. Although he enjoys his liaisons, Wayne concedes that they are not of high value, even to himself.

In addressing his folly, Wayne has lots of options. He could simply cease his promiscuity. He could also tell his wife that he is recklessly unfaithful. Failing that, he could pursue any number of strategies to minimize the threat to his wife (and other partners). For example, he could practice "safe" sex, sleep with fewer other women, or women less at risk. However, Wayne is unwilling to do any of these things, in the latter cases even though he acknowledges that they probably would not make much difference to his own enjoyment, and may make none at all. Instead, he prefers just carrying on as he is, ignoring the wider perils. If asked why, he would simply say that he is used to his life, whatever its flaws, and finds change uncomfortable. If pressed, he may even admit that he "just can't be bothered."

Nevertheless, this is not the end of the story: Wayne is willing to do something. A friend tells him of a scientist who is launching a start-up company dedicated to finding a simple pill that can mask the effects of AIDS. The pill aims to manipulate the body's immune system so as to offset the effects of the virus. The work is highly speculative, and does not offer any kind of solution to the other health threats posed by Wayne's activities (e.g., syphilis). It is also highly plausible to think that it will end up having other harmful side effects. (It is just too early to tell.) Despite this, Wayne decides to invest $10 – a very small amount of his disposable income – in the new company. Indeed, he tells himself that he is responding to a moral emergency, and so is morally required to do this, and has done "the right thing." As a result, he feels better about himself, even though he continues his activities as before. Moreover, he professes that he cannot understand those who criticize his donation: "Can't they see that donating to AIDS research might prevent a catastrophe that we should desperately try to avoid?"; given this, "why can't they stop being fussy about the 'ethics' of the situation, and support a solution that might actually help?"

What should moral philosophy say about Wayne? The obvious answer is that he is deeply immoral.[25] He should stop sleeping around and exposing his wife (and other

[25] Perhaps someone could invent a clever (and presumably convoluted) set of background circumstances that would make his behavior seem less nasty. However, here I will assume that we can take the situation at face value.

partners) to such risks. He has no right to do so, and it is morally vicious for him to continue. This is most of what needs to be said. However, we might add, as a minor side point, that if he is intent on continuing, it is also bad for him to refuse to reduce the risks of his behavior (e.g., by wearing a condom). Indeed, at this point, his recklessness becomes so extreme as to seem callous. We might doubt that he cares for his wife at all, even as an innocent human being, let alone as someone he claims to love. We might even want to say that Wayne shows himself to be not only unjust, but also morally indecent. It is natural to ask the (rhetorical) question: "What kind of person *is* Wayne?"

2.4 *Climate Parallels*

Wayne's folly reflects the abstract case of Agent 1, and the result is just as morally disturbing. There are deep ethical problems with what Wayne does, and with what kind of person he is. The central issue for us is his ethical reasoning. In particular, the thought is that Wayne's argument from moral emergency is deficient, in part because it is ethically shortsighted and morally schizophrenic.

The case of Wayne's folly is intended to be uncontroversial. The interesting questions are not about Wayne, but about whether there are important parallels between his situation and ours with respect to climate change and geoengineering. The resemblance need not be exact to be worrying. After all, analogies are seldom perfect (otherwise they would not be analogies). Instead, the issue is whether the two cases have enough in common to underwrite the same ethical verdict. In short, as the current generation of the world's more affluent slide toward geoengineering, are we collectively vulnerable to the same kinds of moral censure as Wayne?

At first glance, there are some striking parallels. The first set involves the basic ethical problems posed by political inertia on climate change.[26] First, like Wayne's activities, our high emission levels impose risks of severe harm on innocent others (in our case: future people, nonhuman life, and the world's poor). In addition, if anything, our case is worse, as Wayne's victims would in principle at least be able to resist the threat he poses if they became aware of it. This is not so for most of the victims of climate change.

Second, like Wayne's, our behavior violates morally important relationships. Not only does imposing large risks on others undermine our moral stature with other nations and peoples around the world, but it also alienates us from future

[26] Much could be said to elaborate on these parallels, involving serious work in moral and political philosophy. However, we need not delve too deeply here, because the current aim is merely to make good on the analogy. *See* GARDINER et al., *supra* note 1; STEVE VANDERHEIDEN, ATMOSPHERIC JUSTICE (2008) for some entry-points into this literature.

generations of our own communities, and further strains our already fragile relationship with nature.

Third, like Wayne's behavior, much of our emitting seems at least relatively frivolous in the face of the threat it imposes on others. Climate change is projected to cause death, disease, dislocation, and in general widespread suffering. But many carbon emissions in the richer countries support activities that risk triviality by comparison, and objectives that could be achieved in much less damaging ways. For example, carbon emissions are spent on heating houses that are much larger and warmer than people really need them to be, and that are occupied only for a small fraction of the day; they are used for inefficient cars that mostly carry only one person at a time; they are required to manufacture products that are thoughtlessly consumed and quickly disposed of, with little tangible benefit for anyone; they are used for excessive business travel that companies would prefer not to pay for and individuals would rather not take; and so on.

The second set of broad parallels involves the fact that, like Wayne, we have lots of options. At the moment, we are accelerating hard into the climate problem. Global emissions are increasing rapidly, and so are those of most nations. Yet we could do much to combat this. First, we could do things differently using existing technology. Although we probably should not simply stop emitting carbon immediately – because too much of our economies depend on fossil fuels, and because a very large and instantaneous shift would likely cause economic collapse – even fairly aggressive action on cutting emissions seems much less problematic. In the United States, for example, the prospect of gains that would actually save resources is well documented, as is the availability of alternative energy sources (especially if we cease heavy subsidies to those that cause the damage).[27]

Second, we could invest more on innovation, such as in research in alternative energy, on ways to design large-scale infrastructure so as to be less carbon intensive or even carbon neutral, and so on.

Third, we could actually retrench. In the face of the moral problems posed by our actions, we could choose to do less. This may involve absolute sacrifices in quality of life, and may require protecting the more vulnerable from making such sacrifices. However, it is not clear that this would be unwarranted. In addition, there is much to suggest that quality of life may not actually suffer. Not only can we be creative about how to live with less carbon (including being creative in the marketplace), but much research suggests that quality of life is not tightly linked with conventional

[27] For example, the Intergovernmental Panel on Climate Change reports: "studies have consistently found that the total global technical potential for renewable energy is substantially higher than global energy demand." *See* IPCC, *Summary for Policy Makers*, SPECIAL REPORT ON RENEWABLE ENERGY SOURCES 7 (2011). *See also* KRISTEN SHRADER-FRECHETTE, WHAT WILL WORK: FIGHTING CLIMATE CHANGE WITH RENEWABLE ENERGY NOT NUCLEAR POWER (2012).

economic performance beyond some threshold that most developed nations have already substantially exceeded.[28] Under such circumstances, surely we could at least *try* some level of retrenchment, as opposed to acceleration. If we found ourselves going too far – if it were just too painful to bear – we could always turn back.

This concern becomes more vivid if we note the spectacular differences in national levels of per capita emissions across the globe. For example, in 2007 average global emissions per capita were 1.28 metric tons of carbon. However, some countries averaged around five tons of carbon per person (e.g., the United States was at 5.20 tons, Australia at 4.84), others between two and three (e.g., Germany was at 2.61, the UK at 2.41, New Zealand at 2.11), others at around one (e.g., China at 1.35, Argentina at 1.27), and others at much less (e.g., Brazil at 0.52, India at 0.39, and Bangladesh at 0.08).[29] Is it so obvious that (for example) Americans would be dramatically worse off living on emissions levels similar to those of (say) the British, Germans, or New Zealanders (2.11)? More radically, is it clear that any of these countries should insist on maintaining greater per capita emissions levels than (say) Argentines or Brazilians, especially given their greater initial wealth and technological expertise? After all, perhaps Wayne should be content to have sex just with his wife.

This brings us to the third broad parallel between Wayne's folly and the climate case. At the moment, it seems that, despite their availability, we are not yet willing to take up any of these options in a robust way. As a result, the new push within the scientific and policy communities is toward geoengineering. At the time of writing, this push seems likely to succeed. It is plausible that, although resisting substantial emissions cuts, some developed countries will be willing to spend a very small amount of their national budgets (a few hundred million in the case of the UK and United States) pursuing research on climate engineering, even though they know that such engineering may not work, will not fix all the problems even if it does work, and raises serious issues of its own, including those of negative side effects, legitimate governance and irreversibility. However, this is ethically worrying. Is this not a little like Wayne's $10 bet on the AIDS research? If it is, are we not in serious ethical trouble?

3. IMPLICATIONS

The concerns with ethical shortsightedness and moral schizophrenia may account for (and also justify) a fair amount of resistance, even moral outrage, concerning the push toward geoengineering. Hence, potentially these concerns have

[28] Chrisoula Andreou, *A Shallow Route to Environmentally Friendly Happiness: Why Evidence That We Are Shallow Materialists Need Not Be Bad News for the Environment(alist)*, 13 ETHICS, PLACE & ENV'T 1 (2010).

[29] Boden et al., *supra* note 12.

explanatory value. In addition, they explain this resistance without invoking other ethical concerns that are also in play, but which may involve more controversial claims. In particular, neither concern requires assuming that geoengineering is necessarily or inherently bad, or (more specifically) that it constitutes a violation of humanity's relationship to nature. Consider the schizophrenia concern. Nothing in Wayne's folly suggests that AIDS research or vaccines are problematic in themselves. Instead, the natural assumption is quite the opposite: that they are a good thing. Similarly, in drawing the analogy between the cases, we need not presuppose that there is anything bad about geoengineering research or deployment considered in isolation. Of course, this is not to say that proponents of the shortsightedness and schizophrenia arguments must deny that there is something independently bad about geoengineering. Instead, the point is merely that they need take no position on such matters. If claims about the independent badness of geoengineering can be justified on other grounds, they will add to the myopia and schizophrenia concerns.

These points (about explanatory power and ethical presuppositions) are worth noticing for several reasons. First, the fact that both the shortsightedness and schizophrenia concerns can coexist with a wide variety of views about the moral importance of nature (or lack thereof) may provide us with strong pragmatic reasons to highlight it. Perhaps focusing on this ethical worry is a better bet for motivating comprehensive climate action than some more contentious arguments.[30]

Second, at any rate, it seems important that these concerns are not lost amid the efforts of some geoengineering proponents to frame the issue completely in terms of moral emergency.

Third, and more generally, the diagnoses of ethical shortsightedness and moral schizophrenia can play a subsidiary role in supporting the perfect storm hypothesis, as these phenomena appear predictable under the circumstances of the perfect moral storm. If a group (e.g., the current generation of the affluent) is tempted to pass the consequences of its behavior onto others (e.g., future generations), but uncomfortable about admitting that this is what it is doing, then it will be susceptible to forms of argument that appear to justify this, even if they pervert our moral understanding of what is at stake. Arguments about shortsightedness and schizophrenia fit the bill nicely.

4. OBJECTIONS

The analogy between Wayne's folly and our predicament with geoengineering may be attacked in a number of ways. Although a complete defense will not be possible

[30] For a contrary view, see Jamieson, *supra* note 7 (2010). I am agnostic on this question.

within the confines of this chapter, the aim of this section is to cast doubt on some initial reasons for rejecting the analysis.

4.1 *Who Is Wayne?*

The first objection claims that the analogy trades on an ambiguity: it talks of "our" pursuit of geoengineering, but the relevant "we" here is problematic. This threatens the argument because if the people responsible for climate change are not the same as those pursuing geoengineering, then the analogy with Wayne's folly fails. Suppose, for example, that although some political actors and institutions have acted like Wayne, others (e.g., especially climate scientists, environmental organizations) have not. Then one might claim that, as some of these groups have consistently argued for action on climate change (albeit unsuccessfully), there is no ethical shortsightedness or moral schizophrenia in *their* now urging the pursuit of geoengineering. Much as they may lament the ongoing political inertia, they are merely trying to make the best of a bad situation by offering some aid to future generations and the world's poor within that inertia.[31]

My first response is to clear up the ambiguity. I intend the "we" in the analogy to refer to society in general and especially the public bodies that represent it. This is appropriate because a large percentage of the discussion of geoengineering is public argument directed at political institutions, particularly in the form of calls for research programs and action to establish governance structures.[32] In context, many geoengineering advocates are appealing to the same institutions that have (even in their view[33]) been failing adequately to address climate change. Hence, although such advocates may not themselves be subject to the analogy, the bodies that they seek to persuade are. As these are the bodies that will decide, the analogy retains its force.[34]

4.2 *What about a "Pure" Actor?*

The first response is the most important. However, there remains room for debate. In particular, perhaps some of the problems raised by the analogy to Wayne's folly could be sidestepped if other actors stepped in to take up the cause of geoengineering.

[31] I thank Brad McHose for pressing this objection.

[32] *E.g.*, Caldeira & Keith, *supra* note 3; SHEPHERD ET AL., *supra* note 2; Victor et al., *supra* note 3; American Meteorological Society, *Policy Statement on Geoengineering the Climate System* (2009), *available at* http://www.ametsoc.org/policy/draftstatements/2009.

[33] *E.g.*, SHEPHERD ET AL., *supra* note 2.

[34] If the institutions to be persuaded are subject to the analogy, this might have further implications. Most notably, arguments offered to them are subject to some sharp but self-imposed constraints. These constraints are worth investigating, and their moral import should be examined. (See below.)

Suppose, for example, that a new actor were to embark on a geoengineering program, an actor untarnished by the history of the climate problem, and uncompromised by its current relationships and options. Returning to the analogy with Wayne's folly, would this not be like people other than Wayne supporting AIDS research? And what could be wrong with that? Surely such an actor could legitimately forget the contextual question and move straight to the justificatory question?

I have two basic replies. First, in context, finding an agent that has the requisite capabilities to do geoengineering, but is not at all compromised by its past history and future trajectory, may not be so easy. This is especially so given that that the major candidates are likely to remain nation-states and large corporations. Hence, we should not be too quick to assume that a "pure actor" can be found.

Second, there are independent reasons for thinking that the justificatory question cannot be quickly isolated from the contextual question, and these cast doubt on the import of the "pure actor" model. A full defense of this claim would take more space than I have here. So, instead, let me simply gesture toward some relevant concerns.

To begin with, whether we are talking of actual or pure actors, we should beware of any arguments that appear simply to assume or stipulate that the form of geoengineering that ultimately emerges on the international scene is likely to be ethically benign. This worry can be made intuitively plausible simply by listing four examples of ethically worrying forms of geoengineering:

- *Rogue geoengineering* (such as that undertaken by a lone state, corporation, or individual without appropriate consultation with and approval of others)
- *Nonconsensus geoengineering* (such as that carried out by a "coalition of the willing" consisting of (say) the Western powers alone, or China and India alone, or Iran and fellow Islamic states alone, without the approval of the other nations affected)
- *Predatory geoengineering* (such as that which aims to systematically disfavor the interests of some countries in choices between geoengineering schemes or levels of intervention, perhaps in order to secure other strategic advantages)
- *Militarized geoengineering* (such as the weaponization of the climate control system)[35]

These forms of geoengineering are ethically worrying for the same basic reasons. For geoengineering to be morally acceptable any would-be-geoengineer, including an otherwise ethically untainted actor, would be required to act with appropriate authority, and in accordance with adequate norms of global, intergenerational, and

[35] For discussion of historical precedents involving weather modification that give some credence to such worries, see JAMES FLEMING, FIXING THE SKY: THE CHEQUERED HISTORY OF WEATHER MODIFICATION AND CLIMATE CONTROL (2010).

ecological ethics. However, both requirements strongly suggest that there must be due consultation with existing "impure" actors, consideration of their rights and responsibilities, and judgments about their likely future behavior.[36] Given this, it seems unlikely that even a pure actor could simply set aside the contextual question. Its shadow falls over the whole enterprise.

4.3 *Can We Just Get on With Our Science (Please)?*

Perhaps some will accept the problems facing the pure actor model, but nevertheless insist that if scientists want to do research on geoengineering, we should let them. After all, they might say, is free enquiry not one of the basic tenets of scientific research, and indeed of a free society more generally?[37]

This objection is a red herring, based on two mistaken inferences. First, to defend the importance of the contextual question is not to advocate that geoengineering research should not be undertaken, still less that it should be banned absolutely. Indeed, as I have emphasized, the key complaints – ethical shortsightedness and moral schizophrenia – are directed at some emergency arguments for geoengineering, and need not even take a position on the worth of geoengineering itself, or geoengineering research considered in isolation. (Recall that the Wayne's folly example also takes no position on the worth of AIDS research, which I assume to be highly positive.) It is not the projects themselves that are (currently) under scrutiny, but the reasons used to justify their pursuit.

Second, the issue of research is not as "all-or-nothing" as the objection suggests. In particular, there is a large difference between, on the one hand, defending traditional, curiosity-based research, to be published and funded according to normal academic practices (e.g., peer-reviewed journals, National Science Foundation (NSF) applications) in competition with other projects (including other climate projects), and, on the other hand, arguing for the establishment of a new, independent, and targeted research program on grounds of moral emergency. In particular, the latter requires extra defense. This is relevant because the "free enquiry" argument is most naturally at home in defending the first category of research, but most

[36] For example, ethically defensible geoengineering will raise issues of procedural, distributive, and corrective justice. Moreover, these will have implications for the type of geoengineering to be pursued (e.g., types that cause the shutdown of the monsoon in India face a very high burden of proof), how it is researched (e.g., because of the need for trust, and the international and intergenerational stability of the proposal), how it is to be governed (e.g., trigger conditions, salient concerns, review process, etc.), what is to be done about predicable negative side effects (e.g., regional effects that are worse even when global effects are better), and what the support system is when things go wrong (e.g., who is accountable, to what extent, etc.).

[37] Ralph Cicerone. *Geoengineering: Encouraging Research and Overseeing Implementation*, 77 CLIMATIC CHANGE 221 (2006); *see also* Gardiner, *supra* note 7.

of the current advocacy for geoengineering falls into the second. Given this, it seems reasonable to say that the current advocacy does bear special scrutiny.

In addition, there is another side to this point. Normal, curiosity-driven research is subject to scrutiny too, when it competes with other projects for research time and funds, and for publication. So, it is worth emphasizing that putting geoengineering into its own category by appealing to moral emergency actually *privileges* geoengineering research over other kinds of research, in part by shielding it from normal competition. In other words, in context, the issue is not so much that calling for special scrutiny involves discriminating against geoengineering research (as the objection suggests), but rather that strong reasons are needed for privileging geoengineering by *exempting* it from the usual norms. On my reading of the situation, the arguments from moral emergency are an attempt to provide such reasons. Given that this is their aim, they should be scrutinized. My claim is that taking the contextual question seriously is essential to that task. (Note that I do not claim that successful reasons cannot be found, only that there is work to be done.)

4.4 *Do Scientists Not Have a Special Moral Obligation to Pursue Geoengineering?*

A fourth objection suggests that whether or not pure actors need to confront the contextual question, the unique position of scientists means that they have a special obligation to pursue geoengineering. Consider the following argument:

> Scientists have consistently (albeit unsuccessfully) argued for action on climate change. Much as they may lament the ongoing political inertia, there is no schizophrenia in their now urging the pursuit of geoengineering. After all, they are merely trying to make the best of a bad situation by offering some aid within that inertia. Indeed, are they not morally obligated to do this? Surely if it might help, it is worth a try, and the scientists are the only ones in the position to do the trying, because they have the relevant expertise.

In my view, this argument raises a serious ethical question; however, the answer it offers is overly simplistic, and so unhelpful even to scientists. To see why, it is helpful to distinguish the different social roles played by scientists.

One role is that of advisors to decision-making bodies. This seems to be the role that is prominent in the major reports mentioned earlier. However, the worry about moral schizophrenia casts doubt on advocacy for geoengineering within that role. Consider a new variant on Wayne's folly. Suppose that Wayne has a friend who is appalled at his behavior, and so frequently draws Wayne's attention to what is wrong with it, what the alternatives are, and so on. Wayne listens, but ultimately ignores his friend's advice. His friend believes that this is because Wayne does not like the options available, and would prefer a path that is easier on him, even if it is more risky and dangerous for others. What should his friend do? Should he devote himself to exploring the risky options? Does he have an ethical reason to do this?

The main point to make is that this is not obvious. Notice that there are strong considerations that point in the opposite direction. On the one hand, it seems unlikely that the friend has any strong obligation *to Wayne* to do so. He has already given Wayne a number of good options that Wayne refuses to take, and without good reason. On the other hand, the friend may think that he has some obligations to Wayne's wife and the other women involved. However, it is also not obvious that these are best served by the friend's advising Wayne to pursue the donation to AIDS research, or pursuing such research himself. In Wayne's folly, it is highly tempting to argue that the friend would do better to act more directly (for example, by telling Wayne's wife).

These complications are relevant to the pursuit of geoengineering. The fact that existing institutions are failing to act on the good options need not mean that scientists have a strong obligation to commence work on other (inferior) options instead. If they have already suggested good options, they might justly claim that their responsibility is only to work harder promoting or refining the better options. In context, for example, perhaps scientists should try harder to engage the public on the dangers of climate change, or at persuading them that the scientists are telling the truth when they do so. The fact that geoengineering might turn out to be useful is not enough. Other things might be much more useful, and more likely to work (e.g., such as A–X in the Agent 1 example).

Of course, scientists are often not merely in the role of dispassionate advisors. They are sometimes in the role of individual researchers following their own intellectual agendas, advancing their own careers, and so on. This role is perfectly defensible insofar as it goes. However, even here, scientists can ask themselves how their motives and aims interact with those of ethically compromised actors, and also with their role as advisors.

Most obviously, there are difficult issues about conflicts of interest. For instance, some scientists who could contribute to geoengineering research have a personal stake in its pursuit. They stand to win major grants, academic prestige, and in some cases lucrative economic opportunities for their own businesses. Given this, there is the clear potential that their role as advisors might become compromised when they advocate for the pursuit of geoengineering. Hence, safeguards are needed, both for society (to make sure that this does not happen) and for the individual scientists (to protect them from the appearance that it has when it has not).[38]

[38] Such problems are not caused only by careerist motives. They also arise if, to list a few examples, the researchers are actually climate deniers, or believe that conventional responses are unwarranted,

Less obviously, there are live questions that revolve around the background motives and beliefs of scientists, and especially about the role of these in the difficult ethical situation of the perfect moral storm. For example, returning to Wayne's folly, suppose that the AIDS researcher does not try to talk Wayne out of his behavior, but rather encourages it because the researcher wants the funding, or the recognition. Then, his role might be morally problematic. Some have similar fears about geoengineering research. They do not want science to function as an *enabler*, facilitating bad behavior. This, I believe, is one reason some scientists are so worried about the potential "moral hazard" of pursuing geoengineering: that it will itself encourage society at large into further inertia on mitigation and adaptation. It is not just that they are concerned about undermining conventional climate policy per se; they are also concerned about their professional responsibilities to society, and the potential for moral schizophrenia. They do not want to be, or appear to be, complicit in continuing moral failure.

None of this implies that scientists cannot or should not pursue geoengineering, or that ethical geoengineering is impossible. Instead, the discussion serves to highlight the real ethical difficulties that face scientists when they consider geoengineering research in the real world – difficulties that are enhanced when one considers the contextual question. In summary, the worry about the fourth objection – that scientists have a moral obligation to pursue geoengineering given ongoing inertia and the potential for catastrophe – is that it paints too simple a picture of the ethical context within which scientists actually operate. This context is fraught with genuine moral challenges that ought not be dismissed so quickly. Again, the contextual question becomes important, and we lose much if we try to bypass it too quickly with arguments from moral emergency.

4.5 *What Other Options?*

At this point, some might object by introducing a crucial new move.[39] Much of the discussion so far, they will say, depends on the assumption that there are other good or decent options. However, they will continue, this assumption may now be false. Instead, it is possible that climate change has already progressed far enough that

or think that geoengineering is a bad idea but want access to funds to pursue their own basic science projects, and so on. In general, ethical questions are raised when scientists support, or seek to take advantage of, strongly moralized arguments for pursuing a given research strategy but for other reasons that potentially stand in tension with those arguments. In some of these cases, we also see moral schizophrenia. For example, suppose the AIDS researcher urges Wayne to part with his $10 by emphasizing the vulnerability of his wife, but the researcher actually believes that his own research is highly unlikely to help her. Again, there is a split between the values he espouses and what really moves him.

[39] *See also* GARDINER, *supra* note 11,

critical climate thresholds will be breached in the next few decades regardless of what we do about mitigation, and where this breaching is profoundly dangerous for humanity regardless of what we now do about adaptation. Perhaps, the thought goes, "our goose is already cooked"[40] insofar as conventional methods are concerned, and geoengineering – in the form of an offsetting dose of sulfates, for example – is already the only option left.[41] In this case, we are not like Agent 1 in having options A to X still open.

Let me make a preliminary point before offering a more central response: this is a new argument, and there are several contextual worries about it. One is that it involves bold scientific claims about the extent of our current precommitment to climate change, and the inability of conventional policies to overcome it. These claims are controversial, and so require independent defense. For example, the Royal Society's landmark report on geoengineering explicitly declares that "decarbonisation at the magnitude and rate required [to avoid global average temperatures exceeding 2 degrees C above pre-industrial levels this century] *remains technically possible*" and "global failure to make sufficient progress on mitigation of climate change is *largely due to social and political inertia*."[42] Given this, the Royal Society rejects the new framing.

A second contextual concern is that the argument looks likely to be morally schizophrenic unless it really is accompanied by serious efforts at mitigation and adaptation. After all, the claim is not that our goose is *certainly* already cooked (this would be scientifically implausible), but only that it is scientifically credible to say that it *might* be. Given this, there remain strong reasons to engage in conventional ways. In addition, there is no guarantee that the geoengineering research will be successful, or that the situation for its use will arise. Mitigation and adaptation may yet turn out to be the best responses available (even if they are insufficient to ward off some forms of catastrophe).[43]

Let us turn now to a more central response to the objection. Suppose we believe there is a credible threat that we are already precommitted to catastrophe. What would follow about the pursuit of geoengineering? Let us begin by extending Wayne's folly once again. Suppose Wayne concedes that our scruples would be

[40] I borrow the phrase (though not the subsequent argument) from Dale Jamieson.

[41] More weakly, the argument may be that our goose is not yet cooked, but will be before geoengineering is ready to deploy unless we start research now.

[42] SHEPHERD ET AL., *supra* note 2, at 57; emphases added.

[43] Note that the argument also does not sit well in the current political debate, where more mainstream claims about the current extent of climate change and embedded commitments have been subject to vigorous dispute. There is a concern that some political actors who make the precommitment argument may be selectively applying their standards of scientific credibility, insisting on very high levels of proof when substantive action might be required of them (e.g., with mitigation and adaptation), and relaxing these when they seem less liable (e.g., with geoengineering).

well-founded if we assume that merely telling his wife of the danger (so that she can protect herself), or practicing safe sex will be enough. However, he goes on to claim, in fact he has been reckless for so long that there is a credible risk that she has already contracted HIV. If so, then the speculative research may well be the *only* thing that can save her. Of course, he admits, he should own up, or at least practice safe sex, as well as give the $10 (though he shows no sign of being ready to do either). But surely, he insists, the case for the research is unassailable, and he is morally right to invest. Indeed, he asserts (again) that if the risk is realized, this is a much more important thing for him to do than anything else.

Suppose, then, that there is a credible threat that Wayne's wife already has HIV. Is $10 to the start-up really his best and only option? It seems not. Most obviously, he might consider *investing more* than $10. If he really has his wife's welfare at heart, and if he acknowledges his responsibility for her predicament, surely it ought to be a lot more. Less obviously, but much more important, there are *other things he could be doing*. He could be investigating ways of supporting and looking after her if she does contract the illness. He could be looking into the best doctors and health care available. He could be considering what will make her days more bearable. (And so on.) In general, Wayne should not assume that the matter is closed even if some investment in research is necessary.

Similar worries are extremely important in the geoengineering case. Typically, when people think of actually doing something about geoengineering the climate, their focus is predominantly on the idea of throwing a few million dollars at scientific research. However, if the claim is truly that we are already precommitted to severe climate change, this approach seems ethically shortsighted. Surely the state of the planet is worth more than this, and surely the response should be broader. If we really are facing down a global catastrophe, then more is required by way of preparation than merely giving a few scientists funds for modeling and small-scale experiments. For example, issues such as how to protect vulnerable people and infrastructures loom large, as do those of how to compensate those for whom adequate protection is not possible. Talk about geoengineering research dollars can only be a small part of the ethical picture, and wider concerns need to be integrated into any further pursuit of geoengineering. In particular, there needs to be some plan about how to manage a world that is geoengineered in response to looming climate catastrophe, and especially about how to do so in a politically legitimate and morally defensible way. Like Wayne's $10, the call for a limited push on narrow scientific research seems ethically shortsighted and morally schizophrenic. If one is really concerned about already being committed to a climate catastrophe, much more is at stake.

We can phrase the point more abstractly using our earlier example of Agent 1. When this example is applied to climate change, options Y and Z might not be the only ones that include some scientific research on geoengineering. On the one hand, on the positive side, perhaps options S–Z all include this, but S–X include much else besides. For example, perhaps they also include very substantial adaptation funding, geopolitical reform, compensation, and so on, together with major research into geoengineering and other remedial efforts. If one focuses all one's attention on the very restricted geoengineering options (Y and Z) considered completely in isolation, one misses the fact that the wider geoengineering options (S–X) are arbitrarily excluded. Again, there is ethical shortsightedness, and in a way that brings on moral schizophrenia.

On the other hand, on the negative side, the narrowing also obscures other (less ethically welcome) options that also arise in an atmosphere of wider moral corruption. In particular, although Z might be the last option on the list considered by any remotely ethical agent, unethical agents may have more extensive lists. For example, in the climate case, an unethical agent may be happy to consider not only "modest geoengineering research only," but also research on predatory, parochial, and other buck-passing forms of geoengineering as well. If we ignore the contextual question, we may miss the importance of these worries.

4.6 Are We Not Already Advocating for a "Portfolio" Approach?

Some may complain that this response goes too far in suggesting an analogy between Wayne's $10 and current calls for geoengineering research. In practice, they will say, most scientific reports suggest that geoengineering research should be pursued alongside substantial mitigation and adaptation, as part of a "portfolio" approach. Hence, they are not suggesting "modest research only," and indeed are explicitly open to a broader strategy.[44]

I am not so sure. First (as I have just suggested), we have to be concerned not just about how geoengineering is being advocated by scientists, but also about how such arguments are likely to be received and acted on. Given past political inertia, there are realistic worries about whether the other "parts of the portfolio" will be enacted. Moreover, if a robust portfolio were really being enacted, the need for geoengineering research might be at least less pressing, and perhaps nonexistent. Given this, it is far from clear that a few throwaway lines on the need for a comprehensive approach to climate change are really enough. The ethical context of those lines matters.

Second, in any case, it is not obvious that those who currently advocate for a portfolio approach have a full appreciation of the ethical implications of geoengineering. Mitigation and adaptation are not all that is at stake here. As mentioned, ethical

[44] Christopher Preston, *Geoengineering and the Presumptive Argument from Environmental Ethics*, 20 ENVTL. VALUES 457 (2011).

geoengineering would have to address difficult issues of global governance and compensation. However, these would involve deep questions about global legitimacy and international justice that are barely even on the agenda. For example, even when major reports mention governance, they tend to assign it to venues that seem inadequate to the profound issues raised. The Royal Society report, for instance, recommends the United Nations Commission for Sustainable Development, and does not even mention fora such as the UN Security Council, NATO, the G20, or the U.S. Congress.[45] Again, the specter of moral schizophrenia raises its head. If the threat is so profound as to justify the risks of geoengineering, why is there not more emphasis on taking the ethical challenges of actually doing geoengineering more seriously, and preparing to meet them? One answer, of course, is that no one thinks that serious responses to the ethical challenges are really on the table, politically speaking, even if they are morally necessary. This revives the basic worries about moral schizophrenia in the perfect moral storm.

Third, the terms "comprehensive" and "portfolio" are misleading. Many possible climate policies are not even being considered, some of them for ethical reasons. Advocates often claim that the potential for emergency puts geoengineering back on the table, but similar arguments might be offered in favor of other strategies not currently being considered, such as drastic cutbacks in consumption, serious population measures, and so on. I am not advocating such measures. Instead, my point is that most "portfolios" are far from comprehensive, and ethical judgments are already involved in determining their content. Given this, including geoengineering cannot be defended merely on the grounds of the need to be "comprehensive"; instead, other arguments must be offered.[46]

5. CONCLUSION

This chapter distinguishes two questions as central to the ethics of geoengineering. The justificatory question asks "Under what future conditions might geoengineering become justified?," where the conditions to be considered include, for example, the threat to be confronted, background circumstances, governance mechanisms, individual protections, compensation provisions, and so on. The contextual question asks "What is the ethical context of the push toward geoengineering, and what are its implications?" I claimed that early discussions of geoengineering often marginalize both questions because the discussions tend to focus on arguments from emergency that illegitimately brush them aside. Moreover, I argued that discussion of

[45] Stephen M. Gardiner, *Some Early Ethics of Geoengineering: A Commentary on the Values of the Royal Society Report*, 20 ENVTL. VALUES 163, 171 (2011),

[46] Gardiner, *supra* note 48, at 175–76,

the contextual question has an important role to play in explaining what is ethically problematic about some arguments for geoengineering, and that it does so without appealing to (potentially controversial) claims about the moral status of nature or our relationship to it. The key ideas were that some arguments for geoengineering are ethically shortsighted, and morally schizophrenic. These ideas were illustrated through two examples, one abstract (Agent 1), and one more concrete (Wayne's folly). Although both examples were extreme and idealized, even the imperfect analogies provide reasons for concern about our current predicament with geoengineering. Ethically serious discussion of geoengineering should confront such worries, rather than hide behind overly simplistic appeals to moral emergency. As Stocker puts it, "to refuse to do so bespeaks a malady of the spirit."

ACKNOWLEDGMENTS

This chapter significantly extends one part of an argument originally offered in Stephen M. Gardiner, *Is "Arming the Future" with Geoengineering Really the Lesser Evil? Some Doubts about the Ethics of Intentionally Manipulating the Climate System, in* CLIMATE ETHICS: ESSENTIAL READINGS, 284–312 (Stephen M. Gardiner, Simon Caney, Dale Jamieson & and Henry Shue eds., 2010). I thank audiences at Oxford University, University of Bergen, University of Edinburgh, University of Newcastle, University of Oslo, University of Vienna, and University of Washington. I am especially grateful to Derek Bell, Jason Blackstock, Thom Brooks, John Broome, Wil Burns, Simon Caney, Elizabeth Cripps, Roger Crisp, Leslie Francis, Lauren Hartzell, Tim Hayward, Clare Heyward, Angela Kallhoff, David Keith, Graham Long, Catriona McKinnon, Steve Rayner, Phil Rasch, Dominic Roser, Julian Savulescu, Henry Shue, Harald Stelzer, Andrew Strauss, and Allen Thompson. This work was supported by a grant from the Netherlands Institute for Advanced Study in the Humanities and Social Sciences (NIAS), by a visiting fellowship at the Smith School of Environment and Enterprise at Oxford University, and by the Ben Rabinowitz Endowed Professorship in the Human Dimensions of the Environment.

2

The Ethical Foundations of Climate Engineering

Clive Hamilton

The time is coming when the struggle for dominion over the earth will be carried on. It will be carried on in the name of fundamental philosophical doctrines.

Friedrich Nietzsche

1. INTRODUCTION

After years of agonizing among scientists over the dangers of discussing "Plan B," the dam broke with the publication in 2006 of an editorial essay by Nobel Prize–winning atmospheric scientist Paul Crutzen in which he called for serious consideration of geoengineering.[1] Within the expert community, work on geoengineering is now vigorous, with a sharp leap in the number of academic papers published. The debate is poised to move to center stage when the Intergovernmental Panel on Climate Change (IPCC) for the first time includes assessment of geoengineering solutions in its Fifth Assessment Report, due out in stages in 2013 and 2014. Some have been disturbed at the ease with which worries about the morality of openly considering geoengineering seem to have been left behind in favor of a focus on research and governance arrangements. The growing interest in alternatives to mitigation perhaps justifies the fears of those who criticized Crutzen for letting the cat out of the bag, although someone was bound to do it sooner or later.[2]

Whether climate engineering becomes a substitute for carbon abatement, instead of a complement or a backup, remains to be seen. But there can be no doubt that in the wider debate over climate policy technological intervention is everywhere presented as a substitute for social change. Despite the fact that the world's emissions have for some years been tracking at levels higher than the IPCC's

[1] Paul Crutzen, *Albedo Enhancement by Stratospheric Sulfur Injections: A Contribution to Resolve a Policy Dilemma?* 77 CLIMATIC CHANGE (2006).

[2] On criticism of Crutzen, see Mark Lawrence, *The Geoengineering Dilemma: To Speak or Not to Speak*, 77 CLIMATIC CHANGE, 245–48 (2006).

worst-case scenario of the early 2000s, any challenge to the primacy of economic growth is strictly excluded from the official agenda. The whole burden must fall on technology.[3]

Recourse to more and bigger technologies to solve perhaps the most severe threat that modern society has ever faced vitiates any serious reflection on the deeper reasons for humanity's inability to respond to the threat posed by carbon emissions. If the science is only half right – and it is more likely that the scientists have been unduly cautious[4] – then the transformation of the earth's life-support systems, and schemes to engineer the planet's atmosphere, call for sustained moral reflection. Yet work on the ethics of climate change and climate engineering is desultory and rarely seems to recognize the enormity of what is unfolding.[5]

The failure to appreciate the scale of the threat of climate change, or to take in the Promethean nature of geoengineering, is reflected in the question that "climate ethics" believes it must answer, namely, what are the consequences for human well-being of transforming the earth's climate? It is not so much the anthropocentrism of the question that is of interest, but the unrecognized assumption about the kind of *anthropos* that asks such a question – a rational being who gathers evidence on the good and bad consequences, evaluates it, and decides on how to act in a way that most improves human well-being. In short, climate ethics (including geoengineering ethics) is dominated by a consequentialist approach that naturally shies away from questions about *how* we ended up in this mess and what it means for humanity. In so doing, I will argue, it risks entrenching the very ways of thinking that lie at the heart of the climate crisis.

In the consequentialist view, the question of whether it is ethically justified to intentionally shift the planet to a warmer or cooler climate – either by deliberate intervention or by allowing greenhouse gas emissions to reach a target level – depends on an assessment of the costs and benefits of shifting to the new climatic state.[6] This seems to exhaust ethical concerns.[7] Peter Singer defines climate ethics in consequentialist terms: "Climate change is an ethical issue, because it involves

[3] CLIVE HAMILTON, REQUIEM FOR A SPECIES: WHY WE RESIST THE TRUTH ABOUT CLIMATE CHANGE ch. 2 (2010).

[4] James Hansen discusses the natural reticence of the IPCC, leading to more cautious projections, in *Scientific Reticence and Sea Level Rise*, ENVTL. RESEARCH LETTERS, April–June 2007

[5] James Garvey, in THE ETHICS OF CLIMATE CHANGE (2008), is an exception.

[6] A comprehensive overview of the field of climate ethics can be found in Stephen Gardiner, *Ethics and Global Climate Change, in* CLIMATE ETHICS: ESSENTIAL READINGS (Stephen Gardiner et al. eds., 2010).

[7] A number of environmental philosophers have developed alternatives to the consequentialist approach, but their voices are not heard in the public debate. *See, e.g.*, the discussion of eco-phenomenology in Iain Thomson, *Ontology and Ethics at the Intersection of Phenomenology and Environmental Philosophy*, 47 INQUIRY 380–412 (2004).

the distribution of a scarce resource – the capacity of the atmosphere to absorb our waste gases without producing consequences that no one wants."[8]

One immediate implication of this approach is that there is nothing inherently preferable about the natural state. Thus, Powell et al. declare that "there is no prima facie justification for attempting to preserve the current climate, if some other climate might be better for humans and animals."[9] Depending on the assessment of human well-being (Singer would extend it to other sentient beings but this is only an extension of the utilitarian frame), there may be a "better" temperature or climate as a whole. In other words, it is ethically justified for humans to "set the global thermostat" in their interests.

It is apparent that this consequentialist approach to the ethics of climate change, which dominates writing on the topic, is essentially the application of neoclassical economic analysis with a distributional gloss, so that the question of "climate justice" is reduced to the distribution of economic effects. Reflecting the subtle influence of the neoliberal revolution of the last three or four decades, philosophers (even "radical" ones such as Singer) have often adopted uncritically the central tenets of free-market economics. The most obvious, and most revealing, conceptual borrowing from economics is the idea of the atmosphere as a "common resource" or, more properly, a "common property resource."[10] Singer's words quoted above explicitly define climate ethics as a problem of allocating a scarce resource. When we remember that every neoclassical economics textbook defines economics as the analysis of how best to allocate scarce resources, we can see how this kind of "ethics" can become a branch of free-market economics. Among other things, characterizing the atmosphere as a resource implies that it is subject to ownership by humans – it is our property – and that it is available for our use.

In the subsequent sections I describe the consequentialist worldview and the conception of the earth implicit in it, before outlining how the new discipline of earth-system science is destabilizing that view. The argument is made through consideration of the "ethics" of solar radiation management. The advent of the

[8] Peter Singer, *Ethics and Climate Change: A Commentary on MacCracken, Toman and Gardiner*, 15(3) ENVTL. VALUES 415, 415 (2006).

[9] Russell Powell et al., *The Ethics of Geoengineering* (working draft), Oxford Uehiro Centre for Practical Ethics, 2010, at 6. *See* http://www.practicalethics.ox.ac.uk/__data/assets/pdf_file/0013/21325/ (last visited Jan. 15, 2012).

[10] Other concepts imported into climate ethics from economics are "moral hazard," "public goods," and the idea that the social is no more than the aggregation of the individual. And the claim that climate change is an "externality" goes unchallenged. These concepts only make sense when the world is the kind of entity that has as its ideal form a collection of perfect markets, with all of the assumed individualism, self-interest, and equilibrium tendencies embedded in them. In other words, when one begins with perfect markets one begins with a certain conception of the human being, *homo economicus*, the unexamined starting point of free market economics.

Anthropocene heralds a transformation of conditions of life on earth so that the conventional approach to assessing costs and benefits, winners and losers, breaks down. The "system" as a whole is being altered.

2. THE CONSEQUENTIALIST WORLDVIEW

The consequentialism of climate ethics is built on an unstated (and mostly unrecognized) understanding of the natural world, one that grew out of the Scientific Revolution in the seventeenth century and the European Enlightenment philosophy that accompanied it. The transition from an organic conception of nature to a mechanical one in the seventeenth and eighteenth centuries is a history that has been well told.[11] In the modernist view, a human being is a distinct subjective entity that is separate from the world around it, a world on which, guided by its cognitive abilities, it acts to pursue its own individual and collective interests. Through Descartes, and Kant in particular, philosophy responded with the idea of the autonomous subject and the objective external world as a representation. In the Kantian view, the grounding for ethical judgment is the self-legislating moral subject who recognizes no external moral authority. This is a vast topic that cannot be considered here.[12] It is a model in which rational and willing subjects – discrete egos existing inside bodies – exercise control over an inert environment.

This understanding brings to thinking about climate change and geoengineering certain assumptions about the earth, humans, and the relationship between them.

- The earth consists of a collection of resources, that is, materials and energy available for human consumption or other uses.
- This collection of resources can be thought of as a system or collection of systems whose workings can be discovered by inquiring minds.
- Through the mobilization of sufficient intellect and technology the natural world is subject to human control.
- Humans have a right to control the earth, and the only constraint on our dealings with the earth is imposed by enlightened self-interest and perhaps the "interests" of other sentient beings.
- Humans are distinct subjective entities who come to the ethics of climate change as rational calculators whose objective is to pursue their own interests, individual or collective.[13]

[11] By, for example, CAROLYN MERCHANT, THE DEATH OF NATURE: WOMEN, ECOLOGY, AND THE SCIENTIFIC REVOLUTION (2nd ed. 1990); *see also* STEVEN SHAPIN & SIMON SCHAFFER, LEVIATHAN AND THE AIR-PUMP: HOBBES, BOYLE, AND THE EXPERIMENTAL LIFE (1985).

[12] A way into this subject might begin with JERROLD SEIGEL, THE IDEA OF THE SELF (2005).

[13] Although Singer's concern for animal welfare seems to contradict this, in fact his utilitarian approach treats animal welfare in the same way that economists treat human welfare, so that Singer's application

- Because climate ethics is grounded in the self-legislating subject, there is nothing *inherently* desirable about a natural state, and there is no ethical distinction between natural harms and anthropogenic harms.
- There is nothing special about global warming and geoengineering that would prevent the standard ethical framework from being applied.

Although these assumptions have been contested in the past, in this chapter I argue that modern developments in earth-system science, including climate science, are undoing the very conceptions of the earth and the human being constructed in the Scientific Revolution and that now guide ethical thinking about climate change. In particular, I will argue that the meaning of the facts generated by earth-system science challenge climate ethics by suggesting that there is a source of moral authority beyond the self-legislating Kantian subject.

It may seem contradictory that I deploy the results of earth-system science to critique the worldview given to us by the Scientific Revolution. I am suggesting that advances in earth-system science expose the limits and contradictions of the mechanical and systems understanding of the world and the technological thinking that goes with it. The problem is that we have not grasped the implications of earth-system science because we are "too scientific," that is, too habituated to thinking of the world as a systematic totality that we can know and control. If this is so, science itself now challenges the understanding of the world as comprised of resources that are at our disposal and can be grasped with our minds and manipulated with technology. The foundations of climate ethics based on the autonomous moral subject are destabilized.

To make the argument, I will assess the consequentialist approach in the context of one of the several geoengineering technologies being proposed, namely, a program of spraying sulphate aerosols into the upper atmosphere in order to reflect back into space a greater proportion of incoming solar radiation.[14] This form of "solar radiation management" is designed to offset warming by mimicking the cooling effect of large volcanic eruptions. Currently around 23 percent of solar radiation is reflected back into space by the earth's atmosphere.[15] It is estimated that the warming associated with a doubling of CO_2 concentrations could be offset if an additional 2 percent were reflected.[16] Such a program of "global dimming" could be effectuated by a fleet

of economic analysis is even more expansive than that of the economists, so that Singer is more economical than the economists. Whatever the case, extending the field of concern to encompass animals makes no difference to the critique developed in this chapter.

[14] The various technologies are reviewed in the benchmark Royal Society report – ROYAL SOCIETY, GEOENGINEERING THE CLIMATE: SCIENCE, GOVERNANCE AND UNCERTAINTY (2009).

[15] *Id.* at 29. The figure includes the effects of sulphate aerosols due to industry.

[16] *Id.* Or 1.8 percent according to T.M. Lenton & N.E. Vaughan, *The Radiative Forcing Potential of Different Climate Geoengineering Options*, 9 ATMOSPHERIC CHEMISTRY & PHYSICS 5539, 5545 (2009).

of high-flying aircraft fitted with special tanks and spraying devices to inject aerosols into the atmosphere on a continuing basis. Alternatively, a long hose held aloft by balloons could perform the task.[17] There are more benign geoengineering proposals, including some forms of carbon dioxide removal, such as direct air capture that does not transform ecosystems, but sulphate aerosol spraying is currently regarded as the cheapest, most effective and most likely method to be deployed,[18] especially in a so-called climate emergency.

3. CONCEPTION OF THE EARTH

Analysis of the ethics of climate change is built on a particular conception of the earth as a collection of discrete ecosystems or components that can be conceptually grasped. The assumed discreteness and well-defined properties of these systems allow for the idea that technological intervention aimed at manipulation can generate certain defined outcomes. It is a cybernetic conception of the earth as a set of functional systems that are subject to control.

Up to a point, this conception works when applied to particular ecosystems, where an argument can be made that nature can be sufficiently understood and regulated. But earth-system science shows that this conception is especially misleading when trying to understand climate change and planet-wide interventions such as solar radiation management. Several factors come into play.

First, solar radiation management envisages manipulation of the flow of primary energy to the planet as a whole, energy that sustains all living things and ecosystems. The atmosphere acts as the mediator between sun and earth, transferring heat and mass to the biosphere, the hydrosphere (the planet's mass of water), and the cryosphere (the ice masses).[19] By influencing the planet's energy balance, solar radiation management will affect all ecosystems and their interactions. It represents a leap to something entirely new in human history. Beyond deliberate management and exploitation of particular resources or geographical areas, and beyond the

[17] *See* Clive Hamilton, *The Return of Dr Strangelove: The Politics of Climate Engineering as a Response to Global Warming*, June 2010, *available at* http://www.clivehamilton.net.au/cms/media/documents/articles/dr_strangeloves_return.pdf (last visited Aug. 15, 2011).

[18] *See, e.g.*, Scott Barrett, *The Incredible Economics of Geoengineering*, 39 ENVTL & RESOURCE ECON. 45–54 (2008); Martin Bunzl, *Geoengineering Research Reservations*, Presentation to AAAS, Feb. 20, 2010, San Diego, http://sites.google.com/site/mbunzl/GeoengineeringResearchReservations.docx?attredirects=0 (last visited Aug. 15, 2011).

[19] On the relationship to the biosphere, the hydrosphere, and the cryosphere, see, for example, WORKING GROUP I, INTERGOVERNMENTAL PANEL ON CLIMATE CHANGE, IPCC FOURTH ASSESSMENT REPORT: CLIMATE CHANGE 2007: THE PHYSICAL SCIENCE BASIS, UNDERSTANDING AND ATTRIBUTING CLIMATE CHANGE (2007), *available at* http://www.ipcc.ch/publications_and_data/ar4/wg1/en/spmsspm-understanding-and.html.

unintentional degradation of land, rivers, and oceans, solar radiation management seeks to take control of and regulate the atmosphere and climate of the planet as a whole.

On this point, and the five that follow, I ask the reader to take note of the sentiments stimulated by this fact. For some, the Promethean nature of solar radiation management arouses deep misgivings about human capacities.

Second, climate science has shown us that the climate system is extremely complex both in itself, and because changes in it cannot be isolated from changes in the other elements of the earth system.[20] Thus it is well understood (but nowhere answered) that sulphate aerosol injection, although probably effective at suppressing warming, would do nothing to slow the acidification of the oceans. Indeed, if by relieving pressure to reduce emissions, global dimming meant carbon emissions grew more quickly, then it would lead to faster acidification. Ocean acidification, inter alia, interferes with the process of calcification or shell-formation on which a wide array of marine animals – including corals, crustaceans, and molluscs – depend for their survival.

For some, recognizing the mystifying complexity of the earth provokes a sense of trepidation at the thought of interfering with it.

Third, as it is not possible to carry out a test of the effects of sulphate aerosol spraying on the global climate system, any deployment will be embarked upon under conditions of great uncertainty.[21] The risks are multiplied by the fact that scientists will be unable to observe the effects of global dimming for at least ten years into the program because many years of data will be needed in order to separate the effects of aerosol spraying from other influences on global climate.[22] It would matter, for example, if the program were begun during an El Nino event.[23] If, after at least a decade of suppressed warming, it is decided that solar intervention was a bad idea, it would in all likelihood be impossible to stop it for fear of the so-called termination problem, the rapid rebound of global temperatures.[24]

[20] One indicator of this complexity is the huge size of the mathematical models that are being built in an attempt to simulate the global climate. The most powerful computers in the world take several days to complete a single model run.

[21] It is feasible to test aerosol delivery devices but no test can indicate how a deployment program would affect the global climate.

[22] Bunzl, *supra* note 18; Hauke Schmidt, Presentation to a Workshop on Climate Engineering, Institute for Advanced Sustainability Studies, Potsdam, June 9, 2011. Estimates of the cooling caused by a large eruption such as Mount Pinatubo vary widely (from 0.2°C to 0.5°C) because of the difficulty of separating in the data the influence of the eruption from other influences on world temperatures.

[23] Hauke Schmidt has asked: If sulphate aerosol spraying had begun in 1998, a year of record temperatures, would the fall in temperatures in 1999 have been attributed to solar intervention?

[24] Because a program of aerosol spraying would suppress warming within a year or two, any sudden termination of the program would see global warming rise very rapidly to the levels it would otherwise have reached. *See* ROYAL SOCIETY, *supra* note 14, at 35.

For some, the idea of going into solar regulation blind, and the specter of being unable to stop, stirs feelings of horror and an intimation that we would pay dearly for our audacity.

Fourth, the earth system that solar radiation management would seek to control is marked not only by complexity but also by nonlinearities. The "tipping points" that define rapid shifts from one climate state to a quite different one are not well understood, but two facts are known well enough.[25] First, the dangers of tipping points are not theoretical but are of immediate concern. We may have crossed one or two of them already, and we will likely cross two or three more if the temperature reaches 4°C above preindustrial levels,[26] as some now project before the end of the century.[27] In addition, tipping points generate irreversible changes, not just to the climate but to the biosphere. The idea of smooth trade-offs between costs and benefits implicit in the utilitarian framework cannot easily accommodate irreversible impacts. What is a lost species or ecosystem worth? The rate of extinction today is 100 to 1,000 times higher than the natural level, due increasingly to human-induced climatic change.[28] It is expected that up to 30 percent of all mammal, bird, and amphibian species will be threatened with extinction this century.[29]

For some, the abrupt nature of climate change intimates that the earth operates according to its own laws whose unpredictability mocks our plans for control and makes us shrink before the power of natural forces.

Fifth, apart from the uncertainties, unknowns, and threshold effects arising from the complexity and nonlinearity of the earth system, the dominant fact is that CO_2 persists in the atmosphere for many centuries. So it is possible – indeed, likely – that before the larger impacts of warming are felt, humans will have committed future generations to an irreversibly hostile climate lasting a thousand years.

For some, recognizing that what we are doing commits the future inhabitants of the planet to a transformed and less friendly climate rouses a sense of shame for failing to fulfill our responsibilities.

Finally, unless accompanied by sharp reductions in emissions, a continuing program of aerosol spraying would entail the more-or-less permanent transformation of

[25] T.M. Lenton et al., *Tipping Elements in Earth's Climate System*, 105 PROCEEDINGS NAT'L ACAD. SCI. 1786–93 (2008).

[26] *Id.*

[27] *See, e.g.*, K. Anderson & A. Bows, *Reframing the Climate Change Challenge in Light of Post-2000 Emission Trends*, 366 PHI. TRANS. R. SOC'Y. A 3863–82 (2008), and the overview in HAMILTON, *supra* note 3, at ch. 1.

[28] Johan Rockström et al. *A Safe Operating Space for Humanity*, 461 NATURE 472–75 (2009). This is a summary version of Johan Rockström et al., Planetary Boundaries: Exploring the Safe Operating Space for Humanity, Stockholm Resilience Centre, 2009, *available at*, http://www.stockholmresilience.org/download/18.8615c78125078c8d3380002197/ES-2009-3180.pdf

[29] Rockström et al., *supra* note 28.

the chemical composition of the earth's atmosphere, a kind of "chemotherapy" for an ailing atmosphere. For many millions of years the temperature of the earth and the amount of carbon dioxide in the atmosphere have moved together, with rises or falls in one followed by rises or falls in the other.[30] Solar radiation management is an attempt to sunder this primordial link. It is the first conscious formulation of a "planetary technology," a plan to take control of and regulate the earth's climate system as a whole.

For some, the idea that humans in the twenty-first century should make themselves a planetary force of geological scale is supremely reckless and invites retribution.

4. CALCULATION VERSUS HUMILITY

How do these facts emerging from earth-system science change the way we think about geoengineering? For the consequentialist, each of these facts – the attempt to regulate primary energy flow, unfathomable complexity, untestability of solar radiation management, irreversible tipping points, permanence of a changed climate, and interference in geological processes – becomes a "risk"; the science just provides better data to be fed into the calculations that allow ethical conclusions. Those consequentialists more in tune with the zeitgeist of technological hubris respond to the reluctance of the earth to submit to human mastery as a spur to more cunning and greater efforts. So when it is pointed out that sulphate aerosol spraying may suppress warming but will not prevent ocean acidification, some geoengineers immediately look for a solution to this "side effect," proposing that we develop a program of adding lime to the oceans in order to return them to a state more friendly to our interests.[31]

Less-hubristic consequentialists may come to accept that the risks and uncertainties are so pervasive that cost–benefit calculation is unable to reach plausible decisions about what to do. This instrumentalist dumbfounding leaves them applying despairing phrases such as "diabolical" and "a devil's stew."[32] But as long as one cleaves to the cybernetic conception of earth and the authority of

[30] Those rises and falls have sometimes been triggered by other events, such as volcanic eruptions and asteroid impacts, so that at times increases in CO_2 have followed warming. Although climate deniers present this fact as if it disproves warming due to anthropogenic carbon emissions, the reverse conclusion should be drawn – by showing that warming changes the biosphere so as to release more CO_2, we have a positive feedback effect that exacerbates warming caused by humans.

[31] David Victor, *On the Regulation of Geoengineering*, 24(2) OXFORD REV. ECON. POL'Y 327 (2008).

[32] The possibility of catastrophe, including the destruction of civilization, has led Martin Weitzman, an economist with a superior grasp of climate science but limited insight into the worldview in which economics is grounded, to conclude that this factor overwhelms all others in trying to assess the threat of climate change. Martin Weitzman, *On Modelling and Interpreting the Economics of Catastrophic Climate Change*, REStat Final Version, July 7, 2008.

rational calculation there is nowhere for the humble consequentialist to go. The recalibration of risks on the basis of new evidence does not challenge the objectification of the earth or the calculating mode of reaching ethical conclusions by autonomous subjects.

For others, recent developments in earth-system science generate or reinforce a quite different conception of the earth and the ethics of climate change, one that stimulates a very specific moral sense, that of humility in the face of nature. What do we mean by humility in the face of nature? We feel humility when we recognize our own intellectual, physical, and moral limitations and acknowledge a greater power than ourselves.[33] It requires us to temper our self-belief, to acknowledge limits to our ability to control the environment, to accept our insignificance as actors in the cosmos, and to abandon the belief that our future is in our own hands. In the past, the chief grounds for humility has been acceptance of the infinitely greater power of a mysterious and omnipotent god. I am suggesting that earth-system science has revealed that the earth as a whole, our living environment, is vastly more complex, enigmatic, and uncontrollable than we had come to believe, and that taking in these facts causes us to cease thinking we can master the earth and to scale back our ambitions. It means recognizing that the power relation between humans and the earth is the reverse of the one we have assumed for three centuries. In Sections 6 and 8 of this chapter I will argue that this challenges not just our conception of the earth but our understanding of ourselves as moral subjects.

One could contend that my argument is essentially the same as Michael Sandel's criticism of genetic enhancement.[34] Sandel argues that it is the gifted character of human capacities and potentialities that incites a natural regard, and that there is something hubristic and unworthy about attempting to overrule or improve upon this gift. The urge behind both genetic engineering and geoengineering is "a Promethean aspiration to remake nature, including human nature, to serve our purposes and satisfy our desires."[35] However, whereas Sandel finds the source of humility in gratitude for what we have been given, I find the source of humility in acceptance of our limitations in the face of the superior power, complexity, and enigmatic character of the earth.

As consequentialist ethics judges the rightness of an action by the sum of effects on humans (and perhaps other sentient beings), any moral feeling can come about only after the act of calculation. How the numbers turn out tells us how we should feel; if the costs of an action exceed the benefits, then perhaps we are permitted to

[33] "Humility" derives from the Latin *humilis*, which also means grounded or from the earth (related to humus).

[34] Michael Sandel, *The Case against Perfection: What's Wrong with Designer Children, Bionic Athletes, and Genetic Engineering, in Human Enhancement* (Julian Savulescu & Nick Bostom eds., 2009).

[35] *Id.* at 78.

feel indignant.[36] Economists are unapologetic about this; the most ethical course is the best one determined by summing the value of the costs and benefits, perhaps weighted by risks, and maybe with some account of distributional effects. The process of rational calculation is especially attractive to those who feel a greater need for a sense of control.

In the case of solar radiation management, the exclusion of moral feelings toward the object of analysis is easier if the object in question, in this case the earth as a whole, is objectified, that is, regarded as a separate entity – abstract, removed, emotionally distant, and of no ethical concern except insofar as the object can satisfy one's own needs. The objectification of the earth means regarding it as a collection of resources that have instrumental value only, that is, value only as means to human ends. Viewing the earth in instrumental terms, so that the ethics of acting on it are to be judged purely by their effects, requires a certain wonderlessness and estrangement from the earth.

It is important to stress that a non-consequentialist position does not ignore the consequences of an act or a choice. All acts or omissions have consequences. What matters is the mind that is brought to the decision. The consequentialist maintains that geoengineering is (or could be) a "good thing" after all of the positive and negative effects are assessed. This weighing of effects requires that all plans are commensurable so that one can be traded off against another. More precisely, for the consequentialist they can be traded off without feelings of guilt, regret, or anguish.[37] We all at times have to make forced choices, but what makes some choices forced is that the decision entails a moral struggle. For the consequentialist no choices are forced because all effects can be traded off on rational grounds. There are no necessary evils.

Paul Crutzen's call for investigation of geoengineering did not spring from an extreme case of instrumentalism and a cavalier faith in human mastery but from "despair,"[38] a deep anxiety about the failure of the world to act on climate change. He made his call because he is one among a small number of scientists who fully appreciate the implications of humanity's failure to act. For him, geoengineering may be an evil, but it may be the lesser one. What has now become apparent is that the recourse to geoengineering has been taken up by many who do not share Crutzen's understanding of the implications of climate science or his despair. For

[36] Within a worldview built on the idea of mastery over the environment, humility is seen as an expression of weakness. After all, we adopt a position of humbleness before forces we accept are more powerful and beyond our control. Yet if we bring to mind those historical figures renowned for their humility, we recognize that they were also pillars of strength.

[37] On this see Charles Taylor, *Review of The Fragility of Goodness by Martha Nussbaum*, 18(4) CAN. J. PHILOSOPHY 807 (Dec. 1988).

[38] Paul Crutzen, personal communication.

them geoengineering is not the lesser of two evils, but a possible alternative strategy for better meeting human goals. If geoengineering were an evil it could not be turned into a good through cost–benefit analysis.

5. PREFERRING THE NATURAL

As it invests its faith in rational human decision making, consequentialism is intrinsically predisposed to elevate the power of humanity over that of nature. Central to its position is the rejection of the idea that the natural exercises any sort of ethical pull; it must do this because any such ethical pull would be a source of moral authority outside the realm of human calculation. So the popular belief that there is something special about the natural world, perhaps because it is delicately balanced and benevolently configured, is incorrect because natural systems are both inherently unstable (so that human-induced changes are not exceptional) and robust against human interference. Powell et al., for example, refer to research that apparently shows that human interference in natural systems has fairly weak ecological impacts.[39] The point of these claims seems to be to establish the case that, because what humans do cannot disturb the delicate balance of nature, because one does not exist, the risks of intervention are lower than many believe.

Earth-system science shows otherwise. Not only has it destabilized the idea of the earth as a knowable and controllable system but it supports the notion that the natural has a privileged position. How? It is true that over geological time scales the earth's climate system has been highly variable. Yet the last ten thousand years, the epoch known as the Holocene, has been a period of unusual stability for the earth's environment.[40] This time of benevolent constancy has permitted human civilization to flourish.[41] As *Homo sapiens* spread across the continents, settlement was heavily influenced by the climates they found; it is not accidental that deserts and the Antarctic are largely uninhabited and most cities are located near rivers and oceans. The infrastructure for nearly seven billion people to live as they do today has taken several hundred years to develop (a few thousand if we include agriculture), and has been possible because of the relatively stable and sympathetic climate that marks the Holocene.

Nor is it true that this stable and benevolent climate is resilient against human interference. Contrary to claims that the natural world is robust in the face of human interference, geoscientists are now arguing that humans have so transformed the face of the earth as to justify the naming of a new geological epoch to succeed the

[39] Powell et al., *supra* note 9.

[40] For example, W. Dansgaard et al., *Evidence for General Instability of Past Climate from a 250-Kyr Ice-Core Record*, 364 NATURE 218–20 (1993).

[41] Rockström et al., *supra* note 28.

Holocene. The Anthropocene is defined by the fact that the "human imprint on the global environment has now become so large and active that it rivals some of the great forces of Nature in its impact on the functioning of the Earth system."[42] Ellis writes:

> [T]he terrestrial biosphere is now predominantly anthropogenic, fundamentally distinct from the wild biosphere of the Holocene and before.... nature is now human nature; there is no more wild nature to be found, just ecosystems in different states of human interaction, differing in wildness and humanness .... by the latter half of the twentieth century, the terrestrial biosphere made the transition from being shaped primarily by natural biophysical processes to an anthropogenic biosphere in the Anthropocene, shaped primarily by human systems.[43]

The most important features are the huge increase in human numbers, up from eight hundred million in 1750 to nearly seven billion today, and the transformation of the atmosphere due to anthropogenic greenhouse gas emissions.[44]

Although the Holocene was relatively stable, the Anthropocene is likely to be very unstable, depending on decisions made by humans. In a landmark intervention in 2009, twenty-seven experts wrote in *Nature*:

> Many subsystems of Earth react in a nonlinear, often abrupt, way, and are particularly sensitive around threshold levels of certain key variables. If these thresholds are crossed, then important subsystems, such as a monsoon system, could shift into a new state, often with deleterious or potentially even disastrous consequences for humans.[45]

The writers are referring mainly to climate tipping points. Contrary to the comforting conception of robust nature, these scientists believe the upheaval of the Anthropocene "could see human activities push the Earth system outside the stable environmental state of the Holocene," and the focus on past resilience may "lull us into a false sense of security because incremental change can lead to the unexpected crossing of thresholds that drive the Earth System."[46] Abrupt changes are those that happen too quickly for humans and some other species to adapt.[47]

[42] Will Steffen et al., *The Anthropocene: Conceptual and Historical Perspectives*, 369 PHIL. TRANSACTIONS ROYAL SOC'Y 842–67 (2011).

[43] Erle C. Ellis, *Anthropogenic Transformation of the Terrestrial Biosphere*, 369 PHIL. TRANSACTIONS OF THE ROYAL SOCIETY A 1010–35 (2011).

[44] Jan Janzalasiewicz et al., *The New World of the Anthropocene*, 44 ENVTL. SCI. & TECH. 2228–31 (2010).

[45] Rockström et al., *supra* note 28.

[46] *Id.*

[47] *See especially*, National Research Council, *Abrupt Climate Change: Inevitable Surprises*, Committee on Abrupt Climate Change, National Research Council (2002).

6. THE MEANING OF FACTS

How do we react to these startling new facts – the arrival of a new geological epoch under human influence and the dangerous instability of the Anthropocene as compared to the Holocene? Do we attempt to quantify the risks they pose and incorporate them into a cost–benefit analysis, or do they cause us to examine our presuppositions about the relationship of humans to the earth? In other words, should we not reflect on what these new facts *mean*? The conception one has of the world (and one's place in it) carries with it sentiments about the earth beyond utilitarian thinking. Before risks are calculated one asks what is worth risking and whether we have the mandate to take certain sorts of risks.

This is not merely a contrast between people with differing personalities. The difference arises from divergent understandings of the nature of the world and the self – what the earth *is*, and what a human being *is* – so that we are contrasting what might be called ontological arrogance with ontological humility. This idea is reflected, although only indirectly, in the distinction drawn by psychologists between those with an *independent self-construal* – whose conception of self emphasizes individual uniqueness and values autonomy and self-enhancement – and those with a *metapersonal self-construal* – whose self is in some sense inseparably connected to all living things or some wider notion of the earth or cosmos.[48] These are not so much personality characteristics but ways of experiencing the self. Studies show that the kind of self that thinks about the ethics of climate in an instrumentalist way is historically and culturally distinctive.[49]

Here it is vital to understand the way scientific arguments are used to establish an ethical position. Those who argue that the delicate state of nature demands that we "tread lightly on the earth" draw on ecological science not as a form of proof but as a means of evoking a certain sensibility toward nature – one of respect, humility, and even reverence. Images of "nature in the balance," "the majesty of whales," and "the blue planet" are symbols that draw attention to the kind of relationship humans have to the natural world. How the individual arrives at a position on the scale from extreme instrumentalism justifying domination of nature to extreme reverence inviting humility – from, say, Gary Becker (the Nobel Prize–winning economist whose economic analysis of the family defined love as "a non-marketed household commodity) to Mahatma Gandhi – is a complex question not considered here, although we know there are strong cultural factors involved.

[48] *See* the discussion in HAMILTON, *supra* note 3, at ch. 5.

[49] Wendi Gardner, Shira Gabriel & Angela Lee, *"I" Value Freedom, but "We" Value Relationship: Self-Construal Priming Mirrors Cultural Difference in Judgment*, 10 PSYCHOLOGICAL SCI. 4 (July 1999).

The critique of climate ethics is in fact a dispute about what "ethics" means. For those who reject narrow consequentialism, the idea that the ethical can be decided from instrumental calculation is itself unethical. For narrow consequentialists it is always ethically justified to engineer a different climate, even though the process of calculation may show that it is imprudent. The procedure is analogous to resolving the ethics of destroying a sacred site by asking how much money the traditional owners would be willing to accept as "compensation." Simply posing the question this way puts the issue outside genuine ethical judgment. The question itself is morally offensive and, when posed, elicits not a calculative reflex but a sense of outrage.

One of the most commonly expressed ethical concerns about climate engineering arises from the possibility that the availability of an alternative to carbon abatement will reduce the incentive to cut emissions. Borrowed from private market behavior, this "moral hazard" argument is framed in consequentialist terms – as solar radiation management is likely to be environmentally less effective (especially as it does not reduce, and may hasten, ocean acidification), to the extent that political leaders succumb to the temptation to avoid abatement measures and take the easy way out, solar radiation management is ethically dubious.[50] But there is also a non-consequentialist moral hazard objection – geoengineering may facilitate the continuation of bad behavior and is therefore wrong. Bad behavior is bad not only for its harmful consequences but also because of its selfish or malevolent motives. Exxon's environmentally damaging activities and support for climate denial would be rewarded by resorting to climate engineering. If we have a responsibility for the damage we cause, then geoengineering may allow us to opt out of our responsibility for causing climate change. The wrong would be compounded if rich countries with high emissions pursue climate engineering instead of abatement. Solar radiation management would entrench the failure of the North in its duties toward the global South. This is another way of making the case that what matters ethically about geoengineering is not only the outcome but also the human disposition it reveals.

[50] Although moral hazard is usually listed as a potential problem, in my view the pressure to make climate engineering a substitute for abatement will prove irresistible. Already, representatives of the fossil fuel industry have begun to talk of geoengineering as a substitute for carbon abatement. The popular book SUPERFREAKONOMICS insists that the prospect of solar radiation management renders mitigation unnecessary: "For anyone who loves cheap and simple solutions, things don't get much better," STEVEN D. LEVITT & STEPHEN J. DUBNER, SUPERFREAKONOMICS: GLOBAL COOLING, PATRIOTIC PROSTITUTES, AND WHY SUICIDE BOMBERS SHOULD BUY LIFE INSURANCE (2009). For anyone who recognizes the dangers of technological hubris, things do not get much cruder. And Republican presidential candidate and former House Speaker, Newt Gingrich, has declared: "Geoengineering holds forth the promise of addressing global warming concerns for just a few billion dollars a year. Instead of penalizing ordinary Americans, we would have an option to address global warming by rewarding scientific invention … Bring on the American ingenuity."

7. IDENTIFYING "WINNERS AND LOSERS"

When a consequentialist framework is brought to the ethics of geoengineering it is natural to identify winners and losers. For example, it is sometimes claimed that citizens of Canada and Russia will benefit overall from global warming.[51] Those who imagine themselves basking in a more temperate climate are likely to be in for a rude shock because the effects on people of climate change will arise not so much from gradual warming but from extreme events. Russian enthusiasm for warming cooled in 2010 after the unprecedented summer heat wave, drought, and devastating forest fires.[52]

The scientific question is not whether an altered climate would be better or worse (the usual consequentialist frame) but whether it would be safe or dangerous. The objective of the 1992 UN Framework Convention on Climate Change is "to prevent dangerous anthropogenic interference with the climate system."[53] The boundary between safe and dangerous levels of warming is believed to be "what is required to avoid the crossing of critical thresholds that separate *qualitatively different* climate system states."[54] It is in the nature of the climate system that scientists have found it difficult to decide on a safe level of climate change.[55] The European Union adopted the objective of limiting warming to 2°C above preindustrial levels, but many scientists believe that is too dangerous.[56] Some advocate limiting CO_2 concentrations to

[51] *See, e.g.*, Gregg Easterbrook (e.g., http://blogs.reuters.com/gregg-easterbrook/2010/07/23/on-top-secrets-and-climate-change/). Easterbrook's suggestion that readers should ask of global warming "What's in it for me?" may be morally equivalent to urging others to profiteer in a disaster zone. It may be worse because it helps to create the disaster zone.

[52] The scale of the 2010 heat wave is discussed here: http://www.hzg.de/science_and_industrie/klimaberatung/csc_web/010253/index_0010253.html.en (last visited Jan. 15, 2012). Moreover, it is not possible to isolate Canada and Russia from the rest of the world. A financial collapse in the United States or China induced by extreme weather would cause global dislocation. Food price rises because of crop failures in major suppliers cascade through all markets. And a water-war between India and Pakistan could have worldwide fallout.

[53] United Nations, United Nations Framework Convention on Climate Change, Bonn, Germany (1992).

[54] Rockström et al., *supra* note 28.

[55] *See, e.g.*, Stephen H. Schneider & Janica Lane, *An Overview of "Dangerous" Climate Change, in* AVOIDING DANGEROUS CLIMATE CHANGE (Hans Joachim Schellnhuber, Wolfgang Cramer, Nebojsa Nakicenovic, Tom Wigley, and Gary Yohe, eds., 2006), *available at* http://stephenschneider.stanford.edu/Publications/PDF_Papers/Schneider-lane.pdf

[56] "Beyond 2°C, the possibilities for adaptation of society and ecosystems rapidly decline with an increasing risk of social disruption through health impacts, water shortages and food insecurity." K. Richardson et al., *Synthesis Report. Climate Change: Global Risks, Challenges & Decisions. Summary of the Copenhagen Climate Change Congress*, University of Copenhagen, Copenhagen, Mar. 10–12, 2009, at 12. Paleoclimatologists have shown that the Antarctic ice-sheet began to form when CO_2 concentrations fell below 450 ppm, suggesting it would begin perhaps irreversible melting once the concentration rose above that level.

350 ppm, associated with warming of around 1.7°C above the preindustrial average, although even here there are risks, and value judgments are necessary.[57] Despite the difficulties, the idea of a threshold above which warming would be dangerous means that a "safe minimum standard" is a more appropriate decision rule, with cost–effectiveness analysis replacing a cost–benefit framework.

Once again we ask: What is the effect on us of the lack of certainty over what constitutes a safe level of warming in a nonlinear world and the potentially very harmful or even catastrophic consequences of exceeding the safe level? Of course the facts call for greater caution but do they cause us only to calculate differently, to recalibrate risks, or do we reconsider our understanding of the earth and our approach to it? The complex and volatile interactions of earth systems, and our meager understanding of their workings, means that the idea that humans can choose an optimal global average temperature and "set the thermostat" at that level appears increasingly deluded. Any "optimal" degree of warming may prove to be only a temporary way station on a path to more warming, and it is well established that the amount of damage caused by warming is an increasing function of the degree of warming.[58]

8. TECHNOLOGICAL THINKING

> [W] hen we try to see and conceptually come to terms with a certain phenomenon we also have to pay close attention to *how* we approach it ... For there will always be a risk that we let ourselves be guided by a thought model which in the end makes us blind precisely to the phenomenon which we are trying to interpret and understand.[59]

I have argued that the beliefs that there is no prima facie justification for attempting to preserve the current climate and that the optimal temperature should be set through a process of calculation reflect a particular conception of the world and the nature of humans that emerged with the Scientific Revolution and Enlightenment philosophy. It was not just a new conception of the earth that emerged but a new

[57] http://www.nature.com/climate/2009/0912/full/climate.2009.124.html (last visited Jan. 1, 2012); J. Hansen et al., *Dangerous Human-Made Interference with Climate: A GISS ModelE Study*, 7 ATMOSPHERIC CHEMISTRY & PHYSICS 2287–12 (2007). Warming of 1.7°C is an average that may be "safe" for most countries or regions and dangerous for others.

[58] Current climate models "do not include long-term reinforcing feedback processes that further warm the climate, such as decreases in the surface area of ice cover or changes in the distribution of vegetation. If these slow feedbacks are included, doubling CO_2 levels gives an eventual temperature increase of 6°C (with a probable uncertainty range of 4–8°C). This would threaten the ecological life-support systems that have developed in the late Quaternary environment, and would severely challenge the viability of contemporary human societies." Rockström et al., *supra* note 28.

[59] Hans Ruin, *Technology as Destiny in Cassirer and Heidegger, in Form and Technology: Reading Ernst Cassirer from the Present* (Aud Sissel Hoel & Ingvild Folkvord eds., 2010).

conception of the human being as a self-legislating subjective entity, who is distinct from the rest of the world, is guided by its cognitive abilities, and acts to pursue its own interests. The proposed deployment of solar radiation management to offset the effects of anthropogenic global warming is the culmination of the transition to the mechanical conception of nature and the parallel emergence of philosophies built on the idea of the autonomous rational subject exercising control over an inert environment.

The type of thinking embedded in the framework of systems analysis, risk assessment, and cost–benefit analysis can be called "technological thinking." Technological thinking understands the world as a collection of more-or-less useful resources. According to this view technology transforms potentially useful things into useful things without asking about the origins of the world as a collection of potentially useful things. Modern technology therefore challenges nature to supply materials and energy for extraction and storage, to open itself up as possibilities for human progress, providing a path to the fulfillment of human existence. As such, modern technology reveals something essential to the nature of modern humans – the determination to shape the world around us to suit our desires, desires that have no limit.

Plans to engineer the earth through the deployment of contrivances to manipulate the atmosphere represent the fulfillment of three-and-a-half centuries of objectification of nature. The earth as a whole is now represented no longer simply as a collection of objects but as an object in itself, one open to regulation through the "management" of the amount of solar radiation reaching it. Earth-as-object also underlies the idea that we can adjust the volume of greenhouse gases in the atmosphere to a level calculated to be "optimal." Climate engineering represents a conscious attempt to overcome resistance of the natural world to human domination, the last great stride toward total ascendancy. Yet, as I have already suggested, the sheer complexity and unpredictability of the natural world resists attempts at total mastery.

In order to evoke a sense of the way in which climate change and geoengineering prompt a reconsideration of technological thinking, I have pointed to the emerging understandings of the world and our role in it emanating from earth-system science. The idea of the Anthropocene is put forward because humans are now the dominant force in the global biosphere. Yet the earth under the Anthropocene is not mere putty to be shaped at will by humans. We have already seen in the discussion of thresholds, uncertainties, intricate interactions, and unknowns that the earth does not behave obediently according to the systematic, predictable frame we have projected onto it since the Scientific Revolution.

At the same time, climate change is destabilizing this understanding because science itself is pointing toward the inherent uncontrollability, and perhaps the unknowability, of the natural world. We have seen how global warming is affecting the biosphere, the hydrosphere, and the cryosphere. Scientists are now beginning to grasp the way in which human-induced climate change can affect the lithosphere (the outer crust of the earth) and the geosphere (the deeper structures of the planet), including the movement of tectonic plates. It is now emerging that, by shifting the distribution of ice and water over the surface of the earth, human-induced global warming is likely to provoke geological and geomorphological responses, including seismic, volcanic, and landslide activity.[60] Changes in the seasonal snow-load, for example, affect seismic activity in Japan by changing the compression on active faults. According to a recent scientific review of the field, in Iceland and Alaska "melting of ice in volcanic and tectonically active terrains may herald a rise in the frequency of volcanic activity and earthquakes."[61] When glaciers melt the earth "rebounds." The earth's crust may rise by hundreds of meters with a decline in ice load of one kilometer. Moreover, although anthropogenic effects on the climate and biosphere are far more important impacts, the melting of polar ice due to global warming can be expected to alter slightly the earth's rotation speed and its orientation in the solar system.[62]

So we can see that, far from being a phenomenon limited to changes in the weather, human-induced climate change is bringing everything into play in ways that appear increasingly complex and beyond control. Yet plans for solar radiation management challenge the earth as a whole to present itself to us as a system that can be understood, manipulated, and regulated. These facts call not for more calculation of risks but for a radical change in the modern conception of the earth and a repudiation of the idea of the modern subject that founds climate ethics. It is a call for a new kind of subject, the heteronomous subject who recognizes sources of moral authority beyond human calculation in the understanding of the world suggested to us by earth-system science.

Instead we now have governments and scientific societies beginning to deliberate on the "governance" of solar radiation management,[63] that is, the appropriate political institutions for regulating the amount of light reaching the planet. Since the formation of the earth 4.5 billion years ago the amount of solar radiation reaching it has been determined by the sun mediated by the earth's atmosphere. It seems we are no longer happy with the arrangement and want to assume control ourselves. Although individuals may endorse a program of sulphur injections as a regrettable

[60] For a review, see B. McGuire, *Potential for a Hazardous Geospheric Response to Projected Future Climate Changes*, PHILOSOPHICAL TRANSACTIONS OF THE ROYAL SOCIETY A (Mar. 28, 2011).

[61] *Id.*

[62] *How the Japan Earthquake Shortened the Earth Day*, http://www.space.com, (last visited Dec. 25, 2011).

[63] *See* the Royal Society's Solar Radiation Management Governance Initiative (http://www.srmgi.org/).

necessity arising from our previous failure, solar radiation management nevertheless represents the extension of a relentless process of mastery rooted in entrenched social and economic institutions.

Is there any force that can temper this drive for mastery? Are we able today, after all of our astonishing technological accomplishments, to understand the meaning of the ancient Greek stories warning of hubris – the stories of Icarus, of Narcissus, and of Achilles debasing Hector? Most tellingly, are we attuned to the message of the myth of Phaeton who, against all warnings, decided to take control of the forces of the sun, accidentally causing the earth first to freeze then to burn up before he had to be killed? Are the geoengineers modern-day Phaetons, who dare to regulate the sun, and who must be struck down by Zeus before they destroy the earth? Or has the perfection of our rational capabilities forever silenced Nemesis?

ACKNOWLEDGMENTS

This chapter was written while I was a visiting academic in the Department of Philosophy and Centre for the Environment, University of Oxford. I am grateful to Alice Bows, Steve Clarke, Andrew Glikson, Myra Hamilton, Clare Heyward, Andrew Parker, Ingmar Persson, Julian Savulescu, Falk Schmidt, and participants in a seminar at the Oxford Martin School for very helpful comments on a draft of this chapter. Although it has been greatly improved as a result of their comments, all remaining mistakes and misinterpretations remain mine.

3

The Psychological Costs of Geoengineering

Why It May Be Hard to Accept Even If It Works

Gareth Davies

1. INTRODUCTION

When thinking about any controversial policy issue it is helpful to understand the reasons that people have for the stances that they take. If we want to persuade them to change their minds, or even to investigate whether they may have a good point that we have not yet grasped, we have to engage with their actual motivations, fears, aspirations, and interests, and not just the ones attributed to them by those who disagree.

Implicit in this is the idea that the arguments that people may present do not always reveal the entirety of their reasons. When talking about climate change or geoengineering it is common for arguments to revolve around the quantitative and concrete considerations: To what extent is climate change real and caused by human action? What will be the economic and human consequences? What are the risks of geoengineering? What are its costs? Yet, it will be suggested below, these arguments will often be deployed opportunistically. That is to say, the stance that people take on these questions does not in fact explain their attitudes about climate change or geoengineering. Rather, the stances are symptomatic. Most people do not assess the evidence and then use this to found a policy standpoint. Rather, their fundamental attitude is formed by many factors, many fears, and many hopes, and this attitude is then defended with scientific or material arguments that are essentially plucked off-the-shelf from the media or other sources.

The idea that the reasons people may advance for an action may be at best a partial explanation, and that apparently objective and coolheadedly rational explanations may mask more emotional ones, would be banal for a psychologist, or anyone who studies human beings.[1] It is certainly understood by successful politicians,

[1] Amartya Sen, *Rational Fools: A Critique of the Behavioural Foundations of Economic Theory*, 6 PHIL. & PUB. AFF. 317 (1977); Cass Sunstein, *Willingness to Pay versus Welfare*, John M. Olin Law and Economics Working Paper no. 326 (2007) (*available at* ssrn.com).

whose arguments appeal to underlying fears and hopes at least as much as to the quantitative policy-analyst part of our brain. Yet in policy debate this simple observation about human nature is often ignored. Climate change and geoengineering are debated in terms that are often quite divorced from their real psychological importance for most people: the fear, uncertainty, and hope that they may inspire. Such divorced debate has little chance of producing consensus, or even heightened understanding. It bypasses the matters that count.

Rosemary Randall has taken the discussion a step forward by considering climate change and mitigation in terms of loss.[2] She encourages us to understand the real and painful losses that people may suffer as a result of mitigation policies. She points out that there is a Pollyannaish tendency by environmentalists to emphasize the downside of climate change, but to present mitigation as if it is win–win: a move to a world of abundant renewable energy, harmonious sustainable societies, and a happier relationship with nature. These positive visions of a renewable, sustainable world make it truly mysterious why there should be any widespread opposition to them. Yet there clearly is widespread opposition, if only of a rather passive, sullen kind, as such policies are not being adopted at the rates their advocates would like. Clearly, the visions are unconvincing to many, and one of the reasons is that they fail to address the many costs and losses that mitigation will bring. The loss of the freedom to fly long distances and explore the world, to heat one's house to a decadent level, to enjoy the control and privacy of a large and comfortable car – to give a few of the more banal examples – are significant enough to many people that any argument that glosses over them, as if it is self-evident that a world of localism, communality, and warm winter clothes is attractive, loses credibility.

Randall's core insight is that, following the psychology of loss, people need to go through a process of adjustment if they are to be able to accept the losses inherent in mitigation policies, and that facilitating this adjustment process will be at the heart of any successful attempt to persuade the public to adopt climate-friendly policies. This chapter uses that perspective on loss to think about geoengineering, and to analyze attitudes toward it.

The starting point is to borrow some ideas from conventional welfare economics, and policy practices; the idea of cost–benefit analysis (or risk–benefit analysis – essentially cost–benefit analysis taking account of uncertainties) and that of rational self-interest.[3] These economic tools provide a very clear and coherent basis for analysis, especially when stripped of some of the technicality often attached to them and reduced to their underlying insights. First, it is plausible to assume that

[2] Rosemary Randall, *Loss and Climate Change: The Cost of Parallel Narratives*, ECOPSYCH. 118 (2009).

[3] RICHARD WILSON & EDMUND A.C. CROUCH, RISK BENEFIT ANALYSIS (2d ed. 2001); GARY S BECKER, THE ECONOMIC APPROACH TO HUMAN BEHAVIOUR (1976).

people act in a way that they believe serves their interests. Second, as most actions bring advantages and disadvantages, determining what is in one's interest entails balancing of these costs and benefits. The starting point for any policy debate should therefore be to catalog the costs and benefits that result from different choices. This enables us to understand who wins and who loses from these choices, and then to adjust policy to compensate or protect losers, and perhaps to limit gains, according to our sense of justice or according to what is necessary to build a consensus.

This approach has been recently and controversially applied to global policy problems, including climate change, by the Copenhagen Consensus project and its leader Bjorn Lomborg.[4] However, the welfare economics used in the consideration of climate change made only limited use of psychological costs and benefits. This is one of the repeated and fundamental criticisms of modern economics and of ideas such as cost–benefit analysis and rational self-interest: that they exclude important nonmaterial factors, from altruism to intergenerational equity.[5] Yet whether or not these criticisms hold for practical quantitative application of the ideas,[6] they do not hold for the more abstract way in which they are being used in this chapter. At this level they are both highly formal concepts, which in themselves says little about what costs and benefits are involved. There is ample room for interpreting them broadly, if this is appropriate. On such a broad view, altruism can be understood as an attitude that conflates our interest with that of others: we desire to help them, so that their welfare is our happiness.[7] Justice and equity considerations will be costs for an individual if that person cares about these matters: in that case, unjust policies will impose psychological costs upon that individual (will make him or her feel bad). It is entirely realistic that for a given individual the psychological benefits of an altruistic or compassionate policy will outweigh the material costs. Our fundamental interest, it is assumed in this chapter, is our happiness, and it is implausible to think that purely material factors are exclusively relevant, or even dominant, in that.[8] By contrast, it is a psychological truism to suggest that relationships with others, security, self-respect, and such emotional and psychological considerations are the primary influences on happiness, and that any rational self-interested individual will prefer policies benefitting these over policies with purely material benefits.[9]

[4] *See* GLOBAL CRISES, GLOBAL SOLUTIONS (BJORN LOMBORG ed., 2004); SMART SOLUTIONS TO CLIMATE CHANGE (BJORN LOMBORG ed., 2010).

[5] BRUNO FREY, HAPPINESS: A REVOLUTION IN ECONOMICS 15–17 (2008); John Quiggen, *Existence Value and Benefit Cost Analysis: A Third View*, 12(1) J. POL'Y ANALYSIS & MGMT. 195 (1993); John Quiggen, *Altruism and Benefit Cost Analysis*, 36(68) AUSTRALIAN ECON. PAPERS 144 (1997).

[6] *See* Per-Olov Johanssen, *Altruism in Cost-Benefit Analysis*, 2 ENVTL. & RESOURCE ECON. 605 (1992).

[7] *Id.* at 608.

[8] H. MARGOLIS, SELFISHNESS, ALTRUISM AND RATIONALITY (1982).

[9] *Id.*; SUBJECTIVE WELL-BEING: AN INTERDISCIPLINARY PERSPECTIVE (FRITZ STRACK, MICHAEL ARGYLE & NORBERT SCHWARZ eds., 1991).

The focused profit-maximizer who is so often discussed when rational self-interest is mentioned, is not just a myth, but, insofar as we are talking about human beings, a theoretical error. Wealth maximization is not coextensive with rational self-interest. The conflation of the two certainly occurs often in debate and application but this mediocre use of the ideas does not detract from their potential, or from their principled soundness.

The goal of the rest of this chapter is therefore to make a partial cost–benefit analysis of geoengineering, emphasizing in particular the breadth and complexity of the losses that it involves. The suggestion is that attitudes are far better explained by the psychological consequences of climate manipulation than by the economic, climatic, or material ones.

Understanding these psychological consequences enables debate to go forward in two ways. First, it may lead to new evaluations of the cost–benefit balance. Psychological factors should be weighed along with environmental and material ones here. Second, understanding the losses that people are experiencing or are afraid of may help explain those people's policy attitudes, and so may help influence them. Helping those people adapt emotionally to their loss, or simply acknowledging it, or addressing it practically, may result in their changing their stance. Attitudes to geoengineering, as to climate change, may be far more flexible and negotiable than appears if one speaks only about the most obvious and quantifiable climatic and scientific facts.

This approach, described above, to thinking about geoengineering looks very much like what has been described as "shallow ecology."[10] Deep ecology is one of the names for the approach to the environment that values it "for itself," not purely in instrumental or anthropocentric terms.[11] A view that enjoys considerable support in ecological circles, deep ecology regards environmentalism motivated by human interest as shallow in that it sees the nonhuman world as only important insofar as it serves humanity, and as such fails to recognize the "inherent" value of plants, animals, and the environment. Such an anthropocentric approach may justify protection of species or environments where that serves human interests, but the human focus will tend ultimately to lead to exploitation and ecological degradation. Deep ecology also makes a more emotional criticism of instrumental ecology: that it is an impoverished and narrow view of the world. Humanity, it is suggested, cannot realize itself fully if it sees its environment in instrumental terms, as this creates no space to recognize or nurture our essentially "ecological" nature, as one small part of

[10] Arne Naess, *The Deep Ecology Movement: Some Philosophical Aspects*, 8 PHIL. L. INQUIRY 10 (1986); ARNE NAESS, ECOLOGY, COMMUNITY AND LIFESTYLE: OUTLINE OF AN ECOSOPHY (1989); FRITJOF CAPRA, THE WEB OF LIFE 7 (1997).

[11] CAPRA, *supra* note 10; GEORGE SESSIONS & BILL DEVALL, DEEP ECOLOGY 70 (1985).

what is around us.[12] Greater potential happiness lies in according independent value to that environment, and accepting an obligation to respect it. That it misses these deeper points is part of the reason for the label "shallow" accorded to instrumental and anthropocentric environmentalism. A criticism of this chapter could be that it begins from instrumental premises, and so begs at least some of the questions it hopes to examine.

There are two responses to this. One is that this chapter does not take a normative approach to the losses and gains it discusses. If we want to influence people, it helps to understand their reasons, whether or not we agree with them. The presumption that people pursue their interests and value the environment if that serves their interests does not entail approval of that behavior. However, the deep/shallow and inherent/instrumental distinctions are in any case rejected as incoherent. If deep ecology is right in claiming that human happiness lies in valuing the environment for itself (whatever that may mean – a point beyond the scope of this chapter), then it is in our interests to so value it. Rational self-interest may well lead in the direction of a deep ecological world, because rational self-interest is plausibly served by a lifestyle attuned with the world around us and the living things within it, and with our essential emotional needs. The "anti-instrumentalism" argument would be coherent only if the deep ecologists making it claimed that their choice was a painful sacrifice – that they would rather live another kind of life, but dared not contradict some higher moral code. On the contrary, they appear to suggest that their values enrich their life. Deep ecology and rational self-interest are entirely at peace.

2. GEOENGINEERING

"Geoengineering" is the term used to describe intentional manipulation of the climate by human behavior.[13] A number of ways of doing this have been put forward as possible contributions to dealing with climate change. It is suggested that these techniques could cool the world, either by removing carbon dioxide from the atmosphere or by reflecting sunlight and thereby compensating for the warming effect of increasing atmospheric carbon dioxide levels. At the most optimistic, some techniques give reason for hope that they could solve the problem – entirely prevent warming with minimal, or acceptable, side effects – even though establishing

[12] Naess, *supra* note 10, at part 8.

[13] The term is usually traced back to Marchetti: C. Marchetti, *On Geoengineering and the CO2 Problem*, Research Memorandum RM-76-17, International Institute for Applied Systems Analysis, Vienna (1976); *see also* The Royal Society, *Geoengineering the Climate. Science, Governance and Uncertainty*, Report 10/09, RS1636, (2009) at ix. The more precise term "Climate Engineering" is often preferred in recent work: *See* SMART SOLUTIONS TO CLIMATE CHANGE, *supra* note 4, at 6–10.

whether this is actually the case still requires more research.[14] A more widespread view is that some techniques might become necessary or useful as a last resort, or temporary stopgap, preventing catastrophic climatic tipping points being reached, and buying time while humanity adjusts to a low-carbon lifestyle.[15] The conventional wisdom on geoengineering is that it is a very poor second best to conventional mitigation. Reduced carbon emissions are almost universally considered to be a safer and more desirable long-term response to the threat of global warming. Geoengineering, because the climate is so imperfectly understood, might bring very significant, even disastrous side effects, and as such is to be avoided if possible. The cost–benefit analysis only comes out in favor of its use if warming has reached such a point that disastrous consequences are fairly certain without it.[16] The possibility of geoengineering-induced climate catastrophe is unattractive, but more attractive than its global warming–induced probability.

This rhetorical contrasting of the risks of geoengineering against the safety of mitigation encourages a distinct hostility to the former. However, it is apparent that the desirability of geoengineering is at least partly a function of various probabilities: the probability of disastrous side effects if it is deployed, and the probability of disastrous warming if it is not. An attempt to assess these leads to a more sympathetic view. Indeed, if the primary policy concern is protection of the environment in more or less its current state, there appears to be a plausible prima facie case that geoengineering is a desirable policy choice, and should probably be deployed soon.

The techniques of geoengineering that have the most immediate potential, and are the most controversial, are grouped under the acronym SRM, for "Solar Radiation Management."[17] They attempt to reflect back sunlight from the earth, thereby cooling it, just like a sunshade. The two most immediately accessible techniques are the injection of sulfur aerosols into the upper atmosphere, simulating the well-documented cooling effect of volcanic eruptions, and the use of various

[14] *See, e.g.*, Jay Michaelson, *Geoengineering: A Climate Change Manhattan Project* 17 STAN. ENVTL. L.J. 73 (1998); T.M.L. Wrigley, *A Combined Mitigation/Geoengineering Approach to Climate Stabilization* 314 *Sci.* 452 (2006); H.D. Matthews & Ken Caldeira, *Transient Climate-Carbon Simulations of Planetary Geoengineering* 104 PROC. NAT'L ACAD. SCI. USA 9949 (2007); J. Latham et al., *Global Temperature Stabilization via Controlled Albedo Enhancement of Low-Level Maritime Clouds* 366 PHIL. TRANSACTIONS ROYAL SOC'Y 3883 (2008); S. Salter, G. Sortino & J. Latham, *Sea-Going Hardware for the Cloud Albedo Method of Reversing Global Warming* 366 PHIL. TRANSACTIONS ROYAL SOC'Y 3989 (2008).

[15] *See, e.g.*, Royal Society, *supra* note 13, at 44–45; Paul Crutzen, *Albedo Enhancement by Stratospheric Sulfur Injections: A Contribution to Resolve a Policy Dilemma?* 77 CLIMATIC CHANGE 211 (2006); Scott Barrett, *The Incredible Economics of Geoengineering* 39 ENVTL. & RESOURCE ECON. 45 (2008).

[16] Crutzen, *supra* note 15.

[17] *See* Royal Society, *supra* note 13, at 1.

techniques to increase the reflectivity of cloud cover.[18] The latter of these is at a less-established stage of research, and there are still doubts about whether it can be achieved by the methods suggested. However, the striking thing about the sulfur technique is that there is a broad consensus that it is technically feasible, and that the upfront costs would be low, albeit that there would have to be ongoing injections of sulfur to maintain the cooling effect: it would be a continuous process, not a one-off intervention. Suggested methods include squirting the sulfur out of a fleet of ten or so adapted aircraft, and cost estimates vary from a few billion to a few tens of billions dollars per year for the process itself.[19] This is, in the global context, a relatively trivial cost, amounting to a few dollars per capita per year in the developed world. Moreover, there is a scientific consensus that the technique would actually work, and that cooling the earth to pre–Industrial Revolution temperatures would be quite feasible. It is for these reasons that words such as "incredible" have been used in descriptions of the possibilities and economics of SRM.[20]

However, cooling the earth is of little consolation if the side effects are devastating, and the debate about sulfur-based SRM is essentially one about the non-temperature consequences of its use. These are many and various and sometimes frightening, and often uncertain.[21] As the technique would not actually reduce carbon dioxide levels, it would leave the non-temperature consequences of rising carbon dioxide untouched, notably the increased acidification of the oceans. This threatens many aquatic, and ultimately also human, ecosystems, and it is not clear how easy it would be to take countering measures. The technique would undoubtedly result in changes in precipitation patterns, and in local weather systems, so that particular areas might experience more drought, more floods, more extreme weather, or other phenomena that would be particularly threatening on a local scale. There are sophisticated estimates of where and to what extent this is likely to occur, but there is little degree of certainty that they would be accurate, or how the events in question would translate to human or environmental harm. Nor is there much idea at a policy or academic level of how far it might be possible to protect local communities or environments from that harm, this being a question involving considerations of science, politics, law, and social science, and turning on local particularity as much as general commitment.

[18] Latham et al., *supra* note 14; Salter et al., *supra* note 14; Crutzen *supra* note 15; T.M. Lenton & N.E. Vaughan, *The Radiative Forcing Potential of Different Climate Geoengineering Options*, 9 ATMOS. CHEMISTRY & PHYSICS 5539 (2009).

[19] *Policy Implications Of Greenhouse Warming – Mitigation, Adaptation, and the Science Base*, National Academy of Sciences, Committee on Science, Engineering, and Public Policy, ch. 28, at 453 (1992); Alan Robock et al *The Benefits, Risks and Costs of Stratospheric Geoengineering*, 36 GEOPHYSICAL RESEARCH LETTERS L19703 (2009); Royal Society, *supra* note 13, at 32; Crutzen *supra* note 15.

[20] Barrett, *supra* note 15, at 50 (2008).

[21] Royal Society, *supra* note 13, at 28–32 for overview.

As well as this there are what may be called "governance" risks involved in geoengineering generally, and specifically in SRM.[22] Although it would be technically and economically feasible for it to be carried out by a single state, or even a single large corporation, this would create considerable international tensions, and could be interpreted as an aggressive act. It would be unilateral determination of the global climate. As a result it seems clear that it would be desirable for any geoengineering to take place within a fairly robust international governance structure, where consensus could be reached, and within which differences of opinion could be accommodated and accepted. A practical problem is that no such structure exists, at least specifically focused on climatic or environmental issues, and it may be hard to create it. However, the fear more often raised is that if one was created, and geoengineering was carried out under its auspices, this would create such a locus of power and influence that states would be trapped, unable to opt out or walk away, and would inevitably see their autonomy sapped and drained.[23] First, the power to control the global temperature would be of unprecedented import for an international organization, and raise issues of democratic and accountability issues. Second, after starting SRM, it would be dangerous to stop it. Because it does not work by decreasing levels of carbon dioxide, but by countering them, any decision to suddenly stop could result in sudden global warming, which could have more extreme effects than the same warming achieved more gradually.[24] This "rebound" risk would mean that having established an SRM process and framework, states could be held hostage to that framework by fear of the consequences of not maintaining the process. Democratic shortcomings and limitations of sovereignty, as well as nonoptimal decision-making processes, might be locked in. In fact, the fear that SRM would be stopped would perhaps not be the greatest one, as this would harm all parties, and in any case it would be feasible for a state to maintain it unilaterally. However, the degree and type of sulfur injections (or other substances, e.g., nano particles) would inevitably be the subject of ongoing negotiations, and it would not be easy for a state to walk away from these.

All these are good reasons to fear geoengineering, and sulfur-based SRM in particular. Yet there is no reason to think that they represent more frightening risks than those raised by global warming itself. Indeed, there is much similarity. Emission of

[22] David Victor, *On the Regulation of Geoengineering*, 24 OXFORD REV. ECON. POL'Y 322 (2008); John Virgoe, *International Governance of a Possible Geoengineering Intervention to Combat Climate Change*, 95 CLIMATIC CHANGE 103 (2009); Edward A. Parsons, *Reflections on Air Capture: The Political Economic of Active Intervention in the Global Environment* 74 CLIMATIC CHANGE 5 (2006); Thomas Schelling *The Economic Diplomacy of Geoengineering* 33 CLIMATIC CHANGE 303 (1996); Gareth Davies, *Framing the Social, Political and Environmental Risks and Benefits of Geoengineering: Balancing the Hard-to-Imagine against the Hard-to-Measure*, 46(2) TULSA L. REV. 261 (2010).

[23] Victor, *supra* note 22.

[24] Matthews & Caldeira, *supra* note 14.

carbon dioxide is also causing acidification of the oceans, and will result in changes in climate patterns that may have locally, ultimately globally, devastating effects. Addressing this by conventional mitigation will almost inevitably entail constraints on sovereignty, and a degree of international cooperation. It will ultimately trap states within a framework of agreements and supervision that will have to be long term, and from which political and economic considerations will make it hard to walk away.

In fact, on the most important of these issues by far, the environmental consequences, there is broad agreement that although the risks of geoengineering are considerable, they are ultimately less than those of continued global warming: this is inherent in the view of SRM as a last resort.[25] SRM is a less frightening climatic scenario than continued increases in carbon dioxide emissions. This means that whether SRM is desirable, from an immediate environmental point of view, depends on calculations about probabilities. If it is likely that humanity will not, in fact, succeed in reducing emissions in time to prevent disastrous climatic and environmental consequences, then SRM is a desirable choice. If it is likely that humanity (or part of it) is engaging in brinkmanship but will in fact take the measures necessary to prevent such catastrophe, then SRM is an unnecessary risk, and may have the added negative consequence of encouraging delay in emissions reduction, or weakening the case for its necessity.[26] A proper policy decision depends to a large degree on estimates of these chances, alongside factors such as how long it would take to develop and deploy SRM. That is of course far more complex than in the sketch here, as concepts such as "catastrophic climate change" require a more precise and substantial filling in. However, the underlying risk-balancing process involved is quite clear.

For those actively involved in this debate a complicating factor is added by the interdependence of risk and rhetoric: even advocating geoengineering may reduce the chance of mitigation.[27] This is sometimes referred to as the "moral hazard" problem.[28] There is therefore a case to be made for tactically opposing geoengineering in absolute and unconditional terms, in order to raise the sense of impending doom, in the hope that this may increase support for conventional mitigation as the only apparent option.[29] Nevertheless, such a tactic could be quite consistent with awareness that at some point a radical turnaround might be appropriate, to a total embrace of geoengineering, if global warming progressed to such a state that the risk-balance justified

[25] *See supra* note 17.

[26] *See* text to note 30 *infra*.

[27] Alan Robock, *20 Reasons Why Geoengineering May Be a Bad Idea*, 64 BULL. ATOMIC SCI. 14 (2008).

[28] Royal Society, *supra* note 13, at 37–43; Gareth Davies, *Geoengineering–A Critique* 1(3) CLIMATE LAW 429–41 (2010).

[29] Davies, *supra* note 28.

this. Equally, those who do not believe in global warming, or do not find it worrying, might tactically support geoengineering, in the hope that this would distract the public from the relatively expensive and socially disruptive task of reducing emissions.

However, setting these tactical political issues aside, it is suggested that current scenarios on emissions and predicted temperature rise give little reason for confidence that emissions will in fact be reduced to the levels claimed to be necessary to prevent severe climatic disturbances. Given that it is difficult to see that dramatic reductions in global emissions are likely to be achieved in a very short period (if at all), it seems overwhelmingly probable that in the coming decades the earth will experience temperature rises of a sufficient degree that the consequences, in extreme weather events and changed weather patterns, will exceed what is considered likely under a regime of SRM geoengineering.

Of course, to be considered proved such an assertion requires detailed presentation of all kinds of data. That is not the aim of this chapter. However, it is suggested that a more limited assertion can be made on the basis of generally available and accepted knowledge: it is quite possible that – as a result of a failure to reduce emissions sufficiently – the side effects of global warming in the coming decades will be more dramatic than the expected side effects of geoengineering.

The desirability, or not, of geoengineering can be seen to turn on a counterfactual: What will happen without it? How much will humanity reduce emissions of carbon dioxide if we do not geoengineer? That is a difficult question. Political analysis of the likely economic developments in China and India, scientific knowledge of the state of alternative energy sources, and social analysis of public readiness to accept weather and climate consequences are all crucial but complex factors. However, what can be said with reasonable confidence is this: implacable hostility to geoengineering cannot be explained by climatic or environmental concerns. Yet there is great hostility to it among many of those who profess to be environmentalists, or at least to be concerned about the environment and climate change. How can this apparent contradiction be explained?[30]

3. LOSS

Following the rational self-interest idea, the question is how geoengineering may affect the welfare of individuals in ways apart from its direct climatic effects. For convenience this can be looked at in terms of three factors that are relevant to happiness: relative status, security, and hope. This is not a unique or necessary framework for analysis, but it is a possible one: the function of these factors is to order the consequences of geoengineering, not to determine them.

[30] Michaelson, *supra* note 14, at 130–40, offers explanations along similar lines to some of those here.

The use of these factors would not be controversial among social scientists, as research has demonstrated that they are more central to, and more explanatory of, our happiness and welfare than material matters.[31] Within economics, the integration of such emotional considerations into economic theory has been one of the most important and dynamic projects of recent decades. Behavioral economics, now mainstream, looks at the way behavior may be influenced by more complex issues than mere material self-interest, whereas the economics of happiness engages directly with the emotional factors that guide our decisions and preferences.[32]

However, I take the centrality of such psychological factors to welfare as something that hardly needs demonstrating. We can all know it by reflecting upon ourselves, drawing upon our own experience. Moreover, the alternative is almost bizarre. It is extremely unconvincing to claim that bigger televisions or faster cars or even computers and domestic apparatus actually in and of themselves make people happier. Rather, we value these things – if we do – because they enhance our possibilities, status, and security. This offers a better – more plausible and complete – explanation for material aspiration than does an irreducible desire to possess more stuff.

Nevertheless, despite their banality, a few words of explanation about the categories used here may be helpful. Relative status is intended to capture the idea that absolute levels of success matter less than our position in the pecking order. A high level of inequality produces unhappiness among the disadvantaged even if absolute levels of wealth are high throughout the society.[33] The competitive instinct, and the fear of being left behind, mean that keeping up, or getting ahead, count for more than just having a lot. Yet the idea of relative status goes further than just material wealth: it also encompasses the human need for respect and recognition. The granting of respect and recognition to others is a way of treating them as equals, perhaps despite their difference, of removing a sense of diminished relative status and the bitterness and anger that this can lead to.[34] These ideas apply in any of the spheres where status may be relevant, from family life to knowledge and science: the scientist, par excellence, is primarily concerned with discovering something that her peers have not. Better to be a Newton than to be a merely average twenty-first–century scientist who knows far more than Newton ever did, but less than many of her peers.

The notion that security is important to happiness is intended to capture the truism that to have a lot is far less pleasurable if one is aware that it may easily

[31] *See, e.g.*, SUBJECTIVE WELL-BEING, *supra* note 9; FREY, *supra* note 5.

[32] Bruno Frey is the most prominent researcher. *See especially* FREY, *supra* note 5. *See also* GEORGE A. AKERLOF & RACHEL E. KRANTON, IDENTITY ECONOMICS: HOW OUR IDENTITIES SHAPE OUR WORK, WAGES AND WELL-BEING (2010).

[33] Ed Diener, *Subjective Well-Being*, *in* THE SCIENCE OF WELL-BEING 11, 26 (Ed Diener ed., 2009).

[34] CHARLES TAYLOR, MULTICULTURALISM: EXAMINING THE POLITICS OF RECOGNITION (1994).

be lost. Many an individual will trade a lump sum for a guaranteed pension, or a higher salary for job security. Indeed, the idea that happiness lies in having enough, and having faith that this will continue, can be explained in narrowly economic as well as psychological terms. The guarantee of security can be understood as a very valuable insurance policy whose premium is being paid by someone else. Even in financial terms a certain, but lower, income may be more valuable than a higher, but uncertain one. In any case, the modern welfare state and the emphasis on risk avoidance that permeates modern policy making are both examples of a collective agreement that there is a trade-off to be made between immediate profit and freedom, on one hand, and security, on the other.[35] Security, therefore, matters. It gives us the capacity to have a positive view of the future, and releases us from the need to worry. It releases us from the need to engage in anxious psychological processes that are deeply harmful to our happiness – a benefit of immense value. Just as with relative status, this importance is not limited to material security, but to all the things that matter to us. Security about status (what will it count for to have a university degree in twenty years?), about relationships, and about our daily rhythms (will this all have to change, one day?), may make us happier.

"To travel hopefully is better than to arrive" is the traditional expression of the value of hope to welfare. Hope is the sense that changes and developments in one's life may be positive. This is not to say that they need be dramatic or significant, but that they may correspond to what is desired. This matters so much to happiness because its absence is psychologically destructive. Whatever the current state of affairs, if the expectation is that it will all become worse, then even what we have tastes bitter. There may be a certain romantic melancholy pleasure in the sense of "live now, for tomorrow we die," and some may be better than others at translating impending doom into a reason to experience the moment fully. However, for most, a belief that tomorrow may also bring something good is important to the quality of today. Hope is, in this sense, perhaps just a part of security, but it brings a more positive twist to the idea, emphasizing the need not just to maintain the status quo, but the possibility of positive developments. Hope counters not just fear, but boredom and existential despair. One answer to "what is it all for?" is "perhaps tomorrow it will be different." The sense of possibility gives us a license to dream, perhaps the most pleasurable and useful activity that we have.

A policy recognition of the importance of hope is found in the idea of "equal opportunities." Most states no longer aspire to equal distribution of material, or other, goods. Rather, they emphasize that each person should have an equal chance. This is morally founded in the idea that people are responsible for achieving their own

[35] *See, e.g.*, Anthony Giddens, *Risk and Responsibility*, 62(1) MODERN L. REV. 1 (1999).

success. It also has a psychological foundation, however, in the idea that people are better off having a sense of possibility and hope than mere material status without a vision for change or development. Tomorrow may matter more to happiness than today.

4. THE PSYCHOLOGICAL IMPORTANCE OF GEOENGINEERING

4.1 *Relative Status*

Suppose that geoengineering were to be a success, in the sense that it controlled global temperature with minimal or acceptable side effects, at reasonable cost. This is not an unimaginable scenario.[36] It seems likely that in the course of the next decade or so emissions reductions will fail to occur on the scale necessary to stabilize climate, that undesirable environmental consequences will increase in magnitude, that research into geoengineering will continue, and that it will continue to be uncertain whether we are heading for catastrophe or some last-minute escape. That escape could come through a sudden wholehearted investment in emissions reduction, or perhaps through massive advances in sustainable energy generation, but also through acquisition of enough understanding of geoengineering to deploy it in a reasonably safe and predictable way. Many critics of geoengineering contend that such schemes may turn out to be dangerous in ways and degrees beyond what we are able to predict, but it is also true that they may not, or it may even turn out to be more controllable and safer than our current knowledge suggests.

Safe and affordable geoengineering would be good news for many as it would reduce the risks of the currently most accessible and practical source of energy. Not just oil companies, but any society that thinks its energy demands are likely to grow – particularly developing nations – would breathe a sigh of relief. However, for many environmentalists and climate change campaigners this new state of affairs would bring painful psychological and political consequences. They would not just lose an argument, but suffer a blow to their political and moral status.

Losing an argument is of course always disagreeable. In this case, those who are skeptical that climate change is a serious problem (a larger group than those who are skeptical of whether it is occurring at all) would crow delightedly and consider themselves proven correct. As with chlorofluorocarbons a few decades ago, an apocalyptic scenario would be largely dethroned by a technical solution, and the most committed doom-mongers would be left ideationally unclothed. Those who have argued that emissions reduction was the only possible way to prevent massive

[36] *See* text following *supra* note 29.

environmental destruction would be very much diminished in public and political life. Not only would they suffer the political costs of having made false predictions, but they would find it impossible to occupy the moral high ground in public. This would be a heavy blow: a position of moral superiority is one of the most attractive perks of an environmentalist standpoint.

That loss would be all the more bitter for being unfair. It is at least arguable that most of those who resist mitigation do not do so from a sincere belief that carbon dioxide levels are harmless, but because they do not care, or at least the use of fossil fuels serves their interests in ways that outweigh, for them, the global risks of warming. Even if humanity finds a solution, and gets away with it, there is a good argument that these people still deserve to be considered as the "bad guys," whereas even if the emissions reduction camp turns out to have backed the wrong scientific horse, they deserve credit for the fact that they identified a threat, and fought to have it addressed: they did their best to act in a responsible and altruistic way. The negligent company that poisons its workers would not be (entirely) absolved by the fact that a sudden new medical advance means they can unexpectedly all be saved, whereas the campaigners who tried to stop the poisoning should not be laughed at because it turns out that the poison is not as deadly as had been thought. And yet, in politics and public life luck is rewarded: the leader who wins the war will reap the benefits, even if all the generals agree it was thanks to the incompetence of the other side. This is a product of the complexity of the world and human beings. How can we ever really know why people act the way they do, or why things turn out the way they have? Lacking such knowledge, politics tends to be an inductive rather than a deductive practice, in which the public laud those they perceive to have succeeded. As in real science, it is the results that are sacred, and the explanation comes later, and is made to fit them. If one is on the wrong side of history, it rarely matters why.

This fall from grace would be all the worse because of the embedding of climate politics in wider environmental and social issues. If climate change is, as some certainly believe, the inevitable consequence of capitalism, or even colonialism, or a misguided view of humanity's place in nature, then it is easy to argue that it is most effectively combated via justice, or a new economic, ideological, or social order.[37] Removing climate change from that picture risks marginalizing the wider agenda. Will people still listen to the socioeconomic arguments if the most dramatic and appealing piece of empirical support has been taken away? What role is there for preachers of a new world order if problems can in fact be solved individually, if tinkering suffices?

[37] Eileen Crist, *Beyond the Climate Crisis*, 141 TELOS 29, 45–47 (2007); MIKE HULME WHY WE DISAGREE ABOUT CLIMATE CHANGE: UNDERSTANDING CONTROVERSY, INACTION AND OPPORTUNITY 254 (1995).

4.2 *Security*

It has been suggested that ecologists are often apocalyptic: that on some psychological level they embrace global warming as just deserts, and welcome global catastrophe as a sort of ultimate vindication.[38] This is no doubt a vast overstatement at the very least, and probably a distortion of reality. However, geoengineering does undermine certainties and orthodoxies, and will certainly be resisted for this reason. The suggestion that there might be a middle way between emissions reduction and catastrophe turns a straightforward opposition into a complex, dynamic, and unclear balancing process. It is not just that geoengineering may offer an alternative solution, but that it may offer partial solutions, meaning that the optimal position may be an ongoing balance between mitigation and climate management.[39] A comfortingly easy-to-understand choice between a good path and a bad path is replaced by a policy context that is all about technocratic fine-tuning.

This is particularly distressing because it follows a broader trend, in which social and ideational divisions are not only vaguer and more flexible, but often seen more as part of the problem than as demarcating solutions. The zeitgeist encourages contact and acceptance rather than the taking of sides. Divisions such as left and right, employer and employee, Protestant and Catholic, male and female, gay and straight, North and South, are less informative, less legitimate, and less relevant than they once were. For the crusading temperament there are not so many macro-issues left. Geoengineering might take away one of the most satisfying.

Geoengineering also brings uncertainty in more concrete ways. As a less-than-perfect solution it would almost inevitably entail a probabilistic approach to policy, and an ability to commit to taking a reasonable chance. For many that is reckless, irresponsible, or immoral, given what is concerned,[40] but for others it is uncomfortable. Accepting the idea that geoengineering may work, or should be explored, leaves us uncertain about what the best outcome will be, and how it will look: what our world will look like.

Above all, geoengineering entails facing up to the realistic possibility that humanity may not reduce emissions of carbon dioxide. "Reduce emissions or die" is not so much an apocalyptic standpoint as one that refuses to face any scenario other than reduction. As such it offers a certain security. If catastrophe is, deep down, unthinkable (and denial is a common reaction to impending disaster)[41] then many people will hold to the belief that we must be heading for final salvation. Geoengineering,

[38] Deepak Lal, *Eco-Fundamentalism*, UCLA Department of Economics Working Paper no. 732 (1995); Crist, *supra* note 37, at 45–47.

[39] Wrigley, *supra* note 14.

[40] *See, e.g.*, HULME, *supra* note 37, at 353; Crist, *supra* note 37.

[41] ELIZABETH KUBLER-ROSS, ON DEATH AND DYING (1970).

however, introduces the possibility of purgatory, and of a future of endless managed risk, of a fear that never quite goes away.

4.3 *Hope*

If geoengineering removes global warming as a pressing issue, one of the bitterest consequences of this will be the loss of a phenomenon that might have been usable to leverage wider social change. Some activists will see a vision of a better and different world fading as the threat of climate change diminishes.

This is perhaps most so for the deep ecologist who takes global warming as evidence of a misguided relationship with the natural world around us.[42] Geoengineering undermines certain precious ideas – of wilderness, of simplicity, of freedom from the burdens of power, of oneness with nature – and promotes ideas of human control, and of the consequent separateness of humanity from the rest of the natural world.

The end of nature has already been announced, and geoengineering is perhaps the final nail in its coffin.[43] If humanity steers and controls the climate, then not only does an idea of the "natural" state of affairs – the situation without human intervention – become ever more abstract, hypothetical, and even unimaginable, but by choosing the path of conscious control humanity must be taken to have accepted this. The phenomenon of "wilderness," of the area outside of humanity's influence, becomes replaced by the phenomenon of the global garden or global park, where everything is under management to a greater or lesser degree, and what appears wild and untouched is to some extent a simulation – because its climate is steered – and always a human choice.[44]

One reason to fear this state of affairs is concrete – a lack of trust in humanity. Hubris and incompetence may make us unsuited to global management, and short-term successes may merely encourage overextension until we finally make a mess of the world.[45] Those with no faith that humanity can exercise such power responsibly will of course be opposed to any technique that suggests the opposite. From this point of view, the more successful geoengineering is, the greater is the reason to oppose it. Failed attempts to geoengineer might, ironically, be welcome; the immediate environmental harm could be worth the long-term lesson learned about the dangers of global environmental management. The risk here is that geoengineering works.

[42] Crist, *supra* note 37, at 50.

[43] BILL MCKIBBEN, THE END OF NATURE (2006).

[44] *See* PAUL WAPNER, LIVING THROUGH THE END OF NATURE: THE FUTURE OF AMERICAN ENVIRONMENTALISM (2010); William Cronon, *The Trouble with Wilderness; or, Getting Back to the Wrong Nature*, *in* UNCOMMON GROUND: RETHINKING THE HUMAN PLACE IN NATURE 69 (William Cronon ed., 1995).

[45] B. FLYVBERG, N. BRUZELLUS & W. ROTHENDGATTER, MEGAPROJECTS AND RISK: AN ANATOMY OF AMBITION (2003); Davies, *supra* note 28.

More abstractly, the desire to maintain contact with a natural world, to have some sense of how things would be without us, masks many complex ideologies and emotions. There is certainly nostalgia, a desire to return to an imagined simpler past freed from the burden of responsibility that comes with power, to a time before the fall.[46] The icons of modernity – science, control, progress – are the apples into which we have bitten.[47] There may also be an undercurrent of mystical paternalism in the deep ecological desire to be in harmony with nature. It is apparently paradoxical that this desire entails a sense of nature without us – as if our separateness from the rest of nature is an inherent part of our unity with it.[48] One way of resolving this paradox is if our proper place in nature is small, as if there is some higher normativity that goes further than a command not to harm, and entails a command not to dominate or rise above.[49] Some deep ecologists would see this in more pragmatic terms, denying our capacity to be happy if we dominate or rise above the rest of nature, yet even that apparently psychologically realistic perspective raises the question: "why not?" What is it that makes humanity unable to realize itself properly unless it limits its own powers? Is there here a need to believe that there must be something beyond us? Perhaps the most convincing understanding is that we need, for our own self-realization, a relationship with the rest of nature, and that entails giving it enough space to be itself.[50] Dominance, by eliminating the autonomous identity of what is apart from us, entails our own loneliness. Yet the attribution of a self to nature as a whole, even metaphorically, remains obscure and troublingly undefined. The hint of mysticism does not quite go away.

All these visions are dimmed, if not quite shattered, by the sheer technical power that geoengineering would represent if it worked. Even if it solved global warming, it would do so by dominating nature, not submitting to it. If deep ecologists see global warming as a product of an industrial and instrumentalist approach to the globe, then they will see geoengineering as a depressing failure to get to the root of the problem, even a continuance of the same pathology with other symptoms.[51] By contrast, for those who believe in the capacity of humanity to solve its problems with science, geoengineering is a source of great hope.[52] It hints at a liberation from limitation, at the idea that there need be no constraints to creativity of development because there will always be a solution. For others, the political defeat, the fear, and the despair that the prospect of geoengineering promises them derives precisely

[46] Cronon, *supra* note 44, para 17.
[47] Lal, *supra* note 38, at 16–20
[48] Cronon, *supra* note 44, para 18.
[49] Naess, *supra* note 10, part 7.
[50] *Id.*, part 8; Crist, *supra* note 37, at 50–51.
[51] Crist, *supra* note 37, at 50.
[52] *See* Michaelson, *supra* note 14, at 131.

from this prospect, and from the sense that the chance for humanity to collectively choose a different way of life is slipping away.

5. CONCLUSION

The most plausible explanation for the degree of opposition to geoengineering, even to research into it, is the need that it creates for adjustments to views of the world, and – the political, social, and ultimately emotional costs that such adjustment would entail.

However, if geoengineering is a reasonably safe and effective way of limiting global warming, and if humanity lacks the will or organization to reduce emissions sufficiently, both of which will probably become apparent within relatively few years, then to oppose geoengineering on these psychological grounds creates a troubling new dynamic. Broader social and justice agendas, and visions of a better, simpler, more natural life, and even visions of long-term sustainability and prosperity, are pitted against short-to-medium–term global environmental well-being, instead of being synergetic with this, as environmentalists often currently claim.[53] It seems likely that in such a situation the immediate climatic needs will exert a greater grip on most people, and the losses described in this chapter will indeed be imposed upon those who resist. However, as many of those opposing geoengineering are among the more committed and informed environmentalists, they may well have sufficient influence to delay its acceptance, even to a point where considerable climatic harm is done.

The challenge for the policy maker who thinks geoengineering may usefully contribute to preventing global warming is how to ease the adjustments required, and reduce the costs associated with them, so that it becomes possible to achieve the immediate benefits that geoengineering may offer without doing emotional and ideological violence to those who have opposed it. This amounts to separating the narrowly climatic usefulness of geoengineering from the ideological baggage often attached to it.[54]

The themes of status, security, and hope may be brought back into play again here. Even if it appears that the opposition to geoengineering has exceeded what is environmentally defensible, and is based on grounds that to most of the population are either obscure or insufficiently weighty, individuals who have adopted these views can still be recognized and accorded respect for the sincerity and importance of their commitment to finding the best way forward: at least they cared. They can be incorporated in decision making as representatives of a worldview that is probably

[53] *Id. at* 135–37.

[54] *Id.*

only attractive to a minority, but a significant and serious enough minority that they deserve a place at the decision-making table, and to be a recognized and valued voice in the public debate. The other concerns to which climate change – they have argued – is inherently linked may at least be placed on the agenda as independent concerns. Geoengineering may cast doubt on the view of global warming as a result of an inherently destructive process of industrialization, capital exploitation, and colonialism, but that is not to say that there are no destructive side effects of these processes, and a position that should be acceptable to most parties, and which helps prevent the alienation of an important minority, is to compensate for the decoupling of climate change from justice by making justice something that the world cares about in its own right. That, in turn, may ease acceptance of possibly useful geoengineering. Then, ironically, those who said climate change could not be properly addressed without addressing justice will have embodied their own self-fulfilling prophecy, and will have been proved right.

PART II

Law and Governance

4

Geoengineering and Climate Management

From Marginality to Inevitability

Jay Michaelson

1. INTRODUCTION

In 1998, I wrote the first law review article advocating geoengineering as a climate change mitigation strategy; it appeared in the *Stanford Environmental Law Review*.[1] At the time, geoengineering was both unknown and unpopular – a seemingly impossible combination, but as soon as anyone heard of it, they disliked it.

Twelve years later, the political economy of geoengineering – or as I prefer to call it, Climate Management (CM) – has shifted, precisely because the conditions I outlined in 1998 have stayed so strikingly the same. Then, I argued that the lack of political will, the absence and complexity of the climate change problem, and the sheer expense of climate change mitigation made meaningful preventive measures, such as cutting greenhouse gas (GHG) emissions, extremely difficult to undertake.[2] After a decade of obfuscation and misinformation by powerful political actors, this analysis seems stronger than ever.

It was on the basis of this first conclusion – that meaningful GHG reduction would be extremely difficult[3] – that I then proceeded to investigate geoengineering from a political–economic point of view. The causal nexus is crucial, and one I share with most other advocates of CM. Few believe that sulfuric sunscreens and oceanic algae farms are *preferable* to traditional GHG reduction policies. Rather, the claim is that geoengineering is *necessary* because of the unfortunate political economy of GHG

[1] Jay Michaelson, *Geoengineering: A Climate Change Manhattan Project*, 53 STAN. ENVTL. L.J. 73 (1998). The first law review article on a type of geoengineering was James Edward Peterson, *Can Algae Save Civilization? A Look at Technology, Law, and Policy regarding Iron Fertilization of the Ocean to Counteract the Greenhouse Effect*, 6 COLO. J. INT'L ENVTL. L. & POL'Y 61 (1995). As the title implies, it focused exclusively on Ocean Iron Fertilization (OIF).

[2] Michaelson, *supra* note 1, at 81–103.

[3] I discuss these issues infra Part I. *See generally* Tom Wigley, *The Science of Geoengineering*, Powerpoint Presentation to the American Enterprise Institute for Public Policy Research Conference (June 3, 2008), http://www.aei.org/docLib/20080606_WigleyJune3powerpoint.pdf.

reduction, in which the highest costs of mitigation fall precisely on some of the most powerful political actors. Likewise, geoengineering advocates usually do not claim that it should be implemented *in place of* GHG reduction, but in addition to GHG mitigation, and that at the very least serious research is warranted.

Geoengineering was a marginal idea in 1998, and I had only a dozen or so legal, political, and scientific studies to rely upon. Now, much has changed. Although CM remains at the margins of our popular political discourse, there has been an explosion of scientific and policy analyses, particularly since the publication of Paul Crutzen's 2006 editorial on Solar Radiation Management (SRM), probably the most promising potential geoengineering technology.[4] Indeed, one may divide the history of this progress into B.C. – Before Crutzen – and after. This was not some recent law school graduate making a policy recommendation – this was a Nobel laureate, prophet of the ozone crisis, and widely respected leader in climate science. SRM, Ocean Iron Fertilization (OIF: seeding gigantic phytoplankton carbon sinks in the oceans by fertilizing them with iron), and other technologies have since been explored and advanced by credible scientists, scholars, and even entrepreneurs. Such proposals are no longer the stuff of "giant laser space Frisbees," the phrase from Bloom County that I cited in 1998.

CM has also tentatively been explored by conservative think-tanks and pundits. In a sense, this is every environmentalist's worst nightmare: the same conservatives, lobbyists, and business interests who have successfully stymied efforts to pass climate change legislation in the United States are – just as some greens feared – coming around to support engineering the atmosphere instead of reducing our consumption. As I noted in 1998, CM can be postponed; it can be undertaken by making more stuff, rather than less; and it can even be subcontracted out to private actors, without the need of government regulation of ordinary people. Yet the mere fact that conservatives support geoengineering – on the "blue team" in the language of Eli Kintisch's recent book[5] – should not, in itself, cause liberals and greens to join the "red team." CM was not the Right's idea – it has come about, and come to prominence, because of sincere environmental scientists and policy analysts struggling with how to avert massive climate catastrophe.

Yes, it is outrageous, in a way: first, greedy oilmen block GHG reduction by saying that climate change does not exist, and then they support CM because all of a sudden it does. But this nauseating political dynamic is a blessing, not a curse. Are we environmentalists interested in punishing the bad guys, or saving the earth? CM

[4] Paul Crutzen, *Albedo Enhancement by Stratospheric Sulfur Injections: A Contribution to Resolve a Policy Dilemma?*, 77 CLIM. CHANGE 211–19 (2006). *See* ELI KINTISCH, HACK THE PLANET: SCIENCE'S BEST HOPE – OR WORST NIGHTMARE – FOR AVERTING CLIMATE CATASTROPHE 55–58 (2010).

[5] KINTISCH, *supra* note 4. In Kintisch's lingo, the "Blue Team" is pro-geoengineering, the "Red Team" against.

is a climate change strategy that, unlike regulation, might actually stand a chance of becoming reality. It is the only approach to climate change that can act as a compromise between liberals and libertarians, greens and browns. It is the delight of my political opponents, and for that reason, the planet's best hope of survival. As climate change becomes ineluctable, geoengineering becomes inevitable.

This chapter explores this new political dynamic, and reaffirms my conclusion from 1998 that we environmentalists must swallow our pride and our misgivings, and support research into geoengineering – no longer because it is a marginal idea in need of advocates, but because it is the inevitable remedy for climate change that should be carefully researched before it needs to be deployed. Yes, one of our worst nightmares is slowly coming true, but that means an even worse one may be averted.

As a threshold matter, I suggest in Part I that it is time to retire the term "geoengineering." It's too sci-fi to be taken seriously, and it misleadingly suggests that the solution to climate change has to do with bulldozers and dams. For reasons set forth in Part I, I propose "climate management" (CM), which better describes what SRM and OIF are really about, how they differ from GHG reduction, and how they fit within an overall climate risk portfolio.

Following that terminological proposal, Part I explores what has stayed the same since 1998, and what has changed. What has *not* changed is chiefly our inability to do anything about global warming. Perhaps I could have predicted – back in the Clinton administration, pre-Google, pre-9/11 – that the idea of intentionally manipulating the earth's climatic systems would remain a relatively marginal one. But what I did not imagine was that in 2011, politicians would still be arguing whether climate change is a real, anthropogenic phenomenon or not. This is not because the science is uncertain, but because, as former president Clinton recently observed, "we're kind of in a truth-free period right now."[6] The strategy, begun in 1988, to sow uncertainty regarding climate change is now almost as well documented as it is well funded.[7] And so, despite rises in temperatures, a high-grossing documentary by a Nobel laureate, visible changes in glaciers and ice shelves, and widespread understanding of the climate crisis in Europe, the Republican party in Washington still

[6] Bill Clinton, Interview with George Stephanopoulos, ABC News: Good Morning America, Sept. 21, 2010. For a video, see George Stephanopoulous, *Bill Clinton on Palin: "Resilient," Like me, Don't "Underestimate" Her*, George's Bottom Line Blog, Sept. 21, 2010, http://blogs.abcnews.com/george/2010/09/bill-clinton-on-palin-resilient-like-me-dont-underestimate-her.html.

[7] *See* JAMES HOGGAN, CLIMATE COVER-UP: THE CRUSADE TO DENY GLOBAL WARMING (2009); ERIC POOLEY, THE CLIMATE WAR 31–52 (2010); Naomi Oreskes, *Beyond the Ivory Tower: The Scientific Consensus on Climate Change*, 306 SCI. 1686 (Dec. 3, 2004), *available at*: http://www.sciencemag.org/cgi/content/full/306/5702/1686; Peter Jacques, Riley E. Dunlap & Mark Freeman, *The Organisation of Denial: Conservative Think Tanks and Environmental Skepticism*, 17 ENVIRON. POL. 349–85 (2008).

has, as its official position, the view that climate change is either not happening, or is part of some natural cycle, and requires further study.

I rejected such claims in 1998, and I reject them today. Yet if the pseudo-controversy regarding climate change proves anything, it is that my earlier article was correct: we should be very pessimistic about the chances for meaningful GHG reduction, because it would so greatly impact some of the largest and most powerful industrial, commercial, and corporate entities in the world. If anything, I was too optimistic in 1998. Then, I conjectured that these campaigns of deliberate misinformation would end once glaciers melted. Clearly, I was incorrect.

Part I next explores what has changed: first, the scientific and policy discussions of CM, and second, the growing support of CM in conservative quarters. In 1998, there were fewer than half a dozen articles that engaged seriously with geoengineering; now there are hundreds. There have been multiple conferences on the scientific, political, and ethical consequences of geoengineering, and we have a much clearer sense of how OIF and SRM may proceed. There has even been some coverage of geoengineering in the mainstream media, scientific press, and academic sectors. Geoengineering is no longer science fiction, utopian or dystopian fantasy. Whereas it was scarcely polite conversation fifteen years ago, it is now a subject of serious scientific, academic, philosophical, and political discourse.

In Part II, I turn to the normative case for CM, which I believe is stronger than ever. First, I address some of the concerns that have been expressed about CM, focusing on the questions of a "free pass" to polluters, the unintended risks and costs, the equitable considerations, the potential of cataclysm in case of cessation of CM, and the problem of rogue actors. I conclude that the concerns are answerable in every case. Last, as I did in 1998, I turn to the deeper questions of CM for environmentalists. CM is appealing to conservatives not only because it protects economic interests but also because it is ideologically in sync with conservative ideas – it lets the free market be free, uses technology rather than a restraint on behavior, and avoids government regulation. This, in addition to practical concerns, is doubtless why CM appeals to so few environmentalists. It capitulates to precisely those scoundrels who have scuttled our best efforts at a sensible climate strategy, and in so doing, creates an insensible one. Yet not to pursue it, I argue, is to condemn coastal areas, temperate forests, and thousands of species to extinction. What, exactly, is the price of our pride? CM does indeed challenge some of the core assumptions of the contemporary environmental movement. But that may not be a bad thing.

Twelve years ago, I asked whether melting icebergs and record-hot summers would be sufficient to create the political will to act on climate change. Now, we know that they are not. Yet just as CM has become ever more critical, it has also become inevitable. I was a lone voice in 1998, but not today. Whether we like it or not, the inevitability of CM is as much a reality as climate change itself.

2. GEOENGINEERING: FIFTEEN YEARS LATER

The modern theory of anthropogenic climate change dates back to the nineteenth century,[8] and its measurable, visible effects have been with us for at least two decades now. Yet American political discourse remains as if in suspended animation; the two major political parties still differ as to whether climate change is even happening at all. How is this possible? How is it that nothing has changed, despite melting glaciers and record-hot summers?

After a proposal regarding nomenclature, this part explores what has not changed, and what has changed, in the past twelve years of public discourse on climate change on geoengineering. Essentially, because so much has *not* changed in terms of climate change regulation, a great deal has changed in terms of CM. I was a pessimistic prophet of doom in 1998, when I predicted that GHG reduction would not happen soon. In 2010, I am merely reporting the facts.

2.1 *Nomenclature: Let Us Get Rid of "Geoengineering"*

The term "geoengineering" is overbroad, unhelpful, and misleading, and should be set aside. I propose "Climate Management" or "CM" instead.

The Royal Society defines geoengineering as "the deliberate large-scale intervention in the Earth's climate system, in order to moderate global warming."[9] The coinage of the term is attributed to C. Marchetti, who used it in a 1977 article, "On Geoengineering and the CO_2 Problem."[10] Today, it fails the cocktail party test every time: even informed laypeople often ask whether it has to do with open-top coal mining, or carving new canals in Panama. More significantly, there are at least three reasons the term "geoengineering" is negatively impacting the effort to research CM.

[8] *See* Hans von Storch & Nico Stehr, *Towards a History of Ideas on Anthropogenic Climate Change, in* CLIMATE DEVELOPMENT AND HISTORY OF THE NORTH ATLANTIC REALM 17, 17 (Gerold Wefer et al. eds., 2002), *available at* http://coast.gkss.de/staff/storch/pdf/delmenhorst.2002.pdf.

[9] JOHN SHEPHERD ET AL., GEOENGINEERING THE CLIMATE: SCIENCE GOVERNANCE AND UNCERTAINTY (2009), *quoted in* JEFF GOODELL, HOW TO COOL THE PLANET: GEOENGINEERING AND THE AUDACIOUS QUEST TO FIX EARTH'S CLIMATE 16 (2010). This is roughly identical to the definition of "climate engineering" used in the 2010 House Science & Technology Committee report: "the deliberate largescale modification of the earth's climate systems for the purposes of counteracting and mitigating climate change." U.S. House of Representatives Committee on Science and Technology, *Engineering the Climate: Research and Strategies for International Coordination, available at* http://science.house.gov/publications/caucus_detail.aspx?NewsID=2944 (last visited Nov. 4, 2010) [hereinafter U.S. House Report], at ii.

[10] C. Marchetti, *On Geoengineering and the CO2 Problem,* 1 CLIM. CHANGE 59–68 (1977). Marchetti's specific proposal was to inject CO2 into the deep ocean. *See* H. Damon Mathews & Sarah E. Turner, *Of Mongooses and Mitigation: Ecological Analogues to Geoengineering,* 4 ENVIRON. RES. LETTER 045105 (Oct.–Dec. 2009), *available at* http://iopscience.iop.org/1748-9326/4/4/045105/fulltext#erl319524bib35.

First, the category of "geoengineering" has become too broad, and occludes many differences. For example, is OIF more like SRM, or more like afforestation? Yes, OIF involves technological tinkering with areas of ocean, and, like SRM, it could make matters worse rather than better. But the differences are perhaps even greater. The expected oceanic pollution from OIF is relatively limited in spatial and temporal scope; OIF, unlike SRM, could be conducted in a limited area with limited impact on uncooperative nations, and for a limited period of time, at least at first. Although SRM would have secondary effects potentially all around the globe, most people would never even notice OIF. Indeed, with reports that the so-called Great Pacific Garbage Patch has now exceeded the size of Texas, the emphasis on pollution from OIF seems like a highly selective form of criticism. In some ways, OIF is scarcely different from planting trees. Is afforestation also "geoengineering?" (Afforestation also has secondary effects: President Reagan's statement that "trees cause pollution"[11] was based on data that rotting trees may release the carbon dioxide they once absorbed.) The term "geoengineering" confuses the issue.

Second, the term "geoengineering" is bad PR. It connotes science fiction and, to my ears, has a somewhat musty, Arthur C. Clarke air to it. (Arthur C. Clarke was my favorite writer when I was a teenager, but this is 2010.) It rings of arrogance, 1960s-era Space Age fantasy, and a sort of "can-do" spirit that is inappropriate today. Indeed, it evokes the spirit of one of geoengineering's earliest proponents, Dr. Edward Teller, the notorious, arch-conservative nuclear scientist who was a model for Dr. Strangelove.[12] It was Teller who said things such as "we will change the earth's surface to suit us,"[13] and who championed geoengineering at the same time as the "Star Wars" missile defense system.[14] This scientific arrogance is encoded in the term "geoengineering" itself.

Third, "geoengineering" is misleading. Given that the two leading geoengineering proposals have to do with water and air, the "geo" prefix is itself a bit confusing. It connotes bulldozers, dikes, and dams. SRM is not first and foremost an engineering project (although of course it requires sophisticated engineering to accomplish), and neither is OIF. We are not building dams; we are using our limited knowledge of atmospheric science to either increase the albedo and opacity of the stratosphere, or create new carbon sinks in the oceans. Geoengineering is neither geo- nor engineering.

[11] *See* Tim Radford, *Do Trees Pollute the Atmosphere?*, GUARDIAN, May 13, 2004, *available at* http://www.guardian.co.uk/science/2004/may/13/thisweeksciencequestions3.

[12] For a raucous profile of Teller in the context of geoengineering, see GOODELL, *supra* note 9, at 70–88.

[13] *Quoted in id.* at 16.

[14] *Id.* at 85.

My proposed alternative is "climate management" or "CM." Climate management is what SRM and OIF are really about. Unlike "prevention" or "mitigation," which generally refer to GHG reduction, and "adaptation," which generally refers to adapting to the effects of climate change (sea walls, dikes, etc.), "management" can refer to attempting to manage the climate directly. This range of policy options should seem appealingly familiar. (In his recent TED talk, CM advocate David Keith presented a simple continuum of mitigation-geoeng ineering-adaptation[15]; surely mitigation-management-adaptation is even clearer.) Understanding geoengineering as climate management renders comprehensible its positive and negative attributes. We are not talking about a fanciful dream of "hacking the earth." We are talking about Plan B, because Plan A seems so expensive that a few key players remain intent on blocking it. Plan A is best; Plan B may be the best we can do. And of course, as we have said many times, the two are not mutually exclusive. A shift of nomenclature from the technological mechanisms of a climate change strategy to the policy nature of that strategy helps clarify the issue in place, and properly shifts attention from means to ends – or at least from proximate means (technological intervention) to meaningful means (regulation, management, or mitigation).

"Climate Management" preserves the hubris involved in such a policy, as well it should. But it places that arrogance in the context of a long history of similar endeavors – forest management, for example, or wildlife management. We humans have, for centuries, attempted to manage the ecosystems of our planet, and we have a mixed record of doing so. This is not to minimize its potential danger, or potential for folly; indeed, placing CM in the context of other human "management" schemes is meant to capture just that. It is simply to see CM for what it is: a technological effort to manage the scope of climate change.

I also prefer "Climate Management" over the other alternatives that have been proposed of late. Jeff Goodell and Congressman Bart Gordon have lately suggested "climate engineering,"[16] which I admit is better than "geoengineering" but still has an odd, well, "engineering" ring to it. Although SRM, for example, does involve sophisticated devices to deliver sulfur dioxide, the engineering of these devices is only the means to an end; the real strategy is the management of the atmosphere. Mathews and Turner have suggested "direct climate intervention,"[17] which is

[15] David Keith, "David Keith's Unusual Climate Change Idea," *available at* http://www.ted.com/talks/david_keith_s_surprising_ideas_on_climate_change.html (hereinafter Keith, TED Talk) (last visited March 23, 2013).

[16] The suggestion was made at a 2010 conference hosted by the New America Foundation. New America Foundation, *A Future Tense Event: Geoengineering, The Horrifying Idea Whose Time Has Come?*, Sept. 27, 2010, *available at* http://www.newamerica.net/events/2010/geoengineering.

[17] Mathews & Turner, *supra* note 10.

descriptive but perhaps too long, and not descriptive enough. Eli Kintisch's "planet hacking"[18] heightens the absurdity of CM and make it seem like a kind of wonky computer fantasy. (One of the earliest proposed nomenclatures was "climate control" although that term has been coopted by air conditioning systems.[19]) Stephen Dubner and Steven Levitt, of *Freakonomics* fame, dub SRM "Budyko's Blanket," after Belasusian climate scientist Mikhail Budyko, to whom the idea was attributed in the 1992 NAS report,[20] whereas climatologist Alan Robock playfully calls it the "yarmulke solution."[21] Indeed, it is striking how many playful and derogatory terms there are for geoengineering proposals – back in 1998, OIF was still being called the "geritol cure."[22] Perhaps the humor reflects our deep anxiety regarding CM as a climate change methodology. Or maybe it is just a way to ridicule it.

"Climate Management" is less scary than "geoengineering," while still sinister enough to temper our enthusiasm; it is more accurate, and it fits within existing matrices of risk management and risk calculation. For the remainder of this chapter, I will use the terms "geoengineering," "Climate Management," and "CM" interchangeably, and will in all cases attempt to specify whether OIF, SRM, or another particular proposal is the subject of discussion.

2.2 *What Has Not Changed, or: 500 Wrongs Do Make a Right*

What has not changed since 1998? The theory of climate change remains surprisingly intact, with more and more evidence supporting it. Yet as we have learned more about the campaign of misinformation surrounding climate change, the cause for pessimism has increased. It is now clear that some private interests will spend enormous sums on climate change denial, and that such denial will continue to be effective even when the effects of climate change are visible. Now, unlike in 1998, we have detailed studies of the precise ways in which science was manipulated for private gain.[23] The way these campaigns have been prosecuted should lead

[18] KINTISCH, *supra* note 4. Although Kintisch generally supports CM, the term has been used in a derogatory fashion by opponents. *See, e.g.*, Climate Connections, Blog Post, *Geoengineering Moratorium at UN Ministerial in Japan: Risky Climate Techno-Fixes Blocked*, Oct. 28, 2010.

[19] *See* W.W. Kellogg & Stephen Schneider, *Climate Stabilization: For Better or Worse?*, 186 SCI. 1163–72 (1974).

[20] STEVEN D. LEVITT & STEPHEN J. DUBNER, SUPERFREAKONOMICS: GLOBAL COOLING, PATRIOTIC PROSTITUTES, AND WHY SUICIDE BOMBERS SHOULD BUY LIFE INSURANCE 193 (2009).

[21] GOODELL, *supra* note 9, at 115.

[22] *See* Patrick Huyghe, *Geoengineering Our Way Out of Trouble*, 2.1 21STC, *available at* http://www.columbia.edu/cu/21stC/issue-2.1/huyghe.htm (quoting environmental scientist Wallace Broecker's tentative support for "insurance against a bad climate trip" in the form of geoengineering).

[23] *See, e.g.*, NAOMI ORESKES & ERIK M. CONWAY, MERCHANTS OF DOUBT: HOW A HANDFUL OF SCIENTISTS OBSCURED THE TRUTH ON ISSUES FROM TOBACCO SMOKE TO GLOBAL WARMING (2010); HOGGAN, *supra* note 7, at 31–52; Jacques et al., *supra* note 7, at 349–85; Lucia Graves, *Republican Global Warming*

us to be more pessimistic than ever that meaningful GHG reduction will ever be addressed.

Nor is this simply a matter of disseminating information. Several popular books, endless articles in liberal magazines, and two high-profile documentary films (Al Gore's *An Inconvenient Truth* and Leonardo Dicaprio's *The 11th Hour*) have failed to sufficiently mobilize popular opinion: the issue came in dead last in a 2010 Pew Research Center poll of issues that matter to Americans,[24] and the Obama administration has more or less given up on trying to rally public opinion on it.[25] This is a result not of ignorance but of campaigns of misinformation, from the industry-sponsored Global Climate Coalition in the 1990s[26] to the contemporary cadre of "CTTs" (Conservative Think-Tanks) funded by industry groups.[27] Leading "climate sceptics," including Timothy Ball, Frederick Singer, and Steven Milloy, are not even climate scientists.[28] This strategy has been successful: from 1998 to 2002, although the scientific press featured 928 articles supporting or showing evidence for anthropogenic climate change and zero opposing it, 53 percent of major newspaper stories "both sides" of the "scientific debate."[29] Of the 141 books published between 1972 and 2005 denying the seriousness of environmental problems, 130 (92 percent) were published by CTTs, written by authors affiliated with CTTs, or both.[30] And CTTs and their allies have successfully cast doubt on the so-called hockey stick graph of climate change, based on one non-climatologist's discovery of minor mathematical errors,

Deniers Funded by the Energy Industry, HUFFINGTON POST, Oct. 15, 2010, http://www.huffingtonpost.com/2010/10/14/pat-toomey-climate-change-republicans_n_763545.html. *See also* Naomi Oreskes, Video, *You CAN Argue with the Facts*, http://smartenergyshow.wordpress.com/2008/06/03/naomi-oreskes-you-can-argue-with-the-facts/.

[24] *See* Ryan Lizza, *As the World Burns*, NEW YORKER, Oct. 9, 2010.

[25] *Id.* One lobbyist called Obama "the James Buchanan of climate change." *Id.*

[26] GCC's leading funders were ExxonMobil, Royal Dutch Shell, British Petroleum, Texaco, General Motors, Ford, DaimlerChrysler, the Aluminum Association, the National Association of Manufacturers, and the American Petroleum Institute. HOGGAN, *supra* note 7, at 13; Andrew Revkin, *Industry Ignored Its Scientists on Climate*, N.Y. TIMES, Apr. 20, 2009. Per Revkin, in 1997, the year of the Kyoto Protocol, the GCC had an annual budget of $1.68 million.

[27] HOGGAN, *supra* note 7, at 73–85. For example, since 1998, ExxonMobil alone has funded the Competitive Enterprise Institute ($2m), Annapolis Center for Science-Based Public Policy ($1.0m), American Enterprise Institute ($2.8m), Heritage Foundation ($630k), Heartland Institute ($676k) and many others. *See* Exxon Secrets, http://www.exxonsecrets.org/maps.php (last visited Mar. 12, 2011).

[28] HOGGAN, *supra* note 7, at 49–50 (Ball), 156 (Milloy); POOLEY, *supra* note 7, at 33–37 (Singer). Milloy, who regularly appears on Fox News, is a former lobbyist for Exxon, Philip Morris, the Edison Electric Institute, and Monsanto.

[29] Jules Boykoff & Max Boykoff, *Balance as Bias: Global Warming and the U.S. Prestige Press*, 14 GLOBAL ENVTL. CHANGE 125–36 (2004). *See also* HOGGAN, *supra* note 7, at 21–22. This astonishing disparity reveals the extent to which industry-led disinformation campaigns have distorted the public's view of climate science.

[30] Jacques et al., *supra* note 7.

and notwithstanding dozens of scientific studies supporting it.[31] It can no longer be maintained that the public lacks adequate information about climate change. The reality is that huge campaigns of deliberate misinformation have created a false sense of uncertainty that flies in the face of the evidence.

The other fact that has not changed at all since 1998 is that adequate GHG reduction will not be achievable. Current CO_2 levels, for example, are 385 ppm, 100 ppm above the preindustrial level. Yet even the most hopeful CO_2 reduction target is 450 ppm, that is, significantly higher than current levels, and it would require GHG emissions to be reduced 11 percent from current levels by 2030, a reduction that would likely require all new power plants to have zero CO_2 emissions,[32] a wildly uneconomical proposal that would burden the developing world the most. Likewise, the European Union's target of limiting total warming to 3.6 degrees Fahrenheit above preindustrial times would still represent a warmer average temperature than earth has seen in millions of years.[33] And of course, even these targets are based on relatively optimistic projections; if doomsayers such as Richard Lovelock or James Hansen are right, the situation could be far, far worse.[34]

2.3 *What Has Changed*

2.3.1 Geoengineering Comes of Age

Particularly since 2006, there has been an explosion in scientific, legal, and policy discussions of CM. Although proposals for CM to address climate change date back to the 1960s – SRM was first mentioned in a 1965 report[35] – the first major study of geoengineering was published in 1992 by the U.S. National Academy of

[31] Editorial, *Hockey Stick Hokum*, WALL STREET J., July 14, 2006. *See also* Antonio Regalado, *In Climate Debate, the "Hockey Stick" Leads to a Face-Off*, WALL STREET J., Feb 14. 2005, *available at* http://online.wsj.com/public/article/SB110834031507653590-DUadAZBzxHoSiuYH3tOdgUmKXPo_20060207.html?mod=blogs; Geoff Brumfiel, *Academy Affirms Hockey-Stick Graph*, 441 NATURE 1032–33 (June 29, 2006). For a thorough debunking of the "hockey stick hokum" myth, see Jeffrey D. Sachs, *Fiddling while the Planet Burns*, SCI. AM., Sept. 14, 2006, http://www.scientificamerican.com/article.cfm?id=fiddling-while-the-planet; HOGGAN, *supra* note 7, at 109–12.

[32] Tom Wigley, *The Science of Geoengineering*, PowerPoint Presentation to the American Enterprise Institute for Public Policy Research Conference 18 (June 3, 2008), http://www.aei.org/docLib/20080606_WigleyJune3powerpoint.pdf.

[33] KINTISCH, *supra* note 4, at 30.

[34] *See* GOODELL, *supra* note 9, at 101–04; KINTISCH, *supra* note 4, at 27–29; Keith, TED talk, *supra* note 15.

[35] Environmental Pollution Panel of the President's Science Advisory Council (PSAC), *quoted in* NATIONAL RESEARCH COUNCIL, ADVANCING THE SCIENCE OF CLIMATE CHANGE 291–99 (2010) [hereinafter NRC, ADVANCING SCIENCE], *available at* http://americasclimatechoices.org/, at 293. Interestingly, the 1965 PSAC report focused more on albedo enhancement than on GHG reductions.

Sciences,[36] which included a chapter on geoengineering in its assessment of climate change policy, discussing reforestation, OIF, cloud albedo modification, SRM, and the use of space-based reflectors.[37] All of these ideas are still around today, although SRM and OIF have emerged as the most prominent, with albedo enhancement[38] and other methods of CDR (Carbon Dioxide Removal)[39] close behind. SRM in particular has been the subject of numerous scientific studies,[40] legal and policy analyses,[41] philosophical discussions,[42] and prestigious conferences such as the 2007

[36] NATIONAL ACADEMY OF SCIENCES PANEL ON POLICY IMPLICATIONS OF GREENHOUSE WARMING COMMITTEE ON SCIENCE ENGINEERING POLICY, POLICY IMPLICATIONS OF GREENHOUSE WARMING: MITIGATION ADAPTATION AND THE SCIENCE BASE (1992).

[37] This colorful idea has faded from scientific discourse, although it remains a favorite of CM mockers. For the original (short) proposal, see James T. Early, *Space-Based Solar Shield to Offset Greenhouse Effect*, 42 J. BRIT. INTERPLANETARY SOC. 567–69 (1989). Remarkably, this proposal still made it into the U.S. House Report on Geoengineering, although it was dismissed. U.S. House Report, *supra* note 9, at 42.

[38] *See* J. Latham, *Amelioration of Global Warming by Controlled Enhancement of the Albedo and Longevity of Low-Level Maritime Clouds*, 3 ATMOSPHERIC SCI. LETTERS 0–6 (2002); K. Bower, T. Choularton, J. Latham, J. Sahraei & S. Salter, *Computational Assessment of a Proposed Technique for Global Warming Mitigation via Albedo-Enhancement of Marine Stratocumulus Clouds*, 82 ATMOSPHERIC RES. 328–36 (2006). Some have proposed that albedo enhancement may even be accomplished by painting roofs white and other changes to urban environments. *See, e.g.*, H. Akbari, S. Menon & A. Rosenfeld, *Global Cooling: Increasing World-Wide Urban Albedos to Offset CO2*, 94 CLIM. CHANGE 275–86 (2009). For a lay discussion of these proposals, see GOODELL, *supra* note 9, at 163–89.

[39] *See* GOODELL, *supra* note 9, at 25–30 (discussing David Keith's "CO2 Scrubbers"); KINTISCH, *supra* note 4, at 103–25; U.S. House Report, *supra* note 9, at 1.

[40] *See, e.g.*, Crutzen, *supra* note 4; Ken Caldeira & Lowell Wood, *Global and Arctic Climate Engineering: Numerical Model Studies*, 366 PHILOSOPHICAL TRANSACTIONS OF THE ROYAL SOCIETY A: MATHEMATICAL, PHYSICAL, AND ENGINEERING SCIENCES 4039–56 (2008); David Keith, *Geoengineering the Climate: History and Prospect*, 25 ANN. REV. ENERGY & ENV'T 245–84 (2000). *See also* Alan Robock, Allison B. Maquardt, Ben Kravitz & Georgiy Stenchikov, *The Benefits, Risks, and Costs of Stratospheric Geoengineering*, 38(19) GEOPHYSICAL RES. LETTERS 36 (2009); Kevin Bullis, *The Geoengineering Gambit*, MIT TECH. REV. (Jan.–Feb. 2010), *available at* http://www.technologyreview.com/energy/24157/; William J. Broad, *How to Cool a Planet (Maybe)*, N.Y. TIMES, June 26, 2006.

[41] *See, e.g.*, William Daniel Davis, Note, *What Does "Green" Mean?: Anthropogenic Climate Change, Geoengineering, and International Environmental Law*, 43 GA. L. REV. 901 (2009); Alan Carlin, *Global Climate Change Control: Is There a Better Strategy than Reducing Greenhouse Gas Emissions?*, 155 U. PA. L. REV. 1401 (2007) (advocating injections of sulfur into stratosphere to scatter incoming solar radiation in lieu of GHG mitigation); J. Virgoe, *International Governance of a Possible Geoengineering Intervention to Combat Climate Change*, 95 CLIMATIC CHANGE 103–19 (2009).

[42] *See, e.g.*, Rebecca Bendick et al., *Choosing Carbon Mitigation Strategies Using Ethical Deliberation*, 2 WEATHER CLIMATE & SOC., 140–47 (2010). The research grew out of a two-year National Science Foundation grant to study the ethics of SRM. *See* The Center for Ethics, "The Ethics of Geoengineering: Investigating the Moral Challenges of Solar Radiation Management," *available at* http://www.umt.edu/ethics/EthicsGeoengineering/default.aspx *See also* Alyson Kenward, *Scientists Consider Whether to Cause Global Cooling*, Climate Central, Oct. 19, 2010, http://www.climatecentral.org/breaking/news/causing_global_cooling.

gathering at the American Academy of Arts and Sciences[43] and workshops hosted by the Council on Foreign Relations in 2008[44] and 2010.[45]

In addition, the policy conversation has lately spread beyond academic publications. CM has, of late, been the subject of generally positive magazine articles in *Foreign Affairs*,[46] *the Atlantic*,[47] *Salon*,[48] *Slate*,[49] and *Wired*,[50] and has been covered on network news.[51] It received a chapter in the new best-selling sequel book *Superfreakonomics*,[52] and is the subject of at least two mass-market books.[53] CM advocate David Keith has even made it onto the TED Talks.[54]

CM has also begun to be taken seriously at the national policy level. Director of the White House Office of Science and Technology John Holdren has included it as a "possibility" in his presentations on climate change policy,[55] although he later

[43] *See* GOODELL, *supra* note 9, at 190–91; KINTISCH, *supra* note 4, at 3–12. Kintisch credits the Harvard meeting as being a turning point in mainstream acceptance of geoengineering. KINTISCH at 12.

[44] Council on Foreign Relations Geoengineering Blog, *Geoengineering: Workshop on Unilateral Planetary Scale Geoengineering*, http://www.cfr.org/project/1364/geoengineering.html (last visited Nov. 4, 2010).

[45] Council on Foreign Relations, Video, *Developing an International Framework for Geoengineering*, Mar. 10, 2010, http://www.cfr.org/publication/21636/developing_an_international_framework_for_geoengineering_video.html. The panel featured M. Granger Morgan, Head, Department of Engineering and Public Policy, Carnegie Mellon University; John D. Steinbruner, Director, Center for International and Security Studies, University of Maryland; and Ruth Greenspan Bell, Acting U.S. Climate Policy Director, World Resources Institute.

[46] David G. Victor, M. Granger Morgan, Jay Apt, John Steinbruner & Katharine Ricke, *The Geoengineering Option*, FOREIGN AFF., Mar.–Apr. 2009, http://www.foreignaffairs.com/articles/64829/david-g-victor-m-granger-morgan-jay-apt-john-steinbruner-and-kat/the-geoengineering-option.

[47] Graeme Wood, *Re-Engineering the Earth*, ATLANTIC, July–Aug. 2009, http://www.theatlantic.com/magazine/archive/2009/07/re-engineering-the-earth/7552/.

[48] Elizabeth Svoboda, *The Sun Blotted Out from the Sky*, SALON, Apr. 2, 2008, http://www.salon.com/news/feature/2008/04/02/geoengineering; Thomas Rogers, *Can Technology Cool the Planet?*, SALON, Apr. 22, 2010 (reviewing GOODELL, *supra* note 9), http://www.salon.com/books/feature/2010/04/22/how_to_cool_the_planet. The two *Salon* article titles themselves tell the story of how perceptions of CM have evolved in the last two years.

[49] James Rodger Fleming, *Weather as a Weapon*, SLATE, Sept. 23, 2010 (providing a skeptical reading of geoengineering history); Eli Kintisch, *The Politics of Climate Control*, SLATE, Sept. 24, 2010; Rep. Bart Gordon, *Plan B for the Climate*, SLATE, Sept. 24, 2010. Interestingly, *Slate* turned down a proposal of mine to write on geoengineering in 2003, saying that it was "too fringey."

[50] Chris Mooney, *Can a Million Tons of Sulfur Dioxide Combat Climate Change?*, WIRED, June 23, 2008.

[51] NBC Nightly News, Video, Dec. 27, 2009, *available at* http://www.msnbc.msn.com/id/3032619#34596069.

[52] LEVITT & DUBNER, *supra* note 20, at 165–203.

[53] KINTISCH, *supra* note 4; GOODELL, *supra* note 9.

[54] http://www.ted.com/talks/david_keith_s_surprising_ideas_on_climate_change.html

[55] *See* John Holdren, *Climate Change Science and Policy: What Do We Know? What Should We Do?*, Sept. 6, 2010, *available at:* http://www.whitehouse.gov/sites/default/files/microsites/ostp/jph-kavli-9-2010.pdf.

backtracked somewhat after a brief media frenzy.[56] Finally, as this chapter was in final preparation, three major reports from U.S. government agencies were released that together represent the most sustained and important governmental inquiries into CM in history.

First, SRM was thoroughly discussed in a ten-page chapter of the June 2010 report by the National Research Council.[57] Reviewing the current state of SRM research (including not only the usual stratospheric sulfate aerosol proposals but also placing reflective mirrors in space, cloud whitening, and albedo enhancement[58]), the NRC report focused on five questions: whether SRM could reduce climate change, how to reduce undesirable/unintended consequences (the NRC identified four: failure to reduce ocean acidification, uneven regional shifts, possible reduction of stratospheric ozone, and risk of sudden stoppages; these are discussed in Part II[59]), who should decide whether to use SRM (including both practical/legal and ethical considerations[60]), what institutional mechanisms are necessary for monitoring and follow-up, and what kinds of evaluation methods are appropriate.

Second, the House of Representatives' Committee on Science and Technology released in October 2010 its report on "Engineering the Climate: Research and Strategies for International Coordination."[61] The report draws on an eighteen-month inquiry, including three public hearings, and was prepared in concert with the U.K. Royal Society, which released its report, *Geoengineering the Climate: Science, Governance and Uncertainty*, in September 2009.[62] The report, although repeatedly going out of its way not to endorse CM and to recommend GHG mitigation first (I discuss this issue in Section 2.2 below), is nonetheless the most significant U.S. government document ever produced on the subject of geoengineering – or as it calls it, climate engineering – and includes the testimony of nearly every significant scientist who has worked on it (with the notable exception of Lowell Wood). The report focuses on SRM and CRM,[63] proposes no fewer than twenty-six discrete research areas that should be investigated,[64] and calls on the National Science Foundation to lead the research effort[65] while also opining that the National Oceanic

[56] Andrew Revkin, *Science Adviser Lays Out Climate and Energy Plans*, N.Y. TIMES, Apr. 9, 2009, *available at* http://dotearth.blogs.nytimes.com/2009/04/09/science-adviser-lists-goals-on-climate-energy/,

[57] NRC, ADVANCING SCIENCE), *supra* note 35.

[58] *Id.* at 294.

[59] *Id.* at 295–96.

[60] *Id.* at 296–97.

[61] U.S. House Report, *supra* note 9.

[62] JOHN SHEPHERD ET AL., GEOENGINEERING THE CLIMATE: SCIENCE, GOVERNANCE, AND UNCERTAINTY (2009).

[63] *See id.* at 40.

[64] *Id.* at 7–8.

[65] *Id.* at 9.

and Atmospheric Association, Department of Energy, NASA, and EPA should also conduct research.[66] In sum, the report expresses the view that "broad consideration of comprehensive and multi-disciplinary climate engineering research at the federal level begin as soon as possible in order to ensure scientific preparedness for future climate events."[67]

Third, the Government Accountability Office (GAO) released – also in October 2010 – its first ever technology assessment on geoengineering.[68] The report offers both an evaluation of the social, political, and environmental implications of geoengineering and a survey of the state of the science underlying the various CM methodologies. Together, these three reports, along with the 2009 report of the Royal Society, represent a sea change in the official acknowledgment of CM as a climate change strategy. Of course, hortatory documents are not the same as practical actions, and the discourse of denial continues to mold public opinion. Then again, the more denial, the more delay – and the more delay, the more CM becomes inevitable.

3. THE (U.N.) EMPIRE STRIKES BACK

The quick rise of geoengineering has not gone unnoticed by critics of CM, who have struck back at CM with a series of international declarations. The Convention on Biological Diversity (CBD) has been a central locus for this attempt to curtail CM, not because it is the appropriate international body to do so (biodiversity being only one of many issues impacted by CM, and not necessarily the most important) but because it is in this venue that CM opponents have the most ability to enable nonbinding rhetorical statements to be made. First, in 2008, the COP 9 ("Conference of the Parties") meeting decided, in the context of an "integration of climate-change activities within the programmes of work of the Convention"[69] to focus on OIF. COP 9 "(i) endorsed the June 2007 'Statement of Concern regarding iron fertilization of the oceans to sequester CO_2' of their Scientific Groups, (ii) urged States to use the utmost caution when considering proposals for large-scale ocean fertilization operations and (iii) took the view that, given the present state of knowledge regarding ocean fertilization, large-scale operations were currently not justified."[70] Contrary to the claims of some CM opponents, COP 9 did not ban

[66] *Id.* at 10–26.

[67] *Id.* at 38.

[68] U.S. GOV'T ACCOUNTABILITY OFFICE, A COORDINATED STRATEGY COULD FOCUS FEDERAL GEOENGINEERING RESEARCH AND INFORM GOVERNANCE EFFORTS, GAO PUB'N NO. 10–903 (2010).

[69] Convention on Biological Diversity Conference of the Parties, Decision of the Ninth Meeting, COP 9 Decision IX/16, Biodiversity and Climate Change, Sec. A, May 2008, *available at* http://www.cbd.int/decision/cop/?id=11659.

[70] *Id.* at Sec. C.

OIF,[71] but it did express the (nonbinding) opinion of one UN Convention that it should not be pursued.

In October 2010, COP 10 picked up where COP 9 left off. Amid utopian language (I discuss green utopianism in Part II, infra) such as "A new era of living in harmony with Nature is born at the Nagoya Biodiversity Summit,"[72] it issued a second, broader nonbinding declaration regarding CM. The decision submitted by the chair of a COP 10 working group "invites" parties to

> Ensure, in line and consistent with decision IX/16 C, on ocean fertilization and biodiversity and climate change, in the absence of science based, global, transparent and effective control and regulatory mechanisms for geo-engineering, and in accordance with the precautionary approach and Article 14 of the Convention, that no climate-related geo-engineering activities that may affect biodiversity take place, until there is an adequate scientific basis on which to justify such activities and appropriate consideration of the associated risks for the environment and biodiversity and associated social, economic and cultural impacts, with the exception of small scale scientific research studies that would be conducted in a controlled setting in accordance with Article 3 of the Convention, and only if they are justified by the need to gather specific scientific data and are subject to a thorough prior assessment of the potential impacts on the environment[.][73]

As with the COP 9 declaration, it bears repeating that this language is only an "invitation" and bears no enforceable legal weight. One hundred sixty-eight countries are signatories to the CBD treaty, but the treaty has not been ratified by the United States.[74] Practically speaking, the COP 10 language is an effort by a small number of greens to exercise rhetorical power in the grown-up equivalent of a college debate society. Moreover, as Ken Caldeira noted, the language is so expansive as to make "no sense."[75] However, the COP 10 declaration is a clear sign that some

[71] *See, e.g.*, Climate Connections, Blog Post, *Geoengineering Moratorium at UN Ministerial in Japan: Risky Climate Techno-Fixes Blocked*, Oct. 28, 2010, http://climatevoices.wordpress.com/2010/10/28/geoengineering-moratorium-at-un-ministerial-in-japan-risky-climate-techno-fixes-blocked/ (calling the 2008 decision a "moratorium"); Wayne Hall, *Chemtrail Secrets: Strategies against Climate Change?*, SPECTRE MAG., Jan. 2004, *available at* http://www.rense.com/general49/change.htm (calling my article "a masterful attempt to defend the indefensible," a characterization that I do not entirely dispute).

[72] Convention on Biological Diversity Press Release, *A New Era of Living in Harmony with Nature Is Born at the Nagoya Biodiversity Summit*, Oct. 29, 2010, *available at*: http://www.cbd.int/doc/press/2010/pr-2010–10–29-cop-10-en.pdf.

[73] Conference of the Parties to the Convention on Biological Diversity, Tenth Meeting: Biodiversity and Climate Change, *Draft Decision of Working Group I*, Oct. 29, 2010, UN Doc. UNEP/CBD/COP/10/L.36 (footnote removed), *available at*: http://www.cbd.int/doc/meetings/cop/cop-10/in-session/cop-10-L-36-en.doc. (emphasis and footnote in original).

[74] Eli Kintisch, *Proposed Biodiversity Pact Bars "Climate-Related Geoengineering,"* SCI., Oct. 26, 2010, *available at* http://news.sciencemag.org/scienceinsider/2010/10/proposed-biodiversity-pact-bars-.html.

[75] *Id.*

"dark greens" have begun to take geoengineering seriously, and have slammed the door in response.

Finally, it is interesting to note that geoengineering has also been noticed by more than a few conspiracy theorists such as believers in the "chemtrails" conspiracy, which holds that the U.S. government is secretly spraying chemicals on the population, by means of jet contrails. Indeed, this author was one of the earliest sources for this meme, when chemtrails activists discovered my 1998 article, and linked its discussion of SRM to their preexisting suspicions about government chemical trails. Today, there is an entire Web site, geoengineeringwatch.org, which provides exceptionally detailed descriptions of CM science and policy debates, as well as deeply troubling information, such as directions to the offices of leading CM advocates. Of course, such wing nuts remain on the fringes of public discourse – for now.

4. WHAT CAN BROWN DO FOR THE EARTH?

Geoengineering "scrambles old political alliances and carves out new ideological fault lines."[76] Indeed, one of the most intriguing phenomena of recent CM discourse is that CM is beginning to be taken up by some conservative voices. In some ways, this should be regarded as a welcome turn of events, as it represents some evolution from a position of total denial. But "dark greens" may have good reason to worry.

Newt Gingrich has succinctly laid out the conservative case for geoengineering in a blog post that I shall reproduce in its entirety[77]:

> *Can Geoengineering Address Concerns about Global Warming?*
>
> One of the most intriguing and promising areas of scientific innovation today are methodologies to address concerns about global warming by something called geoengineering.
>
> We need to know more about it, but the idea behind geoengineering is to release fine particles in or above the stratosphere that would then block a small fraction of the sunlight and thus reduce atmospheric temperature.
>
> In other words, this is one method that holds the promise of addressing any threat from global warming at a fraction of the cost. Instead of imposing an estimated $1 trillion cost on the economy by Boxer-Warner-Lieberman, geoengineering holds forth the promise of addressing global warming concerns for just a few billion

[76] GOODELL, *supra* note 9, at 15.

[77] Newt Direct Blog, *Stop the Green Pig: Defeat the Boxer-Warner-Lieberman Green Pork Bill Capping American Jobs and Trading America's Future*, June 3, 2008, http://www.newt.org/newt-direct/stop-green-pig-defeat-boxer-warner-lieberman-green-pork-bill-capping-american-jobs-and-t.

> dollars a year. Instead of penalizing ordinary Americans, we would have an option to address global warming by rewarding scientific innovation.
>
> My colleagues at the American Enterprise Institute are taking a closer look at geoengineering, and we should too.
>
> With gas prices already at record highs, the last thing America needs is government regulation that will make gas prices higher, make Americans poorer, and make special interests even richer.
>
> We need innovation, not regulation. We need motivating incentives, not punishing pain.
>
> Our message should be: Bring on the American Ingenuity. Stop the green pig.

Gingrich's analysis neatly sets forth why CM is attractive to conservatives. First, Gingrich's somewhat conditional language leaves open the possibility that climate change is not really taking place; there is no admission here that the threats are real and the concerns are justified – only the claim that geoengineering would help address them. This "cover" is reminiscent of Guido Calabresi's analysis of the role of subterfuge in creating effective public policy.[78] As I noted in 1998, SRM in particular requires so little time to be effective, it is almost like a remedial solution.[79]

Second, the conservative ideological appeal of CM is clearly stated in Gingrich's post. It is striking how CM methodology does fit so squarely within existing conservative and neoconservative thinking. Indeed, not only would CM not cost industry anything, it could actually generate further revenue, at least for those responsible for research, development, and implementation of CM technology. CM requires no new regulations, and allows Gingrich's corporate supporters to continue with business as usual. Indeed, if past experience is any guide, some of them will probably get into the geoengineering business themselves.

Of course, Gingrich's blog post does not mention the many risks associated with CM, some of which are discussed below. It also presents CM as an alternative to regulation, whereas most CM advocates see it as a supplement to GHG reduction. And of course, Gingrich's post moves CM from a desperate last resort to a preferred policy option, which few climate scientists actually believe it to be.

Gingrich's post also refers to a daylong conference held in 2008 at the American Enterprise Institute (AEI), entitled "Geoengineering: A Revolutionary Approach to Climate Change,"[80] and AEI's subsequent endorsement of geoengineering as a

[78] *See* Jay Michaelson, *Rethinking Regulatory Reform: Toxics, Politics, and Ethics*, 105 YALE L.J. 1891, 1923 (1996), *discussing* GUIDO CALABRESI & PHILIP BOBBITT, TRAGIC CHOICES 26, 41 (1978).

[79] Michaelson, *supra* note 1, at 109.

[80] *See* American Enterprise Institute for Public Policy Research Events, *Geoengineering: A Revolutionary Approach to Climate Change*, June 3, 2008, http://www.aei.org/event/1728. *See also* American Enterprise Institute for Public Policy Research, *A Brief History of Geoengineering*, http://www.aei.org/

remedy for "possible" climate change. Curiously, AEI's Web site advertising the conference was greener than anything the CTT has put out in the past:

> For more than twenty years, policymakers have struggled to find ways to reduce greenhouse gas emissions enough to stop global climate change. Congress is likely to enact federal climate legislation in 2009, but many scientists fear that emissions reductions may not occur quickly enough to prevent significant warming. Some scientists also fear that potentially catastrophic effects, such as the melting of the polar ice caps, could happen unexpectedly quickly. If warming proves to be uncontrollable and dangerous, what could we do?
>
> A growing number of climate scientists believe that there may be only one possible answer to that question: change features of the earth's environment in ways that would offset the warming effect of greenhouse gases, a concept known as "geoengineering" (or "climate engineering"). The most plausible way of doing this would be to use very fine particles in (or above) the stratosphere to block a small fraction (roughly 2 percent) of sunlight. While geoengineering science is in its infancy, most scientists who have studied the idea believe it is likely to be feasible and cost-effective.[81]

The conference was legitimate; at least one of the presentations at the 2008 conference insisted that "geoengineering cannot replace mitigation,"[82] and the conference included respectable scientific authorities – not the same old cast of deniers. Given AEI's long-standing opposition to any climate action – and the considerable support ExxonMobil has given to them – this may seem quite surprising. Then again, Jeff Goodell reports one green activist as saying that "combining dire warnings about climate action's economic costs with exaggerated claims about geoengineering's potential is the new climate denialism."[83]

Is this so? No one, of course, can read the minds of AEI's board members. Yet geoengineering is more than practically convenient for conservatives; it is ideologically consistent as well. On a simple level, CM allows polluters to continue polluting and SUV drivers to continue guzzling gas. But it also is consonant with a wider and deeper conservative view that, essentially, the market and human innovation will eventually solve whatever problems they have created, with no need for complex and freedom-abridging government intervention. Geoengineering could well be a market-created, privately implemented solution that could conceivably proceed without any significant government regulation of behavior.

aei-website/managed-content/site-pages/geoengineering/geoengineering-history.html (last accessed March 13, 2011).

[81] American Enterprise Institute Events, Geoengineering, *supra* note 81.

[82] Wigley, *supra* note 3, at 13.

[83] GOODELL, *supra* note 9, at 15, *quoting* Alex Steffen of Worldchanging.

I will return to the conundrums this poses for environmentalists in Section 2.2 below. Clearly, those of us who consider ourselves environmentalists are faced with a question: Must everything my enemy likes also be hateful to me? Or does CM represent an opportunity for an unprecedented cooperation between greens and browns, the former prevailing in ends (yes, the climate is changing, and yes, we need to act), the latter in means?

Summary: Plus C'est la Même Chose, Plus ça Change

What has changed is a result of what has not changed.[84] Twelve years after my initial article, and eighteen years after the 1992 NAS study, at least half of the U.S. Congress still holds the view (publicly at least) that anthropogenic climate change is not even happening. This is a result not of scientific uncertainty but of lobbying and disinformation by the energy industry and others. Nothing has changed, because there is great, concentrated power against change, and a collective action problem on the other side.

Yet precisely because so little has changed in the politics of climate change regulation, much has changed in the science and policy of climate management. The question now is: Has it changed for the better, or the worse?

5. THE CASE FOR PLAN B

As we have seen, there are numerous arguments for investigating the possibility of CM as a climate change strategy. If nothing else, it is not GHG reduction, which has had such a sorry history that it begs for an alternative. Of course, CM is not a panacea, and so I begin the normative case for "Plan B" by addressing the policy objections that have been raised against it. Some of these I first addressed in 1998; others have become known more recently. All, I believe, are answerable. I then conclude in the following section with a reflection on the deeper meaning of CM for environmental advocates: why it is a good idea to surrender to the villains, and what all of us can learn precisely from our resistance to CM.

5.1 *Are We Still Afraid of Giant Laser Space Frisbees?*

In 1998, I identified four primary policy concerns regarding geoengineering: that it simply would not work, that it costs too much, that it is "unnatural," and that it subverts other efforts at regulation. Three of these four concerns remain active

[84] "The more things stay the same, the more they change." This is the inversion of Jean-Baptiste Alphonse Karr's famous aphorism "plus ça change, plus c'est la même chose," the more things change, the more they stay the same.

today. First, CM technologies remain largely untested, and "common sense" as well as humility cause most people to react with skepticism of their efficacy. Second, CM continues to offend the deep sensibilities of environmentalists because it is unnatural – I discuss this concern in Section 2.2. And third, there is no question that as CM grows in legitimacy, it has the potential to undermine efforts at preventive regulation – although my claim is that regulatory efforts are failing well enough on their own.

One policy concern that has faded in importance is that of cost. Indeed, as we have learned more about the true costs of GHG reduction, and the possibilities of relatively low-cost CM, the issue of cost now seems to cut in favor of CM, rather than against it.[85] Perhaps the metaphor of the "Manhattan Project" that I and others have used for a CM R&D effort is misleading; it is possible that after research, CM may be among the least expensive climate change strategies in a policy portfolio. For example, Caldeira, Wood, and Myrhvold estimate the costs of an Arctic-focused SRM process to be only $20 million in start-up costs and $10 million in annual operating costs.[86] In addition to these older concerns, there are three additional ones that were not well-known in 1998, but present serious challenges today. In each case, I find the concerns valid, but addressable. We now turn to these critiques.

5.1.1 Moral Hazard: "Don't Worry, Be Happy"?

Some critics charge that geoengineering gives a "free pass" to polluters, and will undermine efforts to attain meaningful GHG reduction. This is the "moral hazard" argument: that, in David Keith's elegantly simple presentation, "knowledge that geoengineering is possible -> climate impacts look less fearsome ->a weaker commitment to cutting emissions today."[87]

This argument is quite cogent, but my response is simple: the world community has not needed this moral hazard to postpone action on climate change. It has been doing just fine for twenty years.

Even without any real awareness of CM, even with a melting Greenland and shrinking Arctic, American society and government have totally failed to act on climate change. We do not need this moral hazard, because we are already immoral. It is not as if we were *this close* to meaningful action in Copenhagen, or a serious climate bill in the last Congress. And now, as of 2010, "climate is gone," according to

[85] Scott Barrett, *The Incredible Economics of Geoengineering*, 39 ENVTL. RESOURCE ECON. 45–54 (2008).

[86] *Cited in* Levitt & Dubner, *supra* note 20, at 195. A larger, planetary-focused effort is estimated to cost $150 m to begin and $100 m annually to operate. *Id.* at 196.

[87] Keith, TED Talk, *supra* note 15.

GOP strategist Karl Rove.[88] In short, we are nowhere near where we need to be, and we are moving fast in the other direction, if India and China are included. Yes, millions of people now worry about their carbon footprint. But all that is window dressing if we cannot shift our utilities and major industrial bases to less carbon-intensive consumption patterns, and if China and India do not come to the table.

Some have worried that geoengineering's essential political message is "Don't Worry, Be Happy." But this concern, too, is answerable: as CM advocate Nathan Myrhvold has said, blaming geoengineering for complacency is like blaming a heart surgeon for saving the life of someone who does not exercise and who eats too much.[89] Yes, the existence of heart surgery does engender a certain amount of apathy. But how much? Do overeaters really eat too much because a quadruple bypass is available someday? Obviously not – and surely none of us would ban treatments for heart disease because they do not address the "root problem." Likewise here.

5.1.2 Risks: Unk-Unks

As in 1998, one of the leading criticisms of CM today is that there are too many unknowns and too many risks associated with human management of climate. Bad enough are the "known unknowns" – but worse are what Donald Rumsfeld called "unknown unknowns," or, as they are now called, "unk-unks," which in the case of the earth's climate are effectively infinite.[90]

We cannot answer this criticism today because we have not even begun serious research into CM. We do not know whether CM can work, and we will not know until we take it seriously as a policy option. We know that its theoretical basis is sound. But it is premature to object that "we don't know if it will work" because we haven't yet even begun to investigate it.

Some of this criticism, surely, is more based in fear than in rational calculation: "common sense" simply tells us that SRM, OIF, and albedo enhancement are crazy ideas that will not work. So, Al Gore calls CM "nuts."[91] But what is really nuts, as the old cliché holds, is doing the same thing over and over again and expecting a different result. If there is a concern about the feasibility of a particular project, then more, rather than less, research is warranted. Doubtless, the Apollo missions to the moon seemed loony at the time, yet a serious campaign of research and development yielded success. Likewise, perhaps, with climate management.

[88] The Philadelphia Inquirer Blog, *Fracking Karl Rove to Pa.: "Climate is Gone,"* Nov 3, 2010, http://www.philly.com/philly/blogs/attytood/Fracking_Karl_Rove_to_Pa_Climate_is_gone.html.

[89] Levitt & Dubner, *supra* note 20, at 197.

[90] *See* KINTISCH, *supra* note 4, at 25–26.

[91] *Id.* at 200.

These general anxieties are not, of course, the same as specific concerns that climate scientists have raised. For example, it is possible that sulfate aerosol SRM may increase tropospheric sulfate loading and surface deposition or affect cirrus clouds, and there are numerous uncertainties associated with SRM that bear careful and close evaluation.[92] And it is possible that SRM may negatively impact atmospheric ozone levels, and allow ocean acidification to intensify.[93]

Yet all of these potential risks call for more research, rather than less. And because the economics of regulation are so unfavorable, the risks of *not* researching CM are considerable. As such, this a classic risk versus risk dilemma. Indeed, as David Keith has noted,[94] given the feasibility of CM and the temptation to use it as a quick fix, it might be more useful to compare the risk of geoengineering carefully against the risk of doing so hastily. The question is whether we want an eventual CM implementation to be a hasty fix erected amid an emergency, or one that has been carefully planned over time.[95] The climate chickens will come home to roost eventually; will we be ready, or will our concern over risks make those risks more acute?

5.1.3 Equity: The Rain in Spain Falls Mainly...

As we have learned more about the science of geoengineering, questions of equity have become much sharper than they were in 1998, particularly surrounding SRM. It is already known that the effects of climate change will vary across the globe. Some regions will get hotter, others cooler, and some may even experience benefits from climate change. All of this is complicated by uncertainty, as models vary widely regarding the local effects of global climate change. Even if SRM is applied more or less uniformly across the globe, or focused in less-populated areas (such as the Arctic, where it is needed most[96]), the net effects of SRM on local climate patterns will likely vary considerably from place to place, with some areas experiencing more negative effects than others.

[92] *See* Ken Caldeira & Lowell Wood, *Global and Arctic Climate Engineering: Numerical Model Studies*, 366 PHIL. TRANSACTIONS OF THE ROYAL SOC. A: MATHEMATICAL, PHYSICAL & ENGINEERING SCI. 4039–56 (2008).

[93] Wigley, *supra* note 3, at 13.

[94] David Keith, *Geoengineering the Climate: History and Prospect*, 25 ANN. REV. ENERGY & ENV'T 245–84 (2000).

[95] On the possibility of a sudden climatic emergency, see KINTISCH, *supra* note 4, at 39–52.; Richard A. Kerr, *Climate Tipping Points Come in from the Cold*, SCI. 319 (2008). The recent House Science & Technology Committee report suggested that one agenda item warranting immediate attention would be to define the parameters of a "climate emergency" so that policy makers would have a benchmark on which to base a rapid deployment of SRM technology.

[96] According to David Keith, "Geoengineering may be the only tool we have to save certain ecosystems, like the Arctic." *Quoted in* GOODELL, *supra* note 9, at 39.

For example, the effects of SRM's "cooling" may be experienced in radically different ways, far more complex than mere temperature change: as SRM decreases temperature only in the daytime, it reduces the average temperature difference between day and night, which may wreak havoc with local ecosystemic processes, including plant and animal populations, wind, and precipitation. [97] Some areas may have increased precipitation, others decreased precipitation. The severity of these effects likewise will vary from region to region. Some have expressed concern that SRM might lessen rainfall in Africa, Asia, and the Amazon.[98]

A second example of regional variation could arise with regard to albedo enhancement, which may have strong localized effects; if clouds are brighter in one place but unaltered in another, the differential impact could be significant. This, indeed, is one of the advantages of albedo enhancement, as it might be focused on the poles, where warming is worst.[99]

Regional variations in SRM effects are further complicated by the difficulty of measurement. Although climate is essentially predictable, many of the effects of SRM may manifest more in local weather patterns, which, as we all know, are far less predictable. Even if we could imagine a scenario in which countries compensated one another for the undesirable consequences of SRM, it is difficult to imagine how those consequences would be measured, or attributed to SRM as opposed to some other source. History here may be instructive: thirty years into the acid rain crisis, Northeastern states are still bickering with Midwestern ones about the causes of acid rain and what might be done to compensate for them. Surely, such negotiations would be even more fraught on a global scale.

It is also possible that, if SRM takes place in the stratosphere, the sulfates may adversely affect stratospheric ozone levels. In response, David Keith has recently suggested that SRM focus on the mesosphere, the layer above the stratosphere, where such secondary effects would not arise and where, due to differences in atmospheric currents, SRM could be better focused on the polar regions, where it is needed more.[100] I note that a number of private actors (including Virgin's Richard Branson) are developing high-altitude aircraft/spacecraft that may well be able to deliver the sulfates to the mesosphere at extremely low cost, simply as part of their ordinary business.[101]

[97] *See* KINTISCH, *supra* note 4, at 71–72 (citing presentation of Tony Janetos to the National Academy of Sciences, June 2009).

[98] A. Robock, L. Oman, & G.L. Strenchikov, *Regional Climate Responses to Geoengineering with Tropical and Arctic SO2 Injection*, 113 J. GEOPHYSICAL RES.-ATMOSPHERES **15** (2008).

[99] GOODELL, *supra* note 9, at 114.

[100] Keith, TED Talk, *supra* note 15.

[101] *See* Virgin Galactic, Homepage, http://www.virgingalactic.com/overview/environment/ (last accessed March 13, 2011).

Finally, even accounting for unintended regional variations, it is also the case that not all nations have the same incentives regarding climate change. It has been suggested, for example, that Russia may actually stand to gain from global warming; would Russia insist on compensation for a successful SRM scheme (especially one focused on the Arctic) that prevented such a boon from occurring?[102] As Red Team member James Robock has asked, "Whose hand would be on the thermostat? What if India wants it cooler and Russia wants it warmer?"[103] There is no question that SRM is powerful technology, and whose hand is on the levers controlling it, given the differing incentives among nations, will doubtless be among the most contentious of questions as CM becomes discussed internationally.

As before, though, these policy concerns should not be a reason not to invest in research. On the contrary; we do not yet know the nonlocal effects of localized bursts of sulfur dioxide, but this may perhaps be tested in limited trials of SRM technology. As part of any geoengineering agreement, one could envision a presumption of causality if unusual climatic shifts occur, followed by a quick halt to any activities. In addition, if possible climatic side effects are predicted, CM deployment efforts may be carried out in ways that minimize such effects – higher dispersal of SRM particulate matter, for example, or dispersal over the ocean. Most important, the very existence of potentially harmful climatic side effects should cause CM efforts to proceed with a healthy dose of humility and caution. Part of the rationale for the term "climate management" is precisely that it highlights not only the nature of the proposed activity, but its potential for hubris as well.

5.1.4 Future Costs: In the Art of Stopping

One of the most alarming risks that has become better known over the last decade is that of sudden, cataclysmic costs that would accompany a sudden stoppage of an SRM project.[104] In 2007, Ken Caldeira and Damon Matthews showed that, although SRM could return global GHG levels to pre–Industrial Revolution levels within five years, a sudden stoppage could lead to warming rates skyrocketing to twenty times current rates.[105] SRM is like an addiction that cannot be kicked because the withdrawal would kill the addict.

[102] David Pritchard, *quoted in* GOODELL, *supra* note 9, at 192.

[103] *Quoted in* Svoboda, *supra* note 48.

[104] This subtitle is from the great post-punk band Wire's song "In the Art of Stopping" on SEND (2003).

[105] H. Damon Matthews & Ken Caldeira, *Transient Climate–Carbon Simulations of Planetary Geoengineering*, 104 PROC. NAT'L ACAD. SCI., 9949–54 (2007). *See also* Eric Smalley, *Climate Engineering Is Doable, as Long as We Never Stop*, WIRED, July 25, 2007, http://www.wired.com/science/planetearth/news/2007/07/geoengineering#ixzz13aH4Q2G4.

Some have presented scenarios of this stoppage in the context of war or other calamity. In this regard, however, SRM would not be different in kind from nuclear power plants, hazardous waste facilities, and other ultrahazardous sites that require constant monitoring. Indeed, as in the case of nuclear facilities, the consequences are often far more immediate than those of SRM cessation; at least with a geoengineering stoppage, the time frame is a few years, rather than a few minutes.

That being said, the costs of stoppage must be taken into account in any SRM deployment – and provide yet another reason such deployment must be undertaken collectively, rather than by independent actors. Fail-safe systems, multiple redundancies, and secure international locations for SRM facilities are a few of the precautions that must be put in place; there are doubtless hundreds more.

Surely what the cataclysmic consequences of a sudden cessation of SRM teach us is that, contrary to the emerging conservative arguments, CM cannot be the singular policy of climate change mitigation. A more truly conservative use of CM would be to employ it as a stopgap, allowing time for India and China to develop green technologies and to persuade Western governments to implement costly and difficult GHG reduction programs.[106] Perhaps this stopgap lasts twenty years, or perhaps two hundred – but CM should not be a permanent solution to anthropogenic interference with the world's climatic systems.

5.1.5 Monitoring and Institutions: Going Rogue

As we have already seen, geoengineering is particularly susceptible to rogue actors taking matters into their own hands. Now, in one sense, the capacity of geoengineering to be implemented by private actors is an advantage: it resonates with the free market ideologies of conservatives, circumvents the need for costly public action, and could allow for small-scale testing by private entities. Yet as the episode with would-be OIF pioneer Planktos showed, precisely those mavericks who might be most interested in geoengineering are the sort of private actors none of us would want to trust with the well-being of an ecosystem, let alone the planet.[107] As some critics have noted, rogue SRM is like a scheme cooked up by a James Bond villain.[108]

[106] *See* Tom Wigley, *A Combined Mitigation/Geoengineering Approach to Climate Stabilization*, 314 SCI. 452–54 (2006); David Keith, *Engineering the Planet*, IN CLIMATE CHANGE SCIENCE AND POLICY 494–502 (Stephen H. Schneider, Armin Rosencranz, Michael D. Mastrandrea & Kristin Kuntz-Duriseti eds., 2009).

[107] For a profile of Planktos' Russ George, see GOODELL, *supra* note 9, at 144–58.

[108] SRM in particular resembles the "Icarus" project conceived by the villain Gustav Graves in *Die Another Day*, http://www.imdb.com/title/tt0246460/. *See* Eli Kintisch, *Climate Hacking for Profit: A Good Way to Go Broke*, FORTUNE, May 21, 2010, http://money.cnn.com/2010/05/21/news/economy/geoengineering.climos.planktos.fortune/index.htm.

But what did we really learn from the 2007 debacle with the for-profit corporation Planktos, which attempted, on its own initiative, to conduct limited testing of OIF on the high seas? On the one hand, we learned that OIF is so easy and inexpensive that individual actors might experiment with it, perhaps even with profit in mind. On the other hand, we learned (or relearned) that international law is sufficiently plastic to allow concerned nation-states to nip any unwanted experimentation in the bud, using existing international law frameworks such as the Biodiversity Convention; although, as we noted above, these were not binding law, they provided sufficient pretext for nations seeking to threaten force against rogue CM actors.[109]

More generally, what we have learned is that even the relatively modest OIF ambitions of private actors such as Planktos and its saner cousin, Climos, will raise the hackles of environmentalists and public officials. Although this is inconvenient (and clearly represents a double standard when relatively benign research is prevented while far more intrusive oceanic pollution goes unchecked), it shows that the risk of rogue actors is not significant. The international community will not stand for individual rogue actors, and surely even a national project that lacked international assent would also be opposed in the United Nations and elsewhere.

This, of course, is how it must be. Suppose a Branson/Gates/Gustav Graves rogue effort went horribly wrong; would the culpable parties not be liable for thousands of deaths and billions of dollars? What if a Bill Gates–funded project (Gates has given "a few million dollars" to climate scientists Myrhvold, Wood, Keith, and Caldeira[110]) causes crops to fail in Africa? Geoengineering scientists and their backers require a coordinated international framework to avoid liability for failure. No doubt this is why, since 1998, there have been a profusion of law review articles proposing legal frameworks for geoengineering.[111] These are important contributions toward a future CM regime, and no serious CM advocate would deny the need for international cooperation.

In each of these cases, there are well-founded objections to CM in principle, but each objection is answerable – and generally calls for more research and study, which is what CM proponents are advocating. With the exception of the

[109] *See id.* at 157–61.

[110] GOODELL, *supra* note 9, at 114.

[111] *See, e.g.*, William Daniel Davis, Note, *What Does "Green" Mean?: Anthropogenic Climate Change, Geoengineering, and International Environmental Law*, 43 GA. L. REV. 901 (2009); Alan Carlin, *Global Climate Change Control: Is There a Better Strategy than Reducing Greenhouse Gas Emissions?*, 155 U. PA. L. REV. 1401 (2007) (advocating SRM in lieu of GHG mitigation); Alan Carlin, *Why a Different Approach Is Required if Global Climate Change Is to Be Controlled Efficiently or even at All*, 32 WM. & MARY ENVTL. L. & POL'Y REV. 685 (2008) (advocating SRM). *See also* J. Virgoe, *International Governance of a Possible Geoengineering Intervention to Combat Climate Change*, 95 CLIMATIC CHANGE 103–19 (2009).

moral hazard argument, what these Red Team concerns really argue for is the same thing the Blue Team is arguing for: more research and more international discussion.

As in 1998, however, my sense is that these concerns, although significant, are not really what bother the Red Team. Red Teamers are afraid because, well, they are afraid. Geoengineering is frightening to contemplate, and like genetically modified foods or nuclear energy, it is yet another instance of humankind tampering with nature in an instinctually dangerous way. But let us distinguish between the detailed concerns about CM from "concern" more generally. The latter is at once more fundamental and less articulate: that this is simply the wrong way to go about things, that the bad guys want it, and that it is an indulgence of the same human vices that got us into this mess in the first place. I am sympathetic to this sentiment. Yet as I wrote in 1998, "What a Climate Change Manhattan Project asks on a philosophical level is whether the sorts of strategies and norms that have guided thoughtful environmentalism are always applicable, all the time. Many times [...] I have been struck by the ways in which my own proposal flies in the face of what I believe to be the right thing to do environmentally. But the right thing exists in the mind. Climate change is in the atmosphere."[112] It is to these deeper questions that I now turn.

5.2 *The Vicissitudes of Inevitability*

I believe CM to be inevitable, first because it is the lowest-cost option that appeals to the widest range of political actors, and second, because it makes deep sense to those historically most opposed to climate change mitigation efforts. In many ways, the worst fears of environmentalists have come to pass: the most zealous advocates of geoengineering are now no longer climate scientists, but conservative voices such as the *Wall Street Journal* editorial page. Indeed, it is one such conservative voice who I believe has made the clearest case for why CM is inevitable:

> When I talk with people who object to geoengineering, I often say "You don't have to argue with me, and I don't have to argue with you... because I'm going to win." It's just written in the stars. Geoengineering is going to win, because the politicians, when they finally come down to the crunch, are going to ask: What is the cheapest thing that might possibly do the job? They don't care what it is; if it consists of Las Vegas dancers performing in the rotunda of the capital, they'll choose that if it's the cheapest solution. That's the way things work in a democracy. People never pay more than they have to.[113]

[112] Michaelson, *supra* note 1, at 139.
[113] Lowell Wood, e-mail to Jeff Goodell, *quoted in* GOODELL, *supra* note 9, at 125.

Those words were written by Lowell Hood, the father of "Star Wars" (at $60 billion, hardly the "cheapest solution" to anything). But he is right. When the chips go down, as they will, politicians will search for the fastest, cheapest solution to a problem they failed to solve in advance. And Climate Management is just that. It costs nothing to important constituencies, it has a short lead time, and as ugly as it is, it will become the only tenable option. The only real question is whether to implement that option in a rushed, hasty way, or whether to research, plan and test CM in advance. When looked at from a pragmatic point of view, the game is really already over.

In this concluding section, I want to engage with what it means for an environmentalist to support a strategy that indulges precisely those villains who got us into this mess in the first place – and why we should get over ourselves and do so.

5.2.1 Utopian Dreams, Dystopian Nightmares

> The master's tools will never dismantle the master's house. They may allow us temporarily to beat him at his own game, but they will never enable us to bring about genuine change. (Audre Lorde[114])

As described earlier in this chapter, for libertarians and conservatives, geoengineering as a concept is to be embraced enthusiastically, for the practical, financial, and ideological reasons stated previously. For greens, this is a double conundrum. First, research into geoengineering empowers the "Don't Worry, Be Happy" crowd in their debates with sincere activists trying to fight global warming. Second, if CM actually works, the bad guys win: precisely the most heinous twisters of truth to continue profiting from their exploitation of the earth. But we liberals should not burn the planet out of spite. Yes, CM is our worst nightmare – but it also can awaken us from the utopian dream that somehow, American capitalist consumerist society will transform itself, the "people" will rise, and the malefactors of great wealth will be defeated. This is a pipe dream. Do we really believe that Al Gore's Nobel Prize will really defeat ExxonMobil in the halls of the Congress? Even when the American public overwhelmingly supports some action on climate change, the American Congress does not. And so long as climate change mitigation can be said to require "intrusive government regulation" and "job-killing carbon taxes" and other bogeymen of the Fox News Right, it will be a very difficult sell. We need to find a way beyond the partisan battle lines – and geoengineering, precisely because it is so distasteful and abhorrent to dark green liberals such as me, is that way.

[114] Audre Lorde, *The Master's Tools Will Never Dismantle the Master's House, in* SISTER OUTSIDER: ESSAYS AND SPEECHES 112 (1984).

Perhaps we might take a cue from Ken Caldeira and Lowell Wood, two ideological opposites who are now close allies in CM research and development. Wood, as already mentioned, is the father of "Star Wars," the Strategic Defense Initiative, as well as a protégé of Edward Teller, and a long-time conservative Cold Warrior.[115] Caldeira is a former "quasi-socialist" whose car has a bumper sticker saying "JAIL BUSH."[116] Yet they are friends and colleagues who agree on the reality of climate change and the need for CM to address it.

Moreover, we should consider the impact of utopian, idealistic wishes for a carbon-free future might have on "Southern" developing countries and their populations. A serious shift to low-carbon energy sources would take decades, during which time the developing countries would either need to receive some sort of a "pass" (thus further limiting the impact of such a shift) or would be at a severe disadvantage in terms of their own development. Put simply, there is no way to force everyone to live green without harming the developing world.[117] Of course, all this pretends that it is even "our" choice to make, which in fact it is not; India, Brazil, China, and other nations are not asking the North's permission to develop as they wish. Concealed in the utopian insistence on global lifestyle change, then, is a Eurocentrism that blinds us to the needs of billions of poor people.

It is not surprising, in this vein, that environmentalists on the Left, caught in utopian thinking, resort, over and over again, to nonbinding resolutions, aspirational goals, and hopeful summits in which nothing of consequence is actually achieved. En route to COP-15, for example, the United States and several European countries hammered out a meaningless "aspirational goal" for 50 percent GHG reductions by 2050, after (of course) intense negotiation among themselves.[118] Never mind that such targets have no prayer of being achieved in the United States, China, India, Russia, or Brazil; activists pretended, as we often do, that passing a resolution constituted meaningful action. None of that, of course, saved the Copenhagen summit from disaster, and none of it will help Copenhagen the city deal with rising sea levels.

Climate change is real. Perhaps geoengineering bears out the conservative view that human ingenuity can clean up the messes it creates, and perhaps it is caving into precisely the villains who created this mess in the first place. But I for one would gladly swallow my pride in exchange for being able to walk in the forest.

[115] *See* GOODELL, *supra* note 9, at 120–24.

[116] *Id.* at 111.

[117] For a recent example of this dilemma, see Bob Davis, *World Bank Struggles with Coal Power*, WALL ST. J., Nov. 4 2010.

[118] Jonathan Weisman, *Climate Declaration to Get Global Boost*, WALL ST. J., July 3, 2009, http://online.wsj.com/article/SB124656785956688323.html.

Such humility might be a good thing. If utopian dreaming characterizes some dark-green thinking about environmental policy, dystopian nightmares characterize much of the rest of it. The end of the world is always nigh – economically, politically, environmentally. Such predictions are occasionally correct – the global financial system did indeed melt down in 2008, biodiversity has indeed crashed, and species do often become extinct or extirpated – but more often they are not. Surely the astonishing success of the Montreal Protocol to reverse the loss of stratospheric ozone – and the faster-than-expected recovery of the atmosphere – should give us pause. Is there not a grain of truth to the generally conservative sense that human ingenuity can indeed solve serious problems, and that nature can sometimes rebound faster than we suspect? Now, it may well be that such "pause" feels good precisely because the worst-case climate scenarios are so horrifying. Maybe we cannot handle the truth. But as a scholar of religion, as well as of law, I am also mindful that apocalypses tend to be feared excessively – and tend not to happen.

So, yes, I intend the Audre Lorde quotation to be somewhat ironic. I am indeed advocating, in this case, playing by the master's rules – the political arithmetic determined by financial power. It is more than distasteful to do so; it is practically repugnant. But that is politics. If environmentalists are serious about climate change, then we must divest ourselves of our utopian dreams and dystopian nightmares and do the messy work of political compromise. The adage holds that a compromise is a solution that pleases nobody. If nothing else, Climate Management is that.

5.2.2 Killing Mother Nature

"Dark greens," those of us who view nature not as a mere resource or commodity, but as something sacred in itself, obviously must loathe the very idea of "climate management." Perhaps this is the "end of nature," as Bill McKibben prophesied: a giant air conditioner in the sky, whitening the blue sky and creating fake technicolor sunsets. No wonder the likes of Edward Teller – Dr. Strangelove himself, the man who tried to carve new fjords in Alaska with nuclear weapons[119] – was one of geoengineering's earliest proponents. Blue Team veteran David Keith puts it this way:

> This is why geoengineering is so dangerous, and why we need to be careful about how we pursue this ... It's not the end of nature – but it is the end of wildness – or at least our idea of wildness. It means consciously admitting that we're living on a managed planet.[120]

Climate Management *should* feel wrong; in a way, it goes against the very reasons most greens are green to begin with. But there are at least two responses to this critique.

[119] GOODELL, *supra* note 9, at 70–87.
[120] *Quoted in id.* at 45.

First, let us remember that we are talking here not about intellectual, idealist, or social constructions of "nature,"[121] but about actual ice caps, polar bears, estuaries, and trees. Is it really true that the "managed" Arctic will be so different from the present one? Will the polar bears notice? We greens need to get real. We must grant that the human race has made a mess of the planet. The question now is: Can we clean it up? Or do we continue to delude ourselves that somehow we will reform our ways, not be messy anymore, and somehow miraculously avoid the consequences of a century of profligacy? As my intervening decade spent working on questions of law and religion may perhaps evince, I do believe that attitudes and behaviors can change, and that the deepest problems of our society are spiritual (psychological, if you prefer) at root. None of that, however, is likely on the scale or within the time frame needed to avert climate catastrophe. We dark greens need to get over ourselves.

Second, "nature" is already geoengineered beyond recognition. As Steven Levitt and Stephen Dubner put it, "in just a few centuries, we will have burned up most of the fossil fuel that took 300 million years of biological accumulation to make. Compared with that, injecting a bit of sulfur into the sky seems pretty mild."[122] Anthropogenic climate change is, itself, a form of unintentional Climate Management, and although two wrongs do not make a right, they can sometimes get us back to neutral.

5.2.3 One Man Cannot Make a Difference

Stan: Who are the corporations?
Hippie: The corporations run the entire world. And now they've fooled you into working for them.[123]

We are creatures of narrative. Our sacred myths, our everyday lives, and our political minds all are built upon stories, and it has been this way, it would seem, forever. Narratives are about people – good, bad, and in-between – and they imbue a sense of power and moment to our lives. If only Macbeth had chosen differently; if only Moses had not struck the rock. These stories, even when tragic, imbue our own decisions with a sense of importance; our decisions, they say, matter.

[121] On the social construction of "nature," see KLAUS EDER, THE SOCIAL CONSTRUCTION OF NATURE: A SOCIOLOGY OF ECOLOGICAL ENLIGHTENMENT (1996); I.G. SIMMONS, INTERPRETING NATURE: CULTURAL CONSTRUCTIONS OF THE ENVIRONMENT (1993); David Demeritt, *What Is the "Social Construction of Nature"? A Typology and Sympathetic Critique*, 26 PROG. HUM. GEO. 767–90 (2002), http://phg.sagepub.com/content/26/6/767. *But see* Eileen Crist, *Against the Social Construction of Nature and Wildness*, 26 ENVTL. ETHICS 5–24 (2004).

[122] Levitt & Dubner, *supra* note 20, at 197.

[123] *South Park: Die, Hippie, Die!* (Comedy Central 2005), *script available at* http://www.imsdb.com/transcripts/South-Park-Die-Hippie,-Die!.html.

Yet this reliance on narrative misleads us today, in a world of political and economic forces that are not conveyed adequately in tales. If we look for "what each of us can do to solve the climate crisis," we will be looking in the wrong place for solutions. If we really believe that our individual choices, as opposed to our collective political will, make a serious difference, we are deluding ourselves. Individual emissions are ecologically insignificant compared with the emissions from utilities, and the functional difference between my Prius and my neighbor's Hummer SUV, even multiplied across society, is statistically zero. We need collective action to solve this collective problem, with the largest sources of GHG emissions addressed in a systematic, top-down way that prevents free riders and ensures that all of our individual actions actually add up to a meaningful abatement in global GHG emissions. A well-meaning voice in Leonardo DiCaprio's *11th Hour* puts it well: "People need to realize there are things they can do in their everyday lives. Everybody making a change adds up to something meaningful."[124] But this sentiment, however pleasant, is demonstrably false. None of us individually has the power to be a hero.

Nor can we simply hope that the "villains" in climate change will reform their ways. Bashing corporations is, as the *South Park* quote suggests, something of a cliché. The problem is not that corporations are led by Montgomery Burns–like villains. The problem is in the very structure of corporate law, which requires public corporations to maximize profit for shareholders. Imagine if corporate entities really were people, as the legal fiction suggests. What kind of people would they be? Enormously powerful, and richer than the wealthiest billionaire. Nearly omnipresent, with outposts around the globe. And totally, animalistically greedy, with only the single focus of maximizing profits. The villains are not individual "black hats," although those certainly exist. The villain is the system itself.

Climate management does not affect individual behavior, and for that reason it sits poorly with liberals, who cast the problem of, and solution to, climate change in terms of individual choices. Now, of course, it is better to have fluorescent bulbs than not to have them. But doing so confuses ethics with efficacy. Ethically, it may well be wrong to have a large carbon footprint. But practically speaking, it is meaningless, absent collective action. Obviously, if we ever have real climate change regulation, changes in personal behavior will perforce be required. But until then, the focus on individual action is a fallacy that diminishes the importance of collective, pragmatic political action. It is nice to pretend that if each of us acts ethically, the forests will not burn. But given the factual inaccuracy of that view, is it not, in fact, unethical to maintain it?

[124] *The 11th Hour*, directed by Leila and Nadia Conners (Warner Bros., 2007).

6. CONCLUSION, OR, HOW I LEARNED TO KEEP WORRYING AND STILL LOVE CLIMATE MANAGEMENT

Am I still a geoengineering believer? Yes, more than ever. The last twelve years have shown that the vested interests in doing nothing about climate change are stronger than I predicted in 1998, and give rise to a deep pessimism that the United States – and a fortiori China and India – will ever make the GHG reductions necessary to avert dangerous climate change. So what are the next steps?

First, we need an immediate commitment to government subsidized research, on the basis of the October 2010 House Science & Technology Committee Report, the September 2010 GAO Report, and the 2010 NRC report, each of which listed specific areas for further research.[125]

Second, we need a shift in the policy and legal discourse surrounding Climate Management. How we talk matters. Along with the term "geoengineering," I would like to remove the whiff of whimsy that still lingers around many popular, legal, and even scientific discussions of CM. The more this strategy is depicted as being something out of Jules Verne, the less seriously it, and climate change, will be taken. CM has long been too brown for the greens and too green for the browns. But we are clearly at a tipping point. I argued in 1998 that the time for a geoengineering "Manhattan Project" was now. After twelve years, it seems as though "now" has finally come.

Geoengineering is, as the New America Foundation dubbed it, "a horrifying idea whose time has come."[126] As Scott Barrett said, geoengineering's "future application seems more likely than not."[127] It is a matter of simple economics: "the incentives for countries to experiment with geoengineering, especially should climate change prove abrupt or catastrophic, are very strong. It is also because the incentives for countries to reduce their emissions are weaker."[128] Hoping that we can avoid the need for CM is like hoping we can avoid the need for cars and air conditioners, envisioning some utopian future in which everyone will have the personal tastes of a "dark green" and the lack of personal property to match. Not only is this vision hopelessly unrealistic and potentially fascistic, it sacrifices to its ideal eschatological vision the very survival of the biosphere as we know it today.

[125] U.S. GOV'T ACCOUNTABILITY OFFICE, *supra* note 68; NRC, ADVANCING SCIENCE, *supra* note 35, pp. 297–99.

[126] New America Foundation, *supra* note 16.

[127] Barrett, *supra* note 86, at 45.

[128] *Id.*

I return, then, to my concluding paragraph from 1998, which I think still holds true, twelve years on:

> In the end, the debate about geoengineering is largely a debate about what sorts of environmental policies to pursue in an imperfect world. It seems almost preposterous to buck the trends of holistic systems management and suggest running like the Sorcerer's Apprentice from symptom to symptom. It may also seem as though driving less or cutting fewer trees is simpler than scattering dust particles in the stratosphere. It is certainly more elegant. But when the Damocles' sword of massive biotic disruption is hanging over our heads, we should choose what works.[129]

ACKNOWLEDGMENTS

Much of the research for this project was completed when I was a Visiting Assistant Professor at Boston University Law School. I would like to thank Andrew Novak, Jay Wexler, Gerald Leonard, Robert Sloane, and the students in my environmental ethics class for their stimulating conversations and invaluable assistance.

[129] Michaelson, *supra* note 1, at 139.

5

Climate Engineering and the Anthropocene Era

Lee Lane

...Machiavelli discovered the necessity and the autonomy of politics, politics which is beyond moral good and evil, which has its own laws against which it is futile to rebel, which cannot be exorcised and banished from the world with holy water.

Benedetto Croce[1]

1. THE CLIMATE CHANGE PROBLEM AND RESPONSES TO IT

In the last century, and continuing in this one, human wealth and population have risen massively. The speed and scale of the growth surge are unlike anything that had gone before. One of the many side effects has been a substantial rise in greenhouse gas (GHG) levels in the atmosphere. This trend has led to novel threats of climate change and ocean acidification.

Most efforts to counter these threats have sought to lower emissions. These endeavors, though, must steer a course into the force of the gale-force winds of global growth against which, despite twenty years of trying, controls have made scant headway. This chapter delves into the causes of this failure. It concludes that, for at least one or two more generations, the prospects for enhancing human welfare by limiting GHG emissions are slim.

In consequence, other responses merit exploration. One potential option is to take measures to engineer earth's climate. The most promising such strategy would seek to reduce warming even while GHG concentrations keep on rising. Careful research and development (R&D) could narrow the uncertainties about harmful side effects and about the political management problems that might block climate engineering (CE). The greater is the risk posed by very harmful climate change, the greater would be the potential payoff to finding out if CE is a practical option

[1] ISAIAH BERLIN, *The Originality of Machiavelli*, *in* THE PROPER STUDY OF MANKIND: AN ANTHOLOGY OF ESSAYS 269, 298 (Henry Hardy& Roger Hausheer eds., 1997).

Worries about abrupt, very harmful climate change have been on the rise,[2] but many analysts continue to claim that GHG control is inherently superior to CE.[3] Some believe that even R&D on climate engineering endangers the prospects for GHG control, and they have concluded that R&D is, for that reason, a bad idea.[4]

Contrary to these views, I shall maintain here that fairly comparing controls and CE, shows that, although both options may be needed, and neither is perfect, GHG controls may be likely to cause net economic loss. A major advantage of CE would be to allow the process of curbing emissions to unfold more slowly, and, hence, at less cost. This effect, of course, is the outcome most feared by those who regard stringent emission controls to be an urgent moral imperative.

Their attempts to rule CE out of bounds on ethical grounds, or to set impossible terms for its use, may do great harm. An "ethic of responsibility" to use Max Weber's term, suggests a focus on knowing the likely consequences of our policy choices and accepting responsibility for them rather than on more abstract ethical precepts. This more concrete approach might start by exploring the nature of the forces and trends that have created the current climate policy impasse, and to that task we now turn.

1.1 *The Modern Transformation and Its Potential Limits*

Throughout most of history and prehistory, Malthusian forces and destructive rent seeking tended to erode much of the gain in wealth that resulted from technological advance, and, although human population tended to grow over the centuries, it did so only slowly and unevenly.[5] In stark contrast to this story, the twentieth century marked what is, to quote a book title, "Something New under the Sun."[6]

In this age, a vast growth in human population has accompanied an equally vast rise in average per capita wealth. We tend to take the new wealth for granted; it amounts, though, to nothing less than history's most profound revolution.

> The world's economy in the late twentieth century was about 120 times larger than that in 1500. Most of this growth took place after 1820. The fastest growth came in 1950–1973, but the whole period since the World War II saw economic growth at rates entirely unprecedented in human experience.[7]

[2] Richard S.J. Tol, *The Social Cost of Carbon: Trends, Outliers and Catastrophes*, 2 ECON.: THE OPEN-ACCESS, OPEN-ASSESSMENT E-JOURNAL 1, 7 (2008).

[3] DIANA BRONSON, PAT MOONEY & KATHY JO WETTER, RETOOLING THE PLANET? – CLIMATE CHAOS IN THE GEOENGINEERING AGE 34 (2009).

[4] *Id.* at 37.

[5] JOEL MOKYR, THE GIFTS OF ATHENA: HISTORICAL ORIGINS OF THE KNOWLEDGE ECONOMY 31 (2004).

[6] J.R. MCNEILL, SOMETHING NEW UNDER THE SUN: AN ENVIRONMENTAL HISTORY OF THE TWENTIETH CENTURY WORLD (2000).

[7] *Id.* at 6.

Two sets of institutions have called forth this transformation. Institutions are the diverse codes of rules that govern human action in society. The rules that reward discovery and disclosure within the community of scholars have given rise to modern science,[8] and, as those institutions evolved and developed, they sealed the marriage of science and technology.[9] Forging this union has greatly speeded up technologic progress. Traditional industries have been altered beyond recognition, and wave after wave of entirely new techniques have transformed the global economy.

A second trend in institutional change has been equally central. It is the spread of what some call "open order" societies. Open order societies grant relatively unimpeded access to markets and to the political process. The historical record shows that, over the long haul, a high degree of open access to either set of activities seems to require openness to the other as well.[10] Open markets and open politics cause advanced democracies to function very differently from the various autocracies; hence, facing a like challenge, the two types of society often respond in quite unlike ways.[11]

Countries differ greatly in the degree to which they have been able to absorb both the institutions of science and those of open markets and politics. In the case of science, access to knowledge is clearly unevenly distributed.[12] With regard to market and political openness, many states have taken on the trappings of democracy, but there are only about fifty true open order societies in the world.[13]

How well open order society and modern science can thrive in cultures unlike those that gave birth to them remains to be seen.[14] The answer, though, will define the limits of the modern transformation, determine its pace, and greatly affect the ways in which societies will cope with it. It will, therefore, deeply affect the course of manmade climate change, which may be regarded as one manifestation of the modern transformation.

1.2 *The Global Political Economy of Climate Change*

The modern transformation has led to the full dawn of what Paul Crutzen has called the Anthropocene, a new geologic era in which humans exert strong influence over

[8] Paul A. David, *The Historical Origins of "Open Science": An Essay on Patronage, Reputation and Common Agency Contracting in the Scientific Revolution*, 3 CAPITALISM AND SOC'Y 23 (2008).

[9] MOKYR, *supra* note 5, at 18.

[10] DOUGLASS C. NORTH, JOHN JOSEPH WALLIS & BARRY R. WEINGAST, VIOLENCE AND SOCIAL ORDERS: A CONCEPTUAL FRAMEWORK FOR INTERPRETING RECORDED HUMAN HISTORY 20 (2009).

[11] *Id.* at 21.

[12] MOKYR, *supra* note 5, at 2.

[13] Barry R. Weingast, *Why Developing Countries Prove so Resistant to the Rule of Law*, IN GLOBAL PERSPECTIVES ON THE RULE OF LAW 30, 37 (James J. Heckman, Robert L. Nelson & Lee Cabatingan eds., 2010).

[14] NORTH ET AL., *supra* note 10, at 271.

global systems. It would be astonishing if all this change did not come with a price tag, and its darker side is apparent. Greater human wealth and numbers have not only disrupted a host of local or regional ecosystems; they have also upset the balance of several global-scale natural cycles.[15]

Of these, none is a source of greater concern than the disruption of the carbon cycle. Industry, loss of forest cover, and animal husbandry are causing more carbon dioxide and other greenhouse gases to accumulate in the earth's atmosphere. Three concerns have arisen. First, gradual warming will imply climate change that, at some point, is likely to cause costs in excess of its benefits. Second, warming could occasion more abrupt climate change, which, by allowing less time for adaptation to the new conditions, might cause far higher net costs[16] Third, rising levels of CO_2 will acidify the oceans and perhaps thereby lower their productivity.[17]

Richer societies in more temperate climates are better able to adapt to climate change than are poorer ones in tropical climates. Many industrialized countries are, therefore, lukewarm at best to costly plans to curb greenhouse gas (GHG) emissions. Some poorer states are also significant sources of GHG emissions,[18] but all such states lack the wealth to pay for capping their own emissions let alone for inducing richer countries to cut theirs. Compared to gradual warming, abrupt, high impact climate change would be likely to impose large costs even on industrialized states.[19] The likelihood of such change, though, remains in doubt, and the same is also true of ocean acidification.[20]

2. GHG CONTROLS: AN OPTION FRAUGHT WITH CONTRADICTIONS

From the 1970s through the 1990s, experience taught the industrialized countries to focus on abating pollution rather than on finding least-cost means to minimize the harm that it might cause. Their efforts along this line scored a series of seeming successes. Several kinds of pollution were curbed, and damages to public health and property fell. These efforts typically did not use the lowest-cost policy tools to curb emissions, but the gains were visible, and they made emission control seem like the logical first response as the GHG threat came to light.

[15] MCNEILL, *supra* note 6, at 110.

[16] Martin L. Weitzman, *On Modeling and Interpreting the Economics of Catastrophic Climate Change*, 91 REV. ECON. & STAT. 1, 18 (2008).

[17] J.A. KLEYPAS ET AL., IMPACTS OF OCEAN ACIDIFICATION ON CORAL REEFS AND OTHER MARINE CALCIFIERS: A GUIDE FOR FUTURE RESEARCH 1 (2006).

[18] David A. Weisbach, *Responsibility for Climate Change, by the Numbers* 2 (Univ. of Chicago Law & Econ., Working Paper No. 448, 2009), http://www.law.uchicago.edu/files/files/448–255.pdf.

[19] Scott Barrett, *Geoengineering's Role in Climate Policy*, AMERICAN ENTERPRISE INSTITUTE FOR PUBLIC POLICY, Mar. 6, 2009, http://www.aei.org/docLib/Barrett%20Draft.pdf.

[20] KLEYPAS ET AL., *supra* note 17, at 69.

Three main reasons argue against this presumption. First, the net benefits of GHG cuts are small even at best; hence, using wasteful policy tools to curb emissions is nearly certain to yield net costs; yet, as discussed in Section 2.2, the political systems of the major states appear to be strongly biased in favor of the use of just such wasteful tools. Second, for GHG controls to be effective, all major emitting states must adopt them, but, at least for the next several decades, some of those states will be economically better off without GHG controls or with only lax ones. (The less cost-effective are the GHG control options that a country can make use of, the stronger will be its rationale for eschewing action or limiting its severity.) Third, either coercing or bribing reluctant states to adopt controls is likely to cost more to the states taking these actions than they could reap in benefits from emission abatement.

2.1 *GHG Control: Potential Net Benefits Are Small*

In principle GHG control could yield net benefits, but the potential gains appear to be small. By pursuing controls through inefficient policy tools, countries run a high risk of incurring net costs; further, controls are likely to impede progress toward other national policy goals.

2.1.1 Theory Suggests Small Net Benefits

One recent estimate pegged the present discounted value of the potential net gains from GHG control at slightly more than $3 trillion over the next 250 years.[21] Compared to the size of the global economy, which was more than $40 trillion *a year* in 2005, this is not a very big number. Of course, changes in the current state of knowledge about climate science and economics could change expectations on this score, but society can do no better than to act on the best information that it has at any given time.

On this basis, the costs of deep cuts in GHG emissions are high. Globally, proposals by former vice president Gore and British government economist Nicholas Stern have been calculated to entail *net* costs of $17 trillion and $22 trillion, respectively.[22] That is, these proposals are far more expensive than doing nothing at all would be. Even these estimates, though, assume that governments would use the least-expensive possible policy tools in order to meet their targets.[23]

[21] WILLIAM D. NORDHAUS, A QUESTION OF BALANCE: WEIGHING THE OPTIONS ON GLOBAL WARMING POLICIES, Table 5–1 (2008).

[22] *Id.* at 87.

[23] *Id.* at 91.

Some analyses, to be sure, including Stern, reach different answers, but these studies discount the value of future benefits at very low rates.[24] It is worth noting that some economists believe that the Stern analysis cherry-picked studies to find the most pessimistic available estimates of future harm.[25] These narrower problems aside, the main factor that puts Stern at odds with so many other studies is his use of a very low discount rate.[26] Reasons for skepticism about this facet of Stern's analysis will be discussed in Section 4.2.3.

Still other studies take a different path to the conclusion that GHG cuts should be deeper than those suggested by more mainstream analysis. They surmise that the risk of very high-impact climate change demands stronger action than is often assumed.[27] Even though the probability of these outcomes is very low, they would be so damaging that avoiding them might warrant incurring substantial costs.

The scale of this risk remains doubtful. For the remainder of this century, the odds are heavily against these sorts of extremely negative outcomes, and whatever risk does exist, lies far in the future.[28] Then too, unless the tipping point at which abrupt change is triggered falls between business-as-usual emissions and the trajectory that would result from controls, the abatement effort would fail to avoid the threat. As no one knows where the tipping point lies, or even if it exists, the mere existence of the hypothetical risk tells us less than we might wish about the proper course of action.

2.1.2 GHG Control and Other Policy Goals

Climate change, then, is a serious problem. It is, though, not the only such problem facing the world, or, based on the revealed preference of the world's governments, does it seem to be the most important, and GHG control might affect other policy concerns. Some of these side effects might be positive; local air pollution is an example. Some effects, though, are likely to be negative.

For instance, large scale GHG control, were it actually implemented, might well upset the existing global trade regime. The world has benefited greatly from the economic integration produced by the late–twentieth century trade

[24] *Id.* at 168–69.

[25] Richard S.J. Tol, *The Stern Review of the Economics of Climate Change: A Comment*, 17 ENERGY & ENV'T 977, 979 (2006).

[26] NORDHAUS, *supra* note 21, at 160.

[27] Martin L. Weitzman, *On Modeling and Interpreting the Economics of Catastrophic Climate Change*, 91 REV. ECON. & STAT. 1 (2008).

[28] William Nordhaus, *An Analysis of the Dismal Theorem* 21 (Cowles Foundation for Research in Economics at Yale University, Discussion Paper No. 1686, 2009), http://www.dklevine.com/archive/refs4814577000000000116.pdf.

liberalization,[29] but GHG control will raise energy prices and, by doing so, also boost transport costs. This effect would depress trade. Implementing GHG control, if it can be done at all, will probably entail the use of trade sanctions as a coercive measure, and history has shown that conflict can quickly unravel the global trading regime.[30]

GHG controls also depress agricultural production and increase global food prices. Higher energy prices will depress farm output. GHG controls are also causing fuel crops to replace food crops, and forest offset policies can also take land out of food production.[31] Controls are touted as a means of lessening the *future* problems that might arise as climate change subjects the Third World to food shortages; yet food price hikes in Third World cities caused by GHG controls could have the same effect – *today*.

Finally, divergent climate goals among the great powers have already become an irritant in key bilateral relationships. Historically, a rising state's challenge to an existing hegemon has sparked severe conflict.[32] Injecting more discord into the United States/China dyad is not a good thing. More generally, the tone of moralism and opportunism that prevails in the UN climate talks[33] is hardly likely to foster cooperation on other issues.

2.2 *National Institutions and GHG Control*

Thus, even ideal GHG control regimes entail high costs and serious risks. Moreover, in practice, controls will be far from optimal[34]; therefore, their costs could easily exceed their benefits.[35] The International Monetary Fund (IMF) has described the features that are necessary for a GHG control program to be cost-effective. Such a program should impose a price on emissions. This price should be spread uniformly

[29] RONALD FINDLAY & KEVIN H. O'ROURKE, POWER AND PLENTY: TRADE, WAR, AND THE WORLD ECONOMY IN THE SECOND MILLENNIUM 525 (2007).

[30] *Id.* at 535.

[31] Sugandha D. Tuladhar et al., *Effects of Land Use Tradeoffs on the U.S. Agriculture Sector under a Carbon Policy* 2, Oct. 15, 2010, Paper presented to the U.S. Association of Energy Economists, Calgary, Canada, http://www.usaee.org/usaee2010/submissions/OnlineProceedings/USAEE_Calgary_SDT.pdf.

[32] ROBERT GILPIN, *Hegemonic War and International Change*, *in* CONFLICT AFTER THE COLD WAR: ARGUMENTS ON CAUSES OF WAR AND PEACE 75 (Richard K. Betts ed., 2d ed., 2002).

[33] PAUL COLLIER, THE PLUNDERED PLANET: WHY WE MUST – AND HOW WE CAN – MANAGE NATURE FOR GLOBAL PROSPERITY 175 (2010).

[34] Lee Lane & David Montgomery, *Political Institutions and Greenhouse Gas Controls*, AM. ENTERPRISE INST. FOR PUB. POL'Y 8, NOV. 1, 2008, http://www.aei.org/docLib/20081103_LanePoliticalInstitutions.pdf.

[35] Richard S.J. Tol., *Carbon Dioxide Mitigation*, *in* SMART SOLUTIONS TO CLIMATE CHANGE; COMPARING COSTS AND BENEFITS 74, 95 (Bjorn Lomborg ed., 2010).

across all economic sectors and all nations.[36] The price on emissions should not fluctuate unpredictably. Carbon taxes would be the best means of achieving that goal.[37] Actual policies and most proposals now under consideration bear little resemblance to this vision.

2.2.1 Institutional Constraints

Two examples, the United States and China illustrate the point. The two governments' policy processes could hardly be less alike. Yet clearly, GHG controls cannot succeed unless both of these countries adopt strong measures.

In the United States, energy price hikes are unpopular, and central decision makers find it hard to impose their wishes on organized and mobilized interests.[38] The climate policy failures of Presidents Clinton and Obama could serve as textbook illustrations of this point.

In the United States, each branch of government responds strategically to its own unique set of incentives, including the constraints imposed by the other two branches.[39] The separation of power erects a complex array of veto gates at each of which action can be blocked.[40] The climate policy that has emerged is piecemeal and incoherent.

In China, too, higher energy prices are unpopular, and the Chinese Communist Party fears the backlash that they might unleash.[41] Much control of the economy has devolved to regional and local governments, and Beijing's ability to impose its will on them is limited. Incentives for regional and local governments and those built into the money and banking systems push the economy toward high energy-intensity. This fact must further dampen Beijing's very weak appetite for GHG controls.

Neither country's institutions seem likely to lead to cost-effective GHG controls, nor have they.

2.2.2 U.S. Command-and-Control Regulation

The 111th Congress again rejected all GHG cap-and-trade bills. Repeated failures to pass such measures has led the world, as well as most Americans, to believe that the

[36] INTERNATIONAL MONETARY FUND, *Climate and the Global Economy, in* WORLD ECONOMIC OUTLOOK 23, 2 (2008).

[37] *Id.*

[38] Stephen D. Krasner, *US Commercial and Monetary Policy: Unraveling the Paradox of External Strength and Internal Weakness*, IN POWER, THE STATE, AND SOVEREIGNTY: ESSAYS ON INTERNATIONAL RELATIONS 36, 43 (Stephen D. Krasner ed., 2009).

[39] J.P. Rui De Figueiredo, Jr., Tonja Jacobi & Barry R. Weingast, *The New Separation-of-Powers Approach to American Politics, in* THE OXFORD HANDBOOK OF POLITICAL ECONOMY 199, 200 (Barry R. Weingast & Donald A. Wittman eds., 2006).

[40] *Id.* at 219.

[41] SUSAN L. SHIRK, CHINA: FRAGILE SUPERPOWER 53 (2007).

United States is not limiting GHG emissions, but those impressions are at once false and too optimistic. They are false, because, although the United States has spurned cap-and-trade, it has, in fact, been adopting GHG control policies; the impressions are too optimistic because those policies are very likely to cost more than the climate damage that they avoid would have been worth.

In the United States, since 2006 legislative, executive, and judicial actions have put in place a complex array of measures that seek to reduce GHG emissions. Some actions have occurred at the federal level. Thus, the Energy Independence and Security Act of 2007 enacted a series of mandates that were meant in part to lower GHG emissions. They cover lighting and other uses of electricity and natural gas. They also include a national renewable fuel standard, and they tightened corporate average fuel economy standards. The Bush administration launched a series of rulemakings at the Environmental Protection Agency (EPA); this effort sought to promote the use of renewable fuels in transportation. The 2009 stimulus bill added subsidies for specific energy technologies. California and some other states have taken similar steps.

Currently, the EPA plans to regulate under the Clean Air Act CO_2 emissions from large sources. Already though, steps by elected officials, lawsuits in U.S. courts, and regulatory proceedings have effectively halted the construction of new coal-fired power plants that lack the technology to capture CO_2 emissions.

Thus, on the one hand, no one can correctly say that the United States has not acted on GHG control. The marked declines in the Energy Information Administration most recent long-term GHG emission forecasts show the large effects of the existing measures. Even after accounting for the impact of the recession, these forecasts show a notable fall in expected emissions.

Yet, on the other hand, no one should think that the resulting dog's breakfast of measures amounts to *cost-effective* GHG control. To produce the most gain per dollar of cost, standards should prevent all emissions that can be avoided for less than the target cost per ton, and that target cost should be the same across all sectors. In contrast, U.S. GHG control measures compel some sources to pay a lot to avoid an added ton of emissions, they require other sources to pay a little, and they allow still others to pay nothing at all. This patchwork is light-years from the broad uniform system that the IMF called for, and, compared to the IMF model, it will achieve less, cost more, or both. But the United States is not alone in struggling with GHG control.

2.2.3 China's Institutions and GHG Intensity

China has grounds for concern about potential harm from climate change. Water supply is already a problem, and much of the economy is concentrated on the coast.[42] Even so, economic development can reduce these risks, and, for China, it may be a better shield from harm than GHG controls would be.[43]

Consistent with this reasoning, China's GHG policy appears to be mostly a by-product of steps taken for other reasons. The central government seeks to create some new "green" industries, and it appears to be promoting electric vehicles. These steps, though, are likely to have little short-term effect, and they do not fit easily with any commitment to specific emission targets.

Efforts through the Clean Development Mechanism to imitate global cap-and-trade have proven to be highly vulnerable to gaming. CDM offers rewards for emissions cuts. It must, therefore, define a baseline emissions path against which to measure progress, but the need to define baselines creates incentives to overstate emissions. Hence, CDM may appeal most strongly to countries or projects where insiders are best able get away with such overstatements. Also, neither the project's funder nor its manager has an incentive to ask awkward questions about actual performance. Third-party monitoring and detailed accounting rules may limit abuses, but they entail high transactions costs. CDM's record exhibits all of these kinds of problems.[44]

More intriguing is China's pledge to lower its economy's energy intensity.[45] Its stated goal is ambitious, and it may not be met. In any case, given China's blistering growth rate, the intensity target would still leave emissions growing robustly.[46]

The real point, though, is that the intensity target should be juxtaposed with the broad array of other policies that contribute greatly to excess energy use. China's high energy intensity is deeply rooted in its political economy. The officials who control local and regional governments wield a great deal of power over the economy. Whatever other goals Beijing might proclaim, many such officials believe that the Communist Party will reward only those who deliver robust local economic growth.[47] As the Party controls their careers, this belief has a large impact on priorities.

Further, the central government has, in effect, imposed large unfunded mandates on these governments. The latter's ability to carry these burdens depends on a close relationship with business and on the latter's rapid economic growth.[48] The result

[42] David Anthoff & Richard S.J. Tol, *On International Equity Weights and National Decision Making on Climate Change* Section 4 (Hamburg University: Sustainability and Global Change Research Unit, Working Paper FNU-127, 2007).

[43] Thomas C. Schelling, *What Makes Greenhouse Sense*, 38 IND. L. REV. 581, 593 (2005).

[44] Michael Wara, *Measuring the Clean Development Mechanism's Performance and Potential*, 55 UCLA L. REV. 1760, 1797–98 (2008).

[45] Zhang ZhongXiang, *Copenhagen and Beyond: Reflections on China's Stance and Responses* 11 (Foundation Eni Enrico Mattei Sustainable Development Series, Working Paper No. 91, 2010).

[46] *Id.* at 4.

[47] C. FRED BERGSTEN ET AL., CHINA'S RISE: CHALLENGES AND OPPORTUNITIES 78 (2008).

[48] Zhang, *supra* note 45, at 9–10.

has been a symbiotic system. Business owners seek favors (and protection) from local officials, and the latter depend on payments from business to augment state expenditures and personal consumption.[49]

Features of the money and banking systems abet excessive energy use. State-owned banks are still dominant. They pay little or no real interest to depositors, but they also demand little of debtors that have government backing. Thus, state-backed heavy industries can continue to invest well beyond the point of excess capacity; indeed, their inability to earn adequate returns on banked savings may urge them to do so. The product of the resulting excess capacity can be shunted into export markets – thanks to the undervalued yuan.[50] This interlocked complex of institutions has led to excess growth in energy-intensive heavy industries, and that excess growth is, in turn, the main source of China's high energy intensity. [51]

2.3 *The High Costs of Global Agreement on GHG Control*

Over and above these national level problems, another set of challenges exists at the global level, where GHG control must really stand or fall. The crucial task is to build an international "regime." A regime is a set of "implicit or explicit principles, norms, rules, and decision-making procedures around which actors' expectations converge in a given area of international relations."[52] The World Trade Organization and the International Monetary Fund are examples of regimes. Great powers create such structures in order to lower the transaction costs of coordinating their varied interests.[53]

There is a regime for GHG control. It is the United Nations Framework Convention on Climate Change (UNFCCC), and, like some regimes, but unlike others, the UNFCCC has proven to be strikingly feeble. There seems no exit from the endless round of UN-sponsored climate talks. Each year the cycle repeats like one of nature's great herd migrations and with no more prospect of breaking the cycle.

2.3.1 Core Problems of Global GHG Control

The twenty-plus–year record of fruitless yearly UN climate talks offer eloquent evidence of the high transaction costs of reaching an agreement of this kind; yet the

[49] GORDON REDDING & MICHAEL A. WITT, THE FUTURE OF CHINESE CAPITALISM: CHOICES AND CHANCES 95–96 (2010).

[50] BERGSTEN ET AL., *supra* note 47, at 112.

[51] *Id.* at 111.

[52] STEPHEN D. KRASNER, *Structural Causes and Regime Consequences: Regimes as Intervening Variables*, *in* POWER, THE STATE, AND SOVEREIGNTY: ESSAYS ON INTERNATIONAL RELATIONS 113 (Stephen D. Krasner ed., 2009).

[53] ROBERT O. KEOHANE, AFTER HEGEMONY: COOPERATION AND DISCORD IN THE WORLD POLITICAL ECONOMY 52 (1984).

result, or the lack of result, should surprise no one. That so many states are involved, that their values differ, and that trust among many of them is scarce all make striking a deal more costly.[54] Yet those are just the start of the problem. Other factors are more important still.

One core problem is that states' preferences over climate vary widely. Large uncertainties persist, but some states would be better off without controls. Russia, with its cool climate and the prospect of new oil and gas finds from the retreat of Arctic Sea ice, might be an example. So too might other states that are counting on exporting their large reserves of fossil fuels. GHG curbs might harm their economies far sooner and more severely than would climate change.[55]

A second core problem is that the nature of the international institutional process necessary to achieve GHG control confers much power on holdouts. Successful controls will eventually require nearly worldwide cooperation;[56] therefore, the willingness of a few states to incur costs to curb global emissions can have little or no impact on global emission levels. The less concerned states have a large degree of hold-up power. In still other cases, many states might agree with China that GHG controls would slow their economic growth too much to be a good policy. For the United States, its wealth and temperate climate may imply that for several decades adapting to change may be the best policy.[57] The very low cost effectiveness in U.S. GHG control policies highlights the unappealing nature of controls.

A third issue deserves mention. It may be that the international distribution of power has become too diffuse for successful global regime building. Historically, hegemons build regimes. This was certainly the case with Britain in the nineteenth century and the United States in the twentieth, but U.S. power has, in a relative sense, declined.[58]

Political scientist Robert Keohane has suggested, though, that even without a hegemon, it might be possible to build or at least to maintain global regimes that fostered cooperation.[59] In theory, a small number of great powers might solve the collective action problem posed by international cooperation.

Keohane's theory seems to offer a ray of hope for the prospect of building a future GHG control regime. However, that interpretation is open to doubt. Other scholars note that the travails of existing economic regimes have, as a matter of fact, waxed

[54] Gary D. Libecap, *State Regulation of Open-Access, Common-Pool Resources*, *in* HANDBOOK OF NEW INSTITUTIONAL ECONOMICS 568 (Claude Ménard & Mary M. Shirley eds., 2008).

[55] COLLIER, *supra* note 33, at 193–94.

[56] Henry D. Jacoby et al., *Sharing the Burden of GHG Reductions*, MIT JOINT PROGRAM ON THE SCIENCE AND POLICY OF GLOBAL CHANGE, Report No. 167, 11 (2008).

[57] Eric A. Posner & Cass R. Sunstein, *Climate Change Justice* 15–16 (University of Chicago Law & Economics and Public Law and Legal Theory, Working Paper No. 354 (2d series) & 177, 2007).

[58] ROBERT GILPIN, GLOBAL POLITICAL ECONOMY: UNDERSTANDING THE INTERNATIONAL ECONOMIC ORDER 95 (2001).

[59] KEOHANE, *supra* note 53, at 183.

as U.S. hegemony has waned.[60] The failure of the Doha Round only reinforces this observation.

Further, Keohane himself has forthrightly warned that the decline of hegemony makes regime building harder.[61] But GHG control, for reasons stated earlier in this chapter, can ill afford another handicap. Transaction costs are high, rewards for success are meagre, preferences conflict, and the incidence of the costs of controls is ill-matched with that of the benefits. Keohane also points out that success or failure may hinge on the legacy of earlier regimes.[62] Here, too, GHG controls are in severe difficulty of being saddled with the UNFCCC.

2.3.2 Flawed Bargaining Strategy

While all these structural problems persist, nothing that the United States can do will bring about effective GHG control. Nonetheless, at least in Democratic administrations, the United States, Europe, and Japan have gone on indefatigably piling Pelion on Ossa. The object of all this travail has been somehow to arrange to pay reluctant middle-income states to abate their emissions.

Of course, even suggesting that such an offer might be on the table tempts states to display reluctance in hopes of being paid,[63] and so events have played out. The fast-growing middle-income states have demanded a high price for their cooperation. The richer countries have less to fear from climate change than do the middle-income states, and they are now confronted with rising competition from the very countries that are demanding aid. Not surprisingly, a number of key developed countries, including the United States, have balked at the price tag, An agreement along these lines would, after all, impose large costs across the developed world.[64] Compounding the whole scheme's lack of political viability, several of the developed countries have rejected cap-and-trade schemes, a step that makes it much harder to hide from their publics the size of the needed transfer payments to foreigners. The upshot has been that the UNFCCC seems destined to debate this concept forever; yet its main "achievement" has been to create the scandal-plagued CDM (see Section 2.2.3).

2.3.3 Collier's Vision

Paul Collier has recently presented an alternative. His scheme is useful in two regards. First, it illustrates the kind of system that would be required to install effective GHG

[60] GILPIN, *supra* note 58, at 95.
[61] KEOHANE, *supra* note 53.
[62] *Id.* at 184.
[63] COLLIER, *supra* note 33, at 191.
[64] Jacoby et al., *supra* note 56, at 25.

control. And, second, by describing the challenge in a forthright way, it exposes how much quixotic behavior would be required to bring about large-scale GHG cuts.

Collier's pseudo-Bismarckian vision foresees the United States and China as joint prime movers. The scheme would then expand to encompass the EU, Japan, and India. These states, which Collier dubs the G-5, would use the threat of trade sanctions to force the fast-growing middle-income states to adopt controls. The middle-income countries, Collier notes, will host an increasing share of the world's heavy industry, and they are the main line of resistance to a global accord on GHG controls. Yet they are dependent on export markets. For the reasons discussed in this chapter, using trade sanctions to coerce these countries would be more effective than trying to bribe them would be.[65] The really poor countries, he also notes, are dependent on aid, so threatening to withhold it should quickly bring them to heel.[66]

Collier tacitly assumes that his scheme would pass the test of successful regime building in the absence of a hegemon. The validity of this assumption is, to say the least, not obvious. One patent source of doubt lies in the motivation, or more precisely the lack of motivation, of the G-5. Contrary to Collier's hopes, the threat of higher sea levels is not likely to become a salient issue in any U.S. election. The pace of melting of the Himalayan glaciers has, so far, not moved China to do much to curb its own GHG emissions, and it is not clear why that should change. Why, in any case, should these threats suddenly impel both China and the United States to take up the burdens of a joint altruistic *Machtpolitik* in preference to building dikes and reservoirs? Without an impetus from its two prime movers, Collier's blueprint seems doomed to remain on paper.

3. CLIMATE ENGINEERING AS A COMPONENT OF CLIMATE POLICY

Climate engineering (CE) might offer a means of avoiding some of the harm from GHG emissions, but not all of it, and to do so at low cost. Its potential benefits are huge, but its risks may also be high.

3.1 *The Concept of Climate Engineering*

CE might offset some of the warming caused by more GHG in the atmosphere. It would do so by reducing the amount of solar energy absorbed by the earth. GHGs in the atmosphere absorb long-wave radiation (heat) and then radiate it in all directions – including back to the earth's surface; as a result, surface temperatures rise. The more promising form of CE does not lower the GHG concentrations; rather, it

[65] COLLIER, *supra* note 33, at 191.

[66] *Id.* at 192.

seeks to reflect back into space a small amount of the incoming shortwave radiation (sunlight); all else remaining equal, temperatures fall even though GHG levels do not.[67] CE, therefore, lessens at least some of the risks of global warming.

CE is sometimes taken to include a second family of technologies. These other approaches involve carbon dioxide removal (CDR). Several different methods have been proposed. This discussion will exclude CDR, which has characteristics that are almost diametrically opposite the most promising sunlight-based CE systems. The latter have been accurately described as cheap, fast, and imperfect.[68] CDR is, to be sure, imperfect, but it is neither cheap nor fast: "carbon removal can only make a difference if we capture carbon by the gigaton."

The sheer scale of the carbon challenge means that CDR will always be relatively slow and expensive."[69] Most current CDR techniques are certainly very costly compared to their benefits.[70] Some CDR concepts, though, propose to capture carbon by fertilizing the ocean, and some of these techniques may cost less than the other kinds of CDR, but these methods "have only a relatively small capacity to sequester carbon and verification for their carbon sequestration benefits is limited. Furthermore there are likely to be unintended and probably deleterious economical consequences."[71] Such small scale-methods "are only worthwhile if sustained on a millennial time scale."[72] Land-based approaches are less limited in scale but do not have access to the "leverage" that gives ocean fertilization its cost advantage.

Hence, this discussion will exclude CDR. The political economy of CDR differs too much from the potentially more cost-effective forms of CE to do justice to both within the constraints on time and space. Instead, the analysis will focus on the sunlight-based CE concepts.

CE is generating some modest interest. The U.S. National Academy of Science and the British Royal Society have issued reports, and a number of prominent scientists have studied the issue.[73] The U.S. House of Representatives has held hearings[74]

[67] Tim M. Lenton & Nem E. Vaughan, *The Radiative Forcing Potential of Different Climate Geoengineering Options*, 9 ATMOSPHERIC CHEMISTRY & PHYSICS DISCUSSIONS 5339, 5539 (2009).

[68] *Learning to Manage Sunlight: Research Needs for Solar Radiation Management, Testimony Before H. Subcommittee for Energy and Environment of the H. Comm. On Science and Technology* (2009) (testimony of David Keith).

[69] *Id.* at 2.

[70] J. Eric Bickel & Lee Lane, *An Analysis of Climate Engineering as a Response to Climate Change, in* SMART CLIMATE SOLUTIONS 9, 37 (Bjorn Lomborg ed., 2010).

[71] *Geoengineering the Climate: Science, Governance, and Uncertainty*, THE ROYAL SOCIETY, Sept. 2009, RS Policy Document 10/09 at 18.

[72] Lenton & Vaughn, *supra* note 67, at 5539.

[73] *Id.* at 13.

[74] *Geoengineering: Assessing the Implications of Large-Scale Climate Intervention, Hearing Before the H. Comm. on Science and Technology*, 111th Cong. (2009) (statement of Lee Lane, Co-Director, AEI Geoengineering Project).

as had the British House of Commons.[75] The Royal Society in the UK has issued a research report on it,[76] as has a U.S. foundation.[77] The EU has also commenced a new study.[78]

At least two sunlight-based CE concepts may be able to offset the warming expected in this century.[79] One of them contemplates the injection of very fine sulfate particles into the stratosphere.[80] Several delivery methods might be used.[81] After perhaps a year or two, particles would fall to the surface as rain or snow. The quantities that would be needed would be small compared to current sulfate emission levels.[82] The global cooling and other changes that have occurred in the wake of several volcanic eruptions offer an analogue to this concept.[83]

The second approach involves lofting a fine seawater mist into low-level marine clouds. There, the droplets would cause the clouds to "whiten," that is, to reflect more sunlight, and perhaps this would lengthen their lives.[84] Climate models suggest that this approach might cool the planet enough to offset the warming caused by doubling atmospheric GHG levels.[85] The clouds that form in the wakes of ships at sea offer a natural analogue to the concept[86]; one delivery concept is to use a fleet of high-tech, remote-controlled ships to produce the spray.[87]

3.2 *CE: Large Benefits and Risks*

Although the potential benefits of CE are very large, its direct costs appear to be quite small. Doubtless today's very preliminary direct cost estimates will improve

[75] STAFF OF S. COMM. ON SCIENCE AND TECHNOLOGY, 111th CONG., REPORT ON ENGINEERING THE CLIMATE: RESEARCH NEEDS AND STRATEGIES FOR INTERNATIONAL COORDINATION (Comm. Print 2010) (Chairman Bart Gordon).

[76] *See* THE ROYAL SOCIETY, *supra* note 71.

[77] *See* Jason J. Blackstock et al., *Climate Engineering Responses to Climate Emergencies*, NOVIM, July 29, 2009, at iv.

[78] *See* European Commission, Framework-7 Programme, *Implications and Risks of Engineering Solar Radiation to Limit Climate Change*, Coordinated by the Max Planck Institute for Meteorology.

[79] Lenton & Vaughan, *supra* note 67, at 5553.

[80] *Id.* at 5548.

[81] Alan Robock et al., *The Benefits, Risks, and Costs of Stratospheric Geoengineering*, 36 GEOPHYSICAL RESEARCH LETTERS 4–7 (2009).

[82] Paul J. Crutzen, *Albedo Enhancement by Stratospheric Sulfur Injections: A Contribution to Resolve a Policy Dilemma?*, 77 CLIMATIC CHANGE 211, 213 (2006).

[83] THE ROYAL SOCIETY, *supra* note 71, at 31.

[84] John Latham et al., *Global Temperature Stabilization via Controlled Albedo Enhancement of Low-Level Maritime Clouds*, 366 PHIL. TRANSACTIONS ROYAL SOC'Y 3969 (2008).

[85] Keith Bower et al., *Computational Assessment of a Proposed Technique for Global Warming Mitigation via Albedo-Enhancement of Marine Stratocumulus Clouds*, 82 ATMOSPHERIC RESEARCH 328, 329 (2006).

[86] Stephen Salter, Graham Sortino & John Latham, *Sea-Going Hardware for the Cloud Albedo Method of Reversing Global Warming*, 366 PHIL. TRANSACTIONS ROYAL SOC'Y 1, 2 (2008).

[87] *Id.* at 1.

with time, but even if direct costs would be far higher than today's best estimates, they would be unlikely, in themselves, to reverse the prospect of large net benefits.

That, though, is not the whole story. CE may occasion important indirect costs. In fact, most of the controversy that surrounds CE appears to be about the size of the risk of side effects. (It may really be about something else altogether, but, of that, more in Section 4.3.)

3.2.1 Large Net Benefits Compared to Direct Costs

Potentially the benefits of CE are very large compared to the estimated costs of developing and deploying it. CE development costs are likely to start in the tens of millions of dollars per year. Sub-scale experiments may raise costs to one or two billion dollars.[88] Much of the cost at the stage of development and testing may be concentrated in global monitoring to detect the system's full range of global effects.

As responses to climate change go, the annual costs of operating global marine cloud whitening are almost trivial.[89] Stratospheric aerosol system costs appear to be higher, but are minuscule compared with the costs of GHG control. As one recent estimate put the matter:

> At a cost of a few thousands of dollars per ton for aerosol delivery to the stratosphere, the direct cost of offsetting the global mean temperature increase from a doubling of atmospheric CO_2 is estimated to be on the order of $10 billion per year. Of course, the need to operate satellite, atmosphere, and ground-based observation systems to monitor outcomes could increase costs substantially.[90]

Benefits, of course, depend on the scale of the threat from climate change, and the nature of the other policy responses to that threat can also change the expected value of a CE system. The less optimal is the GHG control regime accompanying CE, the greater the net benefits the latter would yield.

Yet another factor is the way in which CE is used. CE, once developed, could be deployed relatively quickly. This feature would allow the option to hold it in reserve to be used in the event of a climate emergency. Alternatively, it might be deployed to forestall warming and to avoid at least some of the harm the latter might cause. Suffice it to say that, even with deployment delayed to 2055, CE could yield present value benefits of $4 to $10 trillion,[91] and direct costs, at least for marine cloud

[88] Bickel & Lane, *supra* note 70, at 41.

[89] *Id.*

[90] Ken Caldeira & David W. Keith, *The Need for Climate Engineering Research*, ISSUES IN SCI. AND TECH. 57, 60 (Fall 2010).

[91] Bickel & Lane, *supra*, note 70, Table 1.7 at 40.

whitening, are only a tiny fraction of benefits.[92] For comparison, recall that, using the same model and the same discount rate, the early deployment of optimal GHG controls, were that possible, might capture about $3 trillion in net benefits.

3.2.2 Unknown Indirect Costs and Possible Remedies

Against this potential benefit must be weighed several kinds of risks. CE, for instance, might change global precipitation patterns. One major concern is a possible impact on the Indian summer monsoon, which plays a large role in the subcontinent's agricultural production. Current models are still too imperfect to quantify this risk.[93]

In addition to possible changes in precipitation patterns, CE may entail other unwanted side effects. Delayed recovery of the damage already present in the ozone layer could result from the use of stratospheric aerosols.[94] Skies might appear to be somewhat whiter, and there would be an increase in acid precipitation from aerosol injections, but this is unlikely to be of more than local import.[95] Stratospheric aerosols might impede surface-based astronomy and further depress the economic appeal of direct solar power.[96] Several other worries also merit research; in general, though, the known risks appear to be small, and the larger fears center on unknown unknowns.[97]

3.2.3 Putting Indirect Costs in Perspective

Four points have a strong bearing on concerns about CE's indirect costs. First, CE is not yet either a fully designed system or a single concept. It is several partial concepts each with distinct virtues and defects. The implications of stratospheric aerosol injection differ a great deal from those of marine cloud whitening, and R&D would be nearly certain to turn up still other approaches.

Second, telling virtues from defects is not always easy. Thus, on the one hand, the somewhat uneven impacts of marine cloud whitening may increase the risks of unwanted regional effects; yet, on the other, this very unevenness may offer a means of fine-tuning regional patterns to avoid harmful effects.[98] Engineered par-

[92] *Id.* Table 1.9 at 41.

[93] Philip J. Rasch et al., *An Overview of Geoengineering of Climate Using Stratospheric Sulfate Aerosols*, 366 PHIL. TRANSACTIONS ROYAL SOC'Y: 4007, 4026 (2008).

[94] *Id.* at 4031.

[95] *Id.* at 4032.

[96] *Id.*

[97] David W. Keith, Edward Parson & M. Granger Morgan, *Research on Global Sun Block Needed Now*, 463 NATURE 426, 426 (2010).

[98] Latham et al., *supra* note 84, at 3985.

ticles, for example, might avoid many problems that some fear from stratospheric aerosol injection.[99]

Third, the side effects have been monetized. Take, for instance, the possible weakening of India's summer monsoon. If it proves to be real and significant, it would be likely to be the biggest or at least one of the biggest risks posed by CE. The real question is how the risk would stack up against the benefits of using CE. The excess of CE's benefits over its direct costs has recently been estimated at $200 billion to $700 billion a year.[100] A 10 percent loss or gain in rain-fed Indian agriculture and forestry in the recent past would have amounted to a change of $4.5 billion to $9 billion a year.[101] It is also worth noting that agriculture in climates such as India's would stand to reap some gains from less warming,[102] and that the monsoon causes damage as well as benefits. That said, the size of the stakes involved, although they warrant concern, do not seem especially large when compared to CE's huge prospective surplus of benefits over direct costs.

Fourth, the relevant comparison is not between a climate with CE and one such as the mid–twentieth century. The world is changing in very large ways. This change will happen with CE or without it. The question at hand is whether the Anthropocene climate is better with CE or in its absence. A return to a climate with far less human impact is not an option.

3.3 *National Institutions and CE*

It may be that R&D on CE would be a good bet; some may disagree even on that issue, but the more troubling question is *will* it be handled competently? The experience with emission control could be cited as grounds for concern. Indeed, based on that experience, one might wonder if the effort will be made at all.

3.3.1 The U.S. Institutions and CE

The U.S. government is often strong abroad but weak at home.[103] This tendency is built into U.S. institutions. Many checks and balances and an array of veto gates makes it difficult for central policy makers to surmount domestic opposition. This pattern may well apply to CE.

[99] Caldeira & Keith, *supra* note 90, at 60.

[100] E-mail from J. Eric Bickel, Assistant professor in the Operations Research/Industrial Engineering Group in the Department of Mechanical Engineering at the University of Texas at Austin to Lee Lane, Resident Fellow American Enterprise Institute (Sept. 16, 2009).

[101] Bickel & Lane, *supra* note 70, at 23.

[102] Robert Mendelsohn, Ariel Dinar & Larry Williams, *The Distributional Impact of Climate Change on Rich and Poor Countries*, 11 ENV'T & DEV. ECON. 159, 161 (2006).

[103] Krasner, *supra* note 38, at 42.

Without R&D, the concept of CE cannot advance. On the one hand, at the R&D stage, CE is not likely to arouse the level of popular opposition that GHG control proposals have stirred up. True, some organized green groups oppose the concept,[104] but others at least claim to be studying it. They are, after all, in a bind. It is difficult to credibly assert that climate change poses an imminent existential threat to mankind while refusing to explore what appears to be a way to ward off its worst effects.

On the other hand, there is not much support for CE either. Much of the political Right disputes climate science, so to them, CE is an answer to a bogus problem. The Left sees CE as threatening to supplant GHG controls, so, to them, it is a bogus answer to a crisis that they long to put to better use. Ideologically, CE is an orphan.

Government R&D spending does not in the main depend on ideological motives, but it gains support from broad popular demand, as in the case of medical research, or it is based on pork barrel politics. The latter thrives on building large "demonstration projects" in the states and districts of influential members of Congress.[105] CE does not, though, seem to offer much "pork"; the concept, in this regard, may be too efficient for its own good. It is also innovative, and some people oppose it on ethical grounds. These features are likely to cause risk-averse government agencies to shy away from taking the lead in fostering CE.

3.3.2 China and CE

China may have stronger motives than the United States to explore CE. The government's parsimoniousness when it comes to climate might well recommend a cheap quick fix to climate change. China's environmental movement is weak,[106] and its government's processes are highly opaque.[107] Domestic resistance would, then, hardly be likely to constrain Beijing should it decide to explore CE.

Having mastered systems such as thermonuclear warheads, ICBMs, cruise missiles, moon rockets, satellites, and satellite killer systems, China could muster the needed technical wherewithal for CE. Some models suggest that China and India have divergent interests with regard to the goals of a CE system, that is, although each might gain from an altered climate, the pattern of temperature and rainfall that would be best for one would be less than ideal for the other.[108] If so, China might have an added motive to seek CE expertise. To wit, doing so might give it an

[104] *See, e.g., Geopiracy: The Case Against Geoengineering*, THE ETC GROUP, Oct. 18, 2010.

[105] LINDA R. COHEN & ROGER G. NOLL, THE TECHNOLOGY PORK BARREL 59 (1991).

[106] PRANAB BARDHAM, AWAKENING GIANTS: FEET OF CLAY 124 (2010).

[107] *Id.* at 126.

[108] Katharine L. Ricke, M. Granger Morgan & Myles R. Allen, *Regional Climate Response to Solar-Radiation Management*, 3 NATURE GEOSCI. LETTERS 537, (2010), *available at* http://www.nature.com/ngeo/journal/v3/n8/full/ngeo915.html.

edge in bargaining about how such a system would be designed. In the case of CE, knowledge, at least private knowledge, is power.

3.4 *Global Institutions and CE*

3.4.1 CE, A Dynamic Option

The world politics of CE are almost the opposite of those of GHG control. CE is easy to start and hard to stop. With CE, holdup power is weak, that is to say, it would be hard to block a state from using CE to cool the planet. Simply refusing to cooperate in the CE is not an effective option: CE's direct costs are too paltry for cost sharing to be essential.

Conventions of international law might, in theory, apply to CE. Some treaties, for example, include broad admonitions against causing cross-border harm.[109] Of course were such theories effective, man-made global climate change would not exist. The UN Convention on the Law of the Sea might limit ocean-based methods, but no such constraint applies to the atmosphere.[110] A recent meeting under aegis of the UN Convention on Biodiversity has advised against CE steps that might harm biodiversity, but experts who took part noted: "At the COP10 in Nagoya, the delegates were not well informed about geoengineering, and negotiations were conducted in haste without proper scientific consideration."[111] It is unlikely that any major country would allow the injunctions of so dubious a body to constrain its actions, and the United States is not even a party to the Convention from which this group claims authority.[112] More broadly, a state that felt its vital interests to be at risk would surely act on former German chancellor Bethmann-Hollweg's precept that "necessity knows no law."

Thus, only two alternatives exist for halting a state that attempts to cool the global climate.

First, another state might try to use economic sanctions or the threat of armed force to end an unwelcome CE effort. (In the case of armed force, far-flung fleets of unmanned cloud-seeding ships might be much more vulnerable to disruption than would stratospheric aerosol injection.) Either force or sanctions would be more likely to work if a great power were wielding it against a lesser state. If the CE is backed by a great power, the prospects for halting it fall dramatically.

[109] THE ROYAL SOCIETY, *supra* note 71, at 40.

[110] *Id.*

[111] Masahiro Sugiyama & Taishi Sugiyama, *Interpretation of CBD COP10 Decision on Geoengineering* ISocio-Econ. Research Ctr., Foundation Central Research Institute of Electric Power Industry SERC, Discussion Paper 10013, 2010).

[112] *Id.* at 9

Coercion or sanctions usually impose high costs on the states that try to use them.[113] Still, a lesser state might be deterred from using CE by a great power's sanctions or threats; coercion sometimes works.[114] Its costs, though, increase dramatically when it is directed against a great power. Not even peers are likely to succeed in coercing other great powers, and the record of such efforts shows that they yield only meager results.[115]

A threat to use armed force against a great power is very dangerous indeed. For that reason, other than in the direst emergency, it is not very credible even for a peer great power to threaten another with force. At least for the remainder of this century, the cooling needed to limit the rise in global mean temperature to two degrees Celsius is relatively modest.[116] The odds are, then, that the potential costs of using armed force will dwarf any likely harm from CE.

Second, states that are located near the poles might be able to counter CE with measures designed to cancel out its cooling effects. Spreading soot on the Arctic or Antarctic ice sheets might be one approach.[117] Such countermeasures, though, entail high risks of unexpected effects. Some of these could be as unwelcome to their author as to the intended target; hence no democracy, or at least not one with a strong environmental movement, is likely to undertake them.

In effect, dueling CE systems would amount to a game of chicken with the global climate. Facing this prospect, a state wishing to cool the planet might choose to await the emergence of a great power consensus before acting; hence, states considering CE would be constrained by the perceived interests and the expected degree of risk-averseness of the great power favoring the least cooling. In today's world, then, Russia might serve as an "engine governor" on other states' wishes to cool the planet.

Hardly any thought, though, has so far gone into the idea of countering global cooling. Other options may exist. A key question is the relative cost of adding an increment of cooling compared to that of adding an equal increment of warming. Of this matter, little is known.

3.4.2 From Disputes about "Whether" to about "How"

As time passes and the net effects of climate change become more negative, disputes about *how* to do CE are likely to supplant those over *whether* to do it. Today,

[113] LLOYD GRUBER, RULING THE WORLD: POWER POLITICS AND THE RISE OF SUPRANATIONAL INSTITUTIONS 37 (2000).

[114] DANIEL W. DREZNER, ALL POLITICS IS GLOBAL: EXPLAINING INTERNATIONAL REGULATORY REGIMES 34 (2007).

[115] *Id.*

[116] J. Eric Bickel, *The Climate Engineering Option: Economics and Policy Implications*, AEI GEOENGINEERING PROJECT, Mar. 3, 2010, at 29.

[117] E-mail from Gregory A. Benford, Professor Emeritus, Physics & Astronomy, School of Physical Sciences, University of California, Irvine to Lee Lane, Resident Fellow, American Enterprise Institute.

though, it is possible to descry only the vaguest outlines of how such disputes might play out. Recent modeling shows that uniform CE systems may be able to lessen GHG-induced changes in temperature and precipitation. They may, though, not affect all regions in the same way and to the same extent, and as the CE intervention rises in scale to offset GHG levels, the regional divergence grows apace.[118]

Hence, a system based on a simple uniform forcing may not be equally satisfactory to all states. It is possible that CE can be fine-tuned to minimize unwanted regional effects. It is uncertain how tightly regional and global climates are coupled, and still more unclear how easy or hard it might prove to be to develop techniques for fine-tuning regional climates. A good deal of trial and error is likely to be called for, and advances in science and technology are likely to extend the process far into the future.

3.4.3 CE and the Logic of Collective Action

It may prove to be relatively easy, in terms of science and technology, to devise a system that leaves all states better off, or it may not. (As with climate change itself, there are also likely to be both winners and losers within nations as well as among them.) Even if all states could gain, disputes are likely to arise over the distribution of the benefits. Side payments might, in theory, resolve such conflicts, but there are sure to be transaction costs of reaching agreement.

Managing such a system through time will require making a long series of bargains. Faced with this kind of task, states often create an international regime. Regimes are costly to build, and building them is more costly still absent a hegemon,[119] but they can also greatly lower the transaction costs of repeatedly striking and revising bargains. How the transaction costs of building a CE regime compare with those of building a GHG control regime remains unclear. What is clear, though, is that the potential net benefits of the former far exceed those of the latter. That fact, combined with CE's more favorable ratio of go-it-alone power to holdup power, suggests that an effective CE regime might emerge where GHG control institutions have failed to take root.

In that case, CE regime structure will likely depend on both the distribution of relative power as well as the need to hold down the transaction costs of managing the system. The goal of controlling transaction costs argues in favor of limiting those with a voice in the regime to the states that, if disgruntled enough, could impose costs on the system managers. The prospect that, in any future regime, knowledge will be a source of power may goad all of the great powers into investing in CE research.

[118] Ricke, Morgan & Allen, *supra* note 108, at 2.

[119] KEOHANE, *supra* note 53, at 50.

In the end, the option of engineering the climate will certainly give rise to disputes. Such disputes are inherent in any policy that would change the global climate. With GHG control, reluctant states can simply avail themselves of their holdup power. With CE, holdup power is weak, so states will have to apply issue linkage and other bargaining strategies.

This fact has led some to fear that unilateral action on CE will unleash destructive conflict.[120] The risk cannot be completely excluded, but there is less here than meets the eye. The prospect of conflict, itself, urges coalition building, and the prospect of significant transaction costs argue for regime building. The outcome is likely to be far from perfect, but a slow, high–transaction-cost bargaining process probably poses a greater threat than does "lone ranger" CE.

3.5 *The Indirect Costs and Benefits of CE*

As with GHG control, attempts to develop or to use CE as a tool of climate policy would affect other national goals. As already discussed, states' preferences over climate differ. Some states are likely to want more cooling or a different trade-off between cooling and precipitation than others do. Currently, climate models' regional resolution is too poor to place much faith in their predictions at this level. The degree to which CE concepts can be refined to match disparate preferences also remains unclear.

The potential conflicts over CE are likely to play out in at least two dimensions. One will be the issue of how close to the global welfare–maximizing Pareto optimal frontier a CE system can come. The transaction costs of striking a bargain and imperfect knowledge will cause actual systems to fall short of the potential net gains. A second, more crucial, issue is likely to center on the distribution of the benefits. States with more bargaining power are likely to set up the system in ways that allow them to capture larger shares of the benefits.[121]

Second, states' preferences over CE and GHG control are likely also to differ. Many OPEC states might prefer cooler climates and less progress on GHG control. For them, CE may be an ideal policy. Russia might be conflicted.

Third, CE will doubtless affect other policies. It is likely to rein in some of the conflicts with economic, energy, and agricultural policy that GHG control has unleashed. Yet CE is likely to occasion frictions of its own. If some states contemplate use of trade sanctions to enforce GHG control, others might consider using

[120] David G. Victor et al., *The Geoengineering Option: A Last Resort against Global Warming?* 88 FOREIGN AFF., Mar.–Apr. 2009, at 64, 67.

[121] Stephen D. Krasner, *Global Communications and National Power: Life on the Pareto Frontier*, *in* POWER, THE STATE, AND SOVEREIGNTY: ESSAYS ON INTERNATIONAL RELATIONS 152 (Stephen D. Krasner ed., 2009).

them in hopes of holding up unwanted CE deployment. With time, if the threat of climate change continues to grow, this risk may lessen, but at least initially CE might be more likely to spark conflict than does GHG control. After all, CE seems likely actually to affect the climate. Belief that the world will adopt GHG controls strict enough to significantly slow climate change demands much willing suspension of disbelief.

4. THE ETHICS OF CE

Some scholars call for more discussion of the ethical aspects of CE.[122] Without doubt, climate change raises ethical issues, and discussing them may cast light on some facets of the issue, but trying to stipulate hard-and-fast ethical rules on which to make climate policy decisions is another matter.

There are many possible sources of ethical guidance on CE. Academic ethicists' views on the matter vary widely, and Section 4.1 shows that no consensus is in the offing.

Welfare economics, as discussed in Section 4.2, offers a coherent ethical system; in it, CE is a matter of costs and benefits. If CE is cheap and effective, it would be a good thing. If not, not. There is much to say in favor of this approach, yet attempting to elevate its rules into a rigorous and all-encompassing system of utilitarian ethics leads to absurdity.

Section 4.3 describes the very different stance of radical environmentalists: to them, GHG control is a moral imperative, and CE tempts mankind to hubris. Few thoughtful people may wish to trod this path to its logical conclusions.

All these views fall afoul of certain hard facts. Human goals are diverse, and the world is rife with conflict. Trade-offs among means and ends are unavoidable. Nowhere are they more necessary than in statecraft, and there is no reason to think that the world politics of CE will be an exception to the rule. Section 4.4 concludes the chapter with a caution against ambitious attempts to base public policy on ethical precepts. It suggests that Max Weber's more modest and worldly notion of an ethic or responsibility may offer limited but useful guidance.

4.1 *Academic Ethics' Limited Guidance*

Academic ethics may be in some respects the least useful resource for dealing with the ethics of CE. It does offer, though, a variety of views. Roger Scruton comes

[122] *Geoengineering: Assessing the Implications of Large-Scale Climate Intervention, Hearing Before the H. Comm. on Science and Technology*, 111th Cong. (2009) (statement of Dr. James Fleming, Professor and Director of Science, Technology and Society, Colby College).

down firmly in favor of exploring the option.[123] Martin Bunzl thinks that researching climate engineering could be ethical[124] but that it might prove to be indecisive.[125] Others reach no final judgment, but find many of the arguments adduced in favor of CE research to be ethically flawed.[126] David Morrow and his colleagues proclaim stringent rules must restrict CE field experiments.[127]

A wider survey would find still other clashing viewpoints. If we put the best face on things, the very diversity of this discourse serves to illuminate diverse facets of CE, but fatwas issued by one school often clash with those of others. This state of affairs reflects the status of philosophy more than it does any special ambiguity of CE.

Nor is there much reason to hope for a convergence of views. Philosophers, like other human beings, embrace countless diverse values,[128] and an impasse has befallen much ethical discourse. Many rival traditions coexist; they rest on seemingly incommensurable principles, and they are apt to reach quite disparate judgments.[129] In this regard, ethics reflects broader trends toward diversity in culture and society.[130]

4.2 *CE and the Ethics of Economics*

Welfare economics can be regarded as one of today's clashing ethical systems. It embodies a utilitarian ethic.[131] Within that system, the ethics of CE comes down to a comparison of costs and benefits.

4.2.1 The Need for Policy Trade-Offs

One virtue of welfare economics as an ethical system is its firm grasp on the need to make trade-offs. This insight suits climate policy well where three approaches (adaptation, GHG control, and CE) are imperfect substitutes. Applying more of any one of them reduces the need for both of the other two. Evidence that CE would work safely would lessen the risk of abrupt climate change, and it might diminish

[123] Roger Scruton, *Climate of Opinion*, AM. SPECTATOR, Mar. 20, 2010, *available at* http://spectator.org/archives/2010/03/06/climate-of-opinion.

[124] Martin Bunzl, *Researching Geoengineering: Should Not or Could Not?*, 4 ENVTL. RES. LETTERS, Oct.–Dec. 2009.

[125] *Id.* at 3.

[126] Stephen M. Gardiner, *Is "Arming the Future" with Geoengineering Really the Lesser Evil? Some Doubts about the Ethics of Intentionally Manipulating the Climate System*, *in* CLIMATE ETHICS (Stephen M. Gardiner et al. eds., 2010).

[127] David R. Morrow, Robert E. Kopp & Michael Oppenheimer, *Toward Ethical Norms and Institutions for Climate Engineering Research*, 4 ENVTL. RES LETTERS, Oct.–Dec. 2009.

[128] BERLIN, *supra* note 1, at 304–05.

[129] ALASDAIR MACINTYRE, AFTER VIRTUE: A STUDY IN MORAL THEORY 5 (2d ed., 1984).

[130] *Id.* at 8.

[131] COLLIER, *supra* note 33, at 10.

the expected costs of gradual climate change as well. Therefore, it would lower the amount of damage expected from emitting another ton of GHG. Less expected damage would, in turn, reduce the value of avoiding the release. Sound climate policy would at the margin substitute some reliance on CE for some reliance on GHG control. It would also lessen the need for adaptation.

The three are, of course, *imperfect* substitutes. Adapting to abrupt or extreme climate change will be very costly. Sunlight-based CE cannot lower the harm from ocean acidification, and GHG control would have little chance of avoiding harm should rapid climate change begin to unfold. Hence, each approach promises to avoid some harm, but an optimal policy is likely to make some use of all three approaches.

4.2.2 Moral Hazard and Hazardous Moralism

Some scientists have claimed that CE presents a "moral hazard." As economists use the term, "moral hazard" arises from asymmetric information between the contracting parties. The insurer or principal knows less than the insured or agent about the latter's behavior or state. The result is often a contract that exposes an insurer or employer to more risk than they would be willing to undertake knowingly.

Moral hazards are a kind of market or policy failure. Moral hazards *raise* risks and costs, and they shift their burden from those who cause them to others. When moral hazard occurs, it lessens total economic welfare. Thus, the term is, in this sense, pejorative.

In this sense of the term, CE is clearly not a case of moral hazard. It involves no contract, let alone one based on asymmetric information. It would allow society to lower harm from GHG emissions, to keep the same level of harm with lower abatement costs, or to do some of both. Should society elect to take some of the resulting efficiency gain in the form of slower and, hence, less costly GHG abatement, that choice would in no way alter the fact that CE that worked would lower total costs and risks – not raise them.

Some scientists disapprove of the option to slow the pace of GHG control, and they have applied the term "moral hazard" as a means of expressing their censure. No great harm would ensure were they to make clear the distinction between the way in which they are using the term and the economists' use of it to name a class of market failure, but few scientists have been at all scrupulous on this point.

4.2.3 Welfare Economics' Limits

The ethics of welfare economics are founded on a simple proposition: "If you want to maximize the sum of human welfare, you should do the following." As discussed above, institutional constraints often prevent states from taking the steps that would enhance welfare. Also, for many reasons, the goal commands only limited support.[132]

The biggest problem stems from adherence to the ethical principle of universalism. Lord Stern is very much a follower of this principle, and it dictates his odd choice of discount rate that was noted back in Section 2.1.1.[133] Universalism dictates that a dollar of loss or gain is of equal value regardless of how near or far from us, in either time or space, is the recipient.

Trying to act on universal welfare maximization would lead quickly to absurdity. The implied very low discount rates would dictate that current generations live at near subsistence levels in order to heap up wealth for future generations that are already likely to be much wealthier than those of today; of course, those generations, too, would be under the same injunction. History, in such a world, would become a treadmill. This hardly amounts to a compelling ethical vision.[134]

Evolution has hardwired people to place greater value on the welfare of those near to them in time and space than on that of those who are distant in either dimension.[135] Stern's universalism, as Collier observes, would suit a colony of ants better than it does a nation of Homo sapiens.[136] Thus, a sense of "the crooked timber of humanity" should temper the strictures of welfare economics. Efforts to increase welfare subject to the discount rates and sense of community that are manifest in actual markets and real polities provide better guides to action than the ethical ideals of the ant world. Some economists see this point, and, seemingly, some do not.

4.3 *Green Ethics and CE*

Neither of these visions, though, holds much appeal to those environmentalists who regard nature as good and mankind as decidedly not.[137] Thus, the ETC Group, a green advocacy group, has attacked the motives and ethics behind what it describes as "geopiracy."[138] To people of this shade of green, reducing GHG emissions is not an instrumental choice. The "natural" state of the world is good, and human influence is certainly highly suspect if not downright bad.[139] Seen in this light,

[132] *Id.*

[133] *Id.* at 26

[134] Kenneth J. Arrow, *Discounting, Morality, and Gaming*, *in* DISCOUNTING AND INTERGENERATIONAL EQUITY 16 (Paul R. Portney & John P. Weyant eds., 1999).

[135] COLLIER, *supra* note 33, at 26.

[136] *Id.* at 27.

[137] Robert H. Nelson, *Environmental Calvinism: The Judeo-Christian Roots of Eco-Theology*, *in* TAKING THE ENVIRONMENT SERIOUSLY 233, 235 (Roger E. Meiners & Bruce Yandle eds., 1993).

[138] *See, e.g.*, *Geopiracy*, *supra* note 104.

[139] Nelson, *supra* note 137, at 240.

curbing pollution is not an instrumental choice to be judged from the standpoint of enhancing human welfare: it is a moral imperative.

The claim that lowering emissions is an ethical duty in itself leads quickly to awkward questions. Was the caveman's wood fire a crime against nature? If new technologies were to make adaptation far more cost-effective, or if climate science found that a new ice age was upon us, would we still be duty bound to go on lowering greenhouse gases? Such views seems hard to credit.

Some environmental scientists insist that climate change should be dealt with by attacking its root cause, not, in other words, by using CE to treat some of its symptoms. Such claims may be either surreptitious attempts to smuggle ethical claims into the discussion, or they may reflect simple confusion.

> Typically in an attempt to find a solution to a problem people look to its causes, or yet more fatuously, to its *root* causes. However, there need be no logical connection between the cause of a problem and appropriate or even feasible solutions.[140]

The false leap from emissions as a cause of the problem to the demand for controls as the sole or the primary response rests on just this simple error.

In any case, mankind is already embarked on massive CE. Greenhouse gas emissions are, themselves, a form of CE; further, current aerosol emissions are offsetting about 40 percent of the warming that man-made GHG emissions would otherwise have caused.[141] Planned CE would extend the scale of the cooling effect and separate it from the harm done by today's aerosols. Substituting a conscious plan for this inadvertent CE would merely lower the costs of the project in which humans are already embarked. Like Wotan in Wagner's *Die Walküre*, mankind may find it easier to renounce the mindfulness of control than it is to divest itself of its substance.

One may, as part of "an ethic of ultimate ends," insist that suppressing emissions should be the only response to climate change. An "ethic of responsibility," though, would take full account of actions' consequences and gauge the latter in light of human nature and extant institutions.[142] In any case, rejecting CE in hopes that stringent GHG control will avoid harm amounts to acting on nothing more than "an ethic of wishful thinking."

[140] COLLIER, *supra* note 33, at 209.

[141] Intergovernmental Panel on Climate Change, *Climate Change 2007: Working Group I The Physical Science Basis: Summary for Policymakers, Contribution of Working Group I to the Fourth Assessment Report of the Intergovernmental Panel on Climate Change*.

[142] MAX WEBER, *Politics as a Vocation*, *in* FROM MAX WEBER: ESSAYS IN SOCIOLOGY 121 (H.H. Gerth & C. Wright Mills eds. & trans., 1958).

4.4 *Philosophy as a Force in Policy Making*

In the end, moralistic objections are likely to have only limited effect. How many divisions, as Stalin famously asked, has the pope? John Rawls cannot even summon the Swiss Guard. Yet ethics may not be entirely without consequences.

4.4.1 Ethics in A Hobbesian World

Preaching moral imperatives in the realm of world politics where force is *ultima ratio regum* is an especially hazardous enterprise. Max Weber put the matter well:

> No ethics in the world can dodge the fact that in numerous instances the attainment of "good" ends is bound to the fact that one must be willing to pay the price of using morally dubious means or at least dangerous ones – and facing the possibility or even the probability of evil ramifications. From no ethics in the world can it be concluded when and to what extent the ethically good purpose "justifies" the ethically dangerous means and ramifications.[143]

Attempts to apply ethical principles to CE encounter just such problems. Take the demand that the governments of all of the people who might be affected by CE experiments must give assent before research can proceed.[144] Such a rule would empower every Third World kleptocracy to extort money by holding up measures to limit the harm from climate change. Needless harm may befall not only the states wishing to engineer the climate but also the subjects of the kleptocrat governments. Not only might they lose or delay the benefits from CE, but foreign aid extorted in this way could make their governments even less responsive to the needs of local taxpayers, a common outcome of foreign aid.[145]

Further, imposing these and other ethical strictures would most hobble the efforts of those open order polities that would be likely to take the greatest care of the interests of other nations. States that have more far-flung global interests, or that prize global lawfulness more highly than less open societies are prone to do, would be the countries most likely to foster quasi-Pareto optimal CE systems. Yet, just because such concerns are more likely to deter such states from forging ahead with research, restrictive rules might retard them in the race for knowledge. If knowledge becomes power over CE system design, restrictive rules may lead to results that are the opposite of those intended.

[143] *Id.* at 121.

[144] Morrow et al., *supra* note 127, sec. 4.1.

[145] ROBERT H. BATES, PROSPERITY AND VIOLENCE: THE POLITICAL ECONOMY OF DEVELOPMENT 64 (2d ed., 2010).

4.4.2 Varieties of Realism and Their Implications for CE

In foreign policy and diplomatic history, realists have often maintained that ethical sentiments have had scant impact on the actual conduct of states and statesmen.[146] In a modern democracy, though, it is not certain that this conclusion is justified.

Earlier realists were perhaps closer to the truth. They believed that ethical notions did sometimes affect even the most important decisions, and that their influence was often baleful.[147] The disastrous "war guilt" provisions of the 1919 Versailles Treaty provide a prime example of the needless harm that can result from injecting moral strictures into serious affairs of state.

Ironically, climate policy seems partially to bear out both the later and the earlier versions of realist theory. As later realists would predict, the moral posturing at the UN climate talks has not led the most puissant of the great powers to harm their own interests by adopting GHG controls. Yet, as their intellectual forbearers might have warned, the posturing impedes agreement on smaller steps that might, in its absence, have been possible.[148]

Similarly, if the course of climate change warrants the use of CE, some state or states will be very likely to use it. Ethical strictures will not alter that part of the outcome, but they may affect the identity of the states managing the CE, and they may change the distribution of bargaining power among them. Nothing guarantees that the impacts will be benign.

[146] JOHN J. MEARSHEIMER, THE TRAGEDY OF GREAT POWER POLITICS 49 (2001).

[147] EDWARD HALLETT CARR, THE THIRTY YEARS CRISIS: 1919–1939 10 (1964).

[148] Lee Lane & David Montgomery. *Organized Hypocrisy as a Tool of Climate Diplomacy*, AEI ENERGY & ENV. OUTLOOK 7 (2009), http://www.aei.org/docLib/05-EEO-Lane-g.pdf.

6

Political Legitimacy in Decisions about Experiments in Solar Radiation Management

David R. Morrow, Robert E. Kopp, and Michael Oppenheimer

1. INTRODUCTION

For better or for worse, geoengineering has moved from the fringes of the climate change debate to the halls of Capitol Hill[1] and Westminster.[2] Of course, a great deal of research remains to be done before the world decides whether to introduce geoengineering as a complement to mitigation and adaptation; academics and policy makers are still wrestling with the scientific, political, legal, social, and ethical questions surrounding the intentional modification of the climate. Here we address the institutional aspects of some of the ethical issues raised by research on geoengineering.

The most ethically challenging form of geoengineering research involves solar radiation management (SRM),[3] which attempts to reduce the earth's absorption of incoming solar radiation. One proposed mechanism for SRM is the injection of aerosols into the stratosphere, which would deflect more solar radiation back into space. In contrast to research into carbon dioxide removal (CDR), which is the other main category of proposed geoengineering activities, SRM research is particularly challenging ethically because studying and testing SRM technologies can require deployment at scales that could have significant regional or global climatic effects.[4] For instance, testing the effects of stratospheric aerosol injection would require lofting enough aerosols into the atmosphere, over a long enough period of

[1] *See* Geoengineering III: Domestic and International Research Governance, 111th Cong. (2010).

[2] *See* Science & Technology Committee, The Regulation of Geoengineering, 2010, H.C. 221 at 3.

[3] Some earlier work, including ours, refers to SRM as "short-wave climate engineering." We regard these two terms as synonymous. *See* David R. Morrow, Robert E. Kopp & Michael Oppenheimer, *Toward Ethical Norms and Institutions for Climate Engineering Research*, 4 ENVTL. RES. LETTERS 045106, 2 (2009). *See also* J.J. BLACKSTOCK ET AL., CLIMATE ENGINEERING RESPONSES TO CLIMATE EMERGENCIES 2 (2009).

[4] BLACKSTOCK et al., *supra* note 3, at 25.

time, to distinguish the effect of the aerosols from normal climatic variation.[5] The consequences of such large-scale testing could cause serious harm to millions of people. For instance, SRM could change regional precipitation patterns, threatening water supplies and agriculture.[6] Moreover, whereas CDR aims to return the atmosphere to an earlier, familiar state, SRM aims to create a new state – one of high greenhouse gas (GHG) concentrations and reduced insolation – about which we know much less.

In an earlier paper we suggested three ethical principles for SRM research based on established principles for biomedical research with human subjects.[7] The analogy between SRM and biomedical research is, like all analogies, imperfect. In this chapter, we consider some of the ethical implications of one limitation of that analogy – namely, the fact that decisions to participate in biomedical experiments are made individually, whereas the decision to "participate" in an SRM experiment is a collective decision. Specifically, we explore the possibility of designing an international institution that would have the moral authority to make collective decisions about SRM experiments. We consider the requisite features of such an institution and examine the characteristics of other global governance institutions as comparable cases.

2. THE BIOMEDICAL MODEL FOR SRM RESEARCH ETHICS

In our earlier paper, we proposed a basic framework for SRM research ethics that derives from principles governing biomedical research with human and animal subjects.[8]

We intend our framework to apply to large-scale SRM experiments. Very roughly, "large-scale SRM experiments" are experiments that are large enough to significantly alter the climate regionally or globally by changing the rate at which the earth absorbs incoming solar radiation, but smaller than would be deployed to counteract the radiative forcing of anthropogenic GHGs on a global basis. For instance, injecting enough aerosols into the stratosphere to distinguish their effect from normal climatic variation[9] constitutes a large-scale SRM experiment; releasing a few tons of

[5] Morrow et al., *supra* note 3, at 6.

[6] Alan Robock et al., *A Test for Geoengineering?* 327 SCI. 530, 531 (2010). *See also* Govindasamy Bala, K. Caldeira & R. Nemani, *Fast versus Slow Response in Climate Change: Implications for the Global Hydrological Cycle*, 35 CLIMATE DYNAMICS **423**, 433 (2010); A. Jones et al., *Geoengineering by Stratospheric SO_2 Injection: Results from the Met Office HadGEM$_2$ Climate Model and Comparison with the Goddard Institute for Space Studies ModelE*. 10 ATMOS. CHEM. PHYS. 5999, 6005 (2010).

[7] Jones et al., *supra* note 6; *Id.* at 1.

[8] Morrow et al., *supra* note 3, at 3–6.

[9] Roughly tens to hundreds of kilotons per year if the injectant is SO_2 as a precursor of sulfate aerosols, based on calculations using previously published significance thresholds and radiative forcing

aerosols from a single airplane to observe their physical and chemical reactions with other particles does not, as such a small quantity of material could not significantly alter the climate. Throughout this chapter, we use "SRM experiment" to refer specifically to large-scale activities. We explicitly exclude climate modeling studies and small-scale field tests of SRM technologies, although we recognize that the line between small-scale field tests and large-scale experiments is fuzzy.

Our ethical framework for SRM experiments includes three basic principles: The Principle of Respect requires that researchers secure the global public's consent, in some appropriate form, before commencing an experiment.[10] The Principle of Beneficence and Justice requires that researchers protect the basic rights of persons affected by their experiments, minimize the risk–benefit ratio of those experiments, and aim to distribute those risks and benefits justly across persons, animals, and ecosystems.[11] The Principle of Minimization requires that experiments should not last longer, cover a greater geographic area, or exert a greater influence on the climate than is necessary to test the specific hypotheses in question.[12]

The analogy between SRM and biomedical research is, like all analogies, imperfect. The key limitation of this analogy is that individuals decide for themselves whether to participate in and face the risks of a biomedical experiment, whereas we must decide collectively whether to subject ourselves to the risks of an SRM experiment. Imagine two people who are considering participating in a trial of an experimental antidepressant. The first person's decision about whether to participate has no effect on the other's decision; it neither precludes nor requires that the other person participate. Thus, the first person's decision does not expose the second to any risks. SRM is different. To "participate" in an SRM experiment, in the relevant sense, is to be subjected to the alteration of the climate. Thus, no one can participate in the experiment unless everyone participates in the experiment. In this respect, an SRM experiment is more like a public health intervention or collective social policy than it is a medical experiment. For example, individuals cannot easily opt out of mandatory vaccination policies, the fluoridation of drinking water, or national pension schemes.

The necessity of "collective participation" in SRM experiments changes the way we think about risk and consent. In the biomedical case, we need to consider only the risks to the individual participant (and, in some cases, his or her family). With SRM, we need to consider both the scale and the distribution of risks. In the biomedical case, we can and should require the informed consent of each participant. If

estimates. For significance thresholds, *see id.* at 7; for radiative forcing estimates, *see* Alan Robock et al., *Tropical and Arctic Geoengineering*, 113 J. GEO. RES. D16101, at 4 (2008).

[10] Morrow et al., *supra* note 3, at 4–5.

[11] *Id.* at 5–6.

[12] *Id.* at 6.

universal informed consent is ethically required for SRM experiments, then ethical SRM experiments are impossible. In general, however, we rarely require unanimous agreement in making collective decisions. Examples of this include democratic governments that sometimes impose military service requirements, change tax rates, institute redistributive social safety nets, protect species or ecosystems, and prohibit or regulate the use of certain technologies, even when significant fractions of the population do not and would not consent to those policies. In discussing consent and SRM experiments, we suggested that some indirect form of consent – such as consent voiced through national representatives – may be ethically sufficient.[13] In what follows, we consider the features that an institution would need in order to serve as a vehicle for such indirect consent.

3. COLLECTIVE DECISIONS, LEGITIMACY, AND GLOBAL GOVERNANCE

We contend that in collective decisions, the central normative concern is the legitimacy of decisions and decision makers rather than universal individual consent. Thus, the ethical conduct of SRM research requires an institution that has the global political legitimacy to make decisions about SRM experiments.

Political philosophers recognize both a normative and a descriptive (i.e., positive) concept of legitimacy. Roughly, an institution is legitimate in the normative sense if it has the *right* to govern, and it is legitimate in the descriptive sense if it is widely *believed* to have the right to govern.[14] Because the ethical conduct of SRM research depends on an institution that has the right to govern SRM research, rather than one that is merely believed to have that right, we focus on the normative sense of legitimacy.

Political philosophers also distinguish between the legitimacy of political institutions and the legitimacy of decisions made by those institutions. To say that an institution is legitimate is to say that with respect to some range of issues, it has the moral authority to make binding decisions for the people within its jurisdiction.[15] To say that a particular decision is legitimate is to say that the institution has the moral right to decide that particular issue in the particular way that it has.[16] The distinction between legitimate institutions and legitimate decisions matters because legitimate institutions can sometimes make illegitimate decisions. A decision might be

[13] *Id.* at 4.

[14] Allen Buchanan & Robert O. Keohane, *The Legitimacy of Global Governance Institutions*, 20 ETHICS & INT'L AFFAIRS 405, 405 (2006).

[15] *Cf.* Allen Buchanan, *Political Legitimacy and Democracy*, 112 ETHICS 689, 689–90; JOHN RAWLS, POLITICAL LIBERALISM 428 (1993).

[16] John Rawls, *Political Liberalism: Reply to Habermas*, 92 J. PHIL. 132, 148 (1995).

illegitimate because it does not result from the proper procedure.[17] If, for instance, a legislative body requires half of its members to be present for a quorum, then a decision is illegitimate if it is made when only a third of the membership is present. Similarly, if a state's legislature enacts a law that violates a right protected by the state's constitution, the law is illegitimate; the only legitimate procedure for abridging that right is to change the constitution. A decision could also be illegitimate if it is grossly unjust.[18] Although states can be legitimate without being perfectly just, not even a legitimate government of a legitimate state has the moral authority to violate the basic rights of its citizens in systematic ways. For instance, procedural propriety presumably would not confer legitimacy on a decision to strip a particular ethnic minority of basic civil rights.

In political (i.e., collective) decision making, legitimacy plays the role that consent plays in individual decision making. Anyone who voluntarily cedes authority over some range of issues to a trade union, a board of directors, a government, or a similar decision-making body thereby acknowledges that legitimacy is an appropriate standard for evaluating collective decisions. As Allen Buchanan puts it, consent, despite its prominence in social contract theorists' accounts of political legitimacy, is "ill-suited to the political world" because "politics seems to be concerned ... with how to get along when consent is lacking."[19]

As "participation" in an SRM experiment is a collective choice, not an individual one, researchers whose experiments have the legitimate approval of an appropriate institution will satisfy the demands of the Principle of Respect. An appropriate institution, in this context, is one with the global political legitimacy to make decisions about SRM experiments.

For the purposes of assessing possible models for a global SRM governance institution, we adopt Allan Buchanan and Robert Keohane's Complex Standard of legitimacy for global governance institutions (GGIs).[20] In broad strokes, the Complex Standard has three parts, each of which we elaborate on below. First, a legitimate institution must enjoy the ongoing consent of democratic states. Second, a legitimate institution must meet certain "substantive" conditions: namely, it must exhibit "minimal moral acceptability," maintain its institutional integrity, and deliver positive benefits relative to alternative feasible institutional arrangements. Third, a legitimate institution must manifest certain "epistemic" or "deliberative virtues," which provide sufficient transparency and accountability to ensure meaningful participation by and due consideration of its stakeholders.

[17] *Id.* at 175.
[18] *Id.* at 176.
[19] Buchanan, *supra* note 15, at 699–700. *See also* RAWLS, *supra* note 15, at 393 and 428.
[20] Buchanan & Keohane, *supra* note 14, at 417–29.

The consent of democratic states is a necessary, but not sufficient, condition for the legitimacy of a GGI. Buchanan and Keohane worry primarily that the "chain of delegation" tying GGIs to the individuals that legitimize the states that legitimize the GGIs may become too long. With such a long leash, the bureaucrats in a GGI may not be appropriately responsive to stakeholders' needs. In the case of SRM, at least, there is a further concern. Several major states – certainly China and arguably Russia – are not democratic in the relevant sense. Indeed, a great deal of the world's population lives in nondemocratic states. Given the potentially broad impact of the decision to be made, we are reluctant to claim that an institution regulating SRM could be legitimate without the consent of at least the larger, less illegitimate nondemocratic states.[21]

The "substantive" conditions for legitimacy combine the need to deliver positive net benefits with the need to avoid gross injustices, corruption, and abuses of power. Buchanan and Keohane explain that GGIs meet the first substantive condition, "minimal morally acceptability," if they do not "persist in committing serious injustices," where a serious injustice consists in violating human rights.[22] They understand "institutional integrity" to mean adherence to a GGI's stated mission and methods. Corruption eroded the integrity of the UN Oil-for-Food Program, for example, because it permitted Saddam Hussein and other government officials to profit from the sale of oil, even though the program aimed to ensure that Iraqi oil revenues would benefit the Iraqi public without further enriching him.[23] Even if a GGI meets these two substantive conditions, it must deliver positive net benefits, as compared with other feasible institutional arrangements. (One feasible arrangement, of course, is the absence of a formal institution.)

The most important part of the Complex Standard, in our view, is the requirement that GGIs manifest the "epistemic virtues" of transparency and accountability.[24] These virtues set democratically legitimate GGIs apart from global bureaucracies staffed by unaccountable technocrats and operating opaquely. Even if the UN General Assembly unanimously voted to establish a GGI to be run by technocratic experts, and even if benevolent experts at the GGI met Buchanan and Keohane's substantive conditions for legitimacy, the GGI would lack legitimacy if the global public had no effective way to monitor and sanction the GGI's activities. A benevolent dictatorship is illegitimate, even if initially installed with public approval, because of the ease with which it can abuse its power; a benevolent but opaque and unaccountable technocracy is illegitimate for the same reason.

[21] *But cf.* Buchanan & Keohane, *supra* note 14, at 412–14.

[22] *Id.* at 419.

[23] *Id.* at 422–23.

[24] *Id.* at 424–33.

Buchanan and Keohane's epistemic virtues serve to overcome the informational asymmetries that enable bureaucracies to subvert the will of their creators. To be transparent and accountable, a GGI must provide information on its goals and behavior in a format that is accessible and intelligible to transnational civil society. Furthermore, there must be mechanisms by which civil society can challenge the GGI's goals, standards, and methods and sanction the GGI for failing to meets its standards or achieve its goals. Manifesting these virtues involves actively engaging with transnational civil society, usually through national governments and international NGOs. Engaging all groups that are significantly affected by an institution may require engaging actors outside the usual circle of governments and NGOs. This is certainly the case with SRM, as those most vulnerable to decisions about SRM experiments may not be well represented by existing NGOs or governments.[25]

In light of Buchanan and Keohane's discussion, we believe that a GGI that met the Complex Standard would have the political legitimacy to make decisions about conducting SRM experiments. We do not claim that such an institution could make decisions about deploying SRM for non-research purposes. Such deployment would involve more serious, longer-term consequences and commitments than an SRM experiment, and so decisions about deployment may require stricter conditions for legitimacy. These stricter conditions may consist merely in more stringent application of the Complex Standard, or they may involve the introduction of further criteria, such as a larger role for the UN General Assembly or other, more directly representative bodies.

4. MODELS FOR AN INSTITUTION TO MANAGE SRM RESEARCH

During the twentieth century, people developed or considered various institutions to govern a wide range of international activities. We examine three of these institutions as possible models for an institution to manage SRM research. None is a perfect analogue because SRM experiments present a new kind of global problem: never before has the world collectively decided whether to conduct experiments that could affect so many people's welfare in such significant ways. Individual states have made momentous decisions, major international organizations have implemented policies with global consequences, and humanity has stumbled collectively into patterns of behavior – such as fossil fuel use – that reshape the globe. None of these decisions, however, constituted an intentional choice by the global public to undertake a risky global experiment for the sake of acquiring new knowledge. Thus, our purpose in reviewing existing GGIs is not to find a single, complete model

[25] *See* Pablo Suarez, Jason Blackstock & Maarten van Aalst, Towards a People-Centered Framework for Geoengineering Governance: A Humanitarian Perspective, 1 GEOENGINEERING Q. 2, 3 (2010).

for managing SRM. Instead, we draw what lessons we can from each case about the ways that an SRM governance body could satisfy the Complex Standard for legitimacy.

4.1 *Institutions for Managing Nuclear Weapons*

SRM would enable humanity to alter the world in a relatively short period of time. Nuclear weapons gave humanity power to alter the world overnight. Given the power of nuclear weapons, the international community has developed a suite of institutions to regulate them. These institutions aim to constrain nuclear testing, curb nuclear proliferation, and reduce the size and danger of existing nuclear arsenals. In this section, we focus mainly on institutions that constrain nuclear testing. We also consider the hypothetical International Atomic Development Agency (IADA), which the United States proposed in 1946 as part of the Baruch Plan.

4.1.1 Nuclear Test Ban Treaties

Between 1963 and 1996, the international community concluded four treaties that constrain the testing of nuclear weapons. These are the 1963 Treaty Banning Nuclear Weapon Tests in the Atmosphere, in Outer Space, and Under Water ("Partial Test Ban Treaty" or PTBT); the 1970 Treaty on the Non-Proliferation of Nuclear Weapons (NPT); the 1973 Treaty on the Limitation of Underground Nuclear Weapon Tests ("Threshold Test Ban Treaty" or TTBT); and the 1996 Comprehensive Nuclear-Test-Ban Treaty (CTBT). We refer to these treaties collectively as the "Test Ban Treaties" (TBTs). The TBTs – especially the NPT – form part of the larger international effort against proliferation and toward disarmament. The history of that larger effort, and of the TBTs in particular, holds important lessons for those interested in forging international agreements about SRM. In other words, the role of transnational civil society,[26] the importance of vested interests at the domestic level,[27] the ways in which nuclear-weapon states promised to protect non–nuclear-weapon states from nuclear aggression,[28] and the various political obstacles confronting diplomats in shaping the TBTs[29] would likely find echoes in the process of shaping SRM treaties. In this chapter, we leave many of those lessons aside to focus narrowly on the question of the legitimacy of the TBTs' constraints on nuclear weapons tests.

[26] *See* Rebecca Johnson, Unfinished Business: The Negotiation of the CTBT and the End of Nuclear Testing 25 (2009).

[27] *See* JOHNSON, *supra* note 26, at 32, 41, 47.

[28] *See* S.C. Res. 255, U.N. Doc. S/RES/255 (June 19, 1968).

[29] *See* JOHNSON, *supra* note 26, at 9–172.

Nuclear weapons tests share important features with SRM experiments. Like SRM experiments, nuclear weapons tests threaten the global public directly, through exposure to radioactive fallout, and indirectly, by contributing to the development of dangerous technologies. Furthermore, states conduct nuclear tests in part because they believe that the development or maintenance of nuclear weapons may be vital to their national interests in the future – a view that some states may one day adopt with respect to SRM technologies.

The PTBT bans all nuclear explosions except those conducted underground.[30] (The treaty exempted subterranean tests partly because of technical difficulties in distinguishing such tests from earthquakes.) The treaty's purpose was to curtail the testing of nuclear weapons in order to slow the nuclear arms race and protect the public from radioactive fallout. Beginning in 1955, small multilateral conferences of major powers struggled for eight years to negotiate a ban on nuclear testing. Frustrated by the failure of these negotiations, the United States, the USSR, and the UK hammered out a treaty over the course of ten days in Moscow in 1963.[31] This is not to say that the multilateral negotiations were fruitless. They laid the groundwork for the final negotiations, helping to ensure that the negotiations in Moscow generated a treaty to which most states consented. One hundred and eight parties signed the treaty that fall. The treaty has 124 parties, including all of the nuclear-armed states except China, France, and North Korea, none of which signed the treaty.[32] The PTBT did not involve the creation of a separate bureaucracy; the treaty implicitly relies on state parties to detect violations of the treaty.

In the years following the PTBT, the NPT emerged from bilateral and multilateral negotiations, including discussion in the UN General Assembly. The treaty prohibits the transfer of nuclear weapons–related technologies from nuclear-weapon states to any other State, and it prohibits non–nuclear-weapon states from developing or acquiring nuclear explosives.[33] Rather than prohibiting certain kinds of nuclear tests, it prohibits tests by certain actors – namely, states that had not already detonated a nuclear explosive prior to January 1, 1967.[34] The treaty opened for signature in 1968 and entered into force in 1970. It currently has 190 parties, indicating broad global consent.[35] The most prominent dissenters are India, Israel, and Pakistan.

30 *See* Treaty Banning Nuclear Weapon Tests in the Atmosphere, in Outer Space and Under Water, Aug. 5, 1963, 14 UST 1313, 480 UNTS 4, at Art. I.

31 U.S. Department of State, Treaty Banning Nuclear Weapon Tests in the Atmosphere, in Outer Space and Under Water (n.d.), http://www.state.gov/t/isn/4797.htm.

32 *See id.*

33 Treaty on the Non-Proliferation of Nuclear Weapons, July 1, 1968, 21 UST 483, 729 UNTS 169 at Art. I–II.

34 *Id.* at Art. IX.

35 United Nations Office for Disarmament Affairs, "Non-Proliferation of Nuclear Weapons (NPT)" (n.d.), http://www.un.org/disarmament/WMD/Nuclear/NPT.shtml.

India has been a particularly vocal critic of the treaty, which it sees as entrenching international power differentials.[36]

The NPT relies heavily on the International Atomic Energy Association (IAEA) to monitor compliance with the treaty's provisions. Although this makes the IAEA central to the global nonproliferation regime, the IAEA plays only an indirect role by helping to enforce the regimes prohibiting nuclear weapons tests.

In 1974, the United States and the USSR negotiated the TTBT. The treaty restricts underground tests to yields of less than 150 kilotons.[37] Concerns about verification stalled ratification for sixteen years until highly technical protocols were devised through bilateral meetings in the late 1980s. Following the adoption of these protocols, both parties ratified the treaty in 1990. The United States and Russia remain the sole parties to the treaty.[38] Despite this delay, both parties announced in 1976 that they would abide by the 150-kiloton limit,[39] and according to the officially stated yields of their nuclear tests, both have done so.[40] Critics condemn the treaty as a ruse by which the superpowers could claim progress on disarmament without imposing meaningful limits on the development of their nuclear arsenals.[41]

The CTBT, which would ban all nuclear explosions of any kind, was opened for signature in September 1996 but has yet to enter into force. As of 2010, the treaty is awaiting the ratification of nine key states: China, Egypt, Indonesia, India, Iran, Israel, North Korea, Pakistan, and the United States. Some of these states have signed the treaty; others have not.[42] Even without the CTBT in force, however, nuclear tests have all but ceased. None of the major nuclear-weapon states have conducted tests since 1996; India, Pakistan, and North Korea have each conducted two tests since then.[43] The CTBT arguably played a causal role in this reduction in testing. China and France stated that their final tests in 1996 were meant to avoid the need for further testing once they had signed the CTBT. The largest reduction, though, came earlier from the end of the Cold War, which enabled Russia, the UK, and the United States to cease testing in the early 1990s.

[36] JOHNSON, *supra* note 26, at 19.

[37] Treaty on the Limitation of Underground Nuclear Weapon Tests, July 3, 1974, 13 ILM 906 (1974) at Art. I.

[38] U.S. Department of State. "Threshold Test Ban Treaty" (n.d.), http://www.state.gov/t/isn/5204.htm.

[39] *Id.*

[40] U.S. Department of Energy, "United States Nuclear Tests: July 1945 through September 1992" (2000), http://www.nv.doe.gov/library/publications/historical/DOENV_209_REV15.pdf at 71–89; Ministry of the Russian Federation for Atomic Energy, "USSR Nuclear Weapons Tests and Peaceful Nuclear Explosions: 1949 through 1990" (1996), http://npc.sarov.ru/english/issues/peaceful/peaceful_e.pdf at 30–48.

[41] JOHNSON, *supra* note 30, at 20.

[42] *Id.* at 3–4.

[43] Comprehensive Nuclear Test-Ban Treaty Organization Preparatory Committee, "Nuclear Testing: 1945–2009" (n.d.), http://www.ctbto.org/nuclear-testing/history-of-nuclear-testing/nuclear-testing-1945–2009/page-7-nuclear-testing-1945–2009.

Like the PTBT, the CTBT emerged from a long series of discussions in various forums. Among the most important of these forums was the Group of Scientific Experts (GSE), which had collaborated since 1976 to develop the technical knowledge needed to monitor and verify compliance with a test ban.[44] Decades of discussion about a CTBT culminated in two years of negotiations in the Conference on Disarmament (CD) from 1994 to 1996. Despite substantial progress in those two years, Indian opposition still threatened to scuttle the treaty. Only Belgian and Australian parliamentary maneuvering brought the draft treaty out of the CD and into the UN General Assembly. The General Assembly endorsed the draft by an overwhelming majority.[45]

The CTBT calls for a dedicated international organization to monitor compliance and implement the treaty; a Preparatory Commission has worked since 1997 to lay the groundwork for implementation. The central task of the Comprehensive Nuclear-Test-Ban Treaty Organization (CTBTO) is to monitor compliance with the treaty. This involves operating the International Monitoring System (IMS), which monitors for physical and chemical signs of a nuclear explosion, and if necessary, conducting on-site inspections after suspected nuclear tests. The IMS consists of 337 facilities around the world. These facilities monitor seismic events, hydroacoustic activity, atmospheric infrasonic waves, and airborne radionuclides.[46] The data they gather is channeled to the International Data Centre (IDC) in Vienna and made available for civilian research. The IDC provides both raw data and quality-controlled data bulletins to member states, along with software and training to help member states interpret the data.[47] This arrangement grew out of negotiators' insistence that the IDC make its data transparent to member states that lack the resources to interpret raw data.[48]

Collectively, the TBTs provide a framework by which the international community has forbidden various classes of dangerous experiments. What lessons do the TBTs hold for those looking to create a legitimate SRM GGI?

First, the TBTs suggest a way to achieve some progress, ethically speaking, with respect to SRM experiments, even if the international community cannot reach perfect agreement on whether or how to conduct such experiments. The TBTs do not constitute the global community's consent to any particular test; they merely

[44] *See* JOHNSON, *supra* note 26, at 149.

[45] *Id.* at 46–142.

[46] Comprehensive Nuclear-Test-Ban Treaty Organization Preparatory Commission, *The CTBT Verification Regime: Monitoring the Earth for Nuclear Explosions* 2 (2009), http://www.ctbto.org/fileadmin/user_upload/public_information/2009/Verification_Regime_final_web.pdf [hereinafter CTBTO Preparatory Commission].

[47] *Id.* at 5.

[48] JOHNSON *supra* note 26, at 149.

express, through a legitimate GGI, a refusal to "participate" in certain kinds of experiments. The decision to perform nuclear tests or not remains in the hands of those states that are authorized to do so under the NPT, or have refused to join the NPT. If the international community cannot agree on which SRM experiments they would like to perform, they might at least agree on what kinds of SRM experiments they will not tolerate. Conducting an SRM experiment in the gaps left by a partial test ban would still be deeply ethically problematic, as it still amounts to human experimentation without consent. However, even if it only had the power to restrict the kinds of experiments that may be performed, a politically legitimate SRM GGI would increase the global public's control over the climate and might help deter the most dangerous experiments. Although it would not satisfy the Principle of Respect, it would be ethically better than nothing.

Second, the history of the TBTs demonstrates that meaningful treaties that enjoy widespread international support – and thus satisfy one of the criteria in the Complex Standard for legitimacy – can emerge from small multilateral negotiations. The PTBT, for instance, was ultimately negotiated by just three states, and yet it covers over one hundred states. The lesson for SRM is that, if wider negotiations falter, a relatively small working group may be able to produce a treaty that the broader international community finds acceptable.

Admittedly, none of the TBTs have attained universal support. Each lacks the support of at least one major power, including at least one major democracy: China and France declined to sign the PTBT, although both have signed the CTBT; India, Israel, and Pakistan reject the NPT; China, India, Indonesia, Iran, and the United States, among others, have yet to ratify the CTBT.

Some of the differences between nuclear weapons and SRM, however, give reason to hope that multilateral negotiations about SRM may be more productive than negotiations over nuclear test bans. The TBTs were negotiated in a context in which some states already had developed, tested, and deployed nuclear weapons; no one has yet tested or deployed SRM technologies. Thus, a treaty that prohibits the further development of SRM technologies would not institutionalize existing inequalities in the way that the NPT did. Furthermore, nuclear weapons pose a greater technological challenge than (some forms of) SRM. Thus, many states could develop SRM technologies, whereas fewer are capable of producing nuclear weapons. If any state were to deploy such technologies, all states would face the consequences of an altered climate. This increases each state's incentive to seek genuine international consensus relative to the nuclear weapons case, where each state had to worry mainly about unfriendly states with high technological capacity. These differences may facilitate agreements that enjoy even broader international support than the TBTs do. Conversely, the relative ease of conducting SRM experiments provides many more opportunities for political maneuvering. In principle, even small states could threaten to conduct SRM experiments or withhold support from a treaty in order to extract concessions during the treaty negotiations. To the extent that larger states can link SRM to other issues on which they have leverage over smaller states, however, such threats would not be particularly credible.

Third, any SRM GGI will need the capacity to collect and analyze massive quantities of data. The epistemic criteria of the Complex Standard requires that an SRM GGI include an international organization that can relay this data to interested parties in a comprehensible format. The IMS and IDC provide a useful model for collecting and disseminating that data in an epistemically virtuous way. The dual military–civilian use of the IMS also suggests that SRM monitoring could piggyback on existing facilities.

One of the limitations of the analogy between the TBTs and SRM is particularly instructive, too. Continued nuclear weapons tests provided no global benefit. Thus, the TBTs provide a net benefit to the global public, as required by the Complex Standard, although their benefit might not be as great as that of some alternative institution (e.g., a CTBT that is more likely to enter into force). SRM experiments might provide a global benefit, either by preparing the global public to deploy SRM effectively or by revealing that SRM is unwise. In the event that SRM experiments turn out to be beneficial, an SRM GGI modeled on the TBTs would prove to be detrimental to the global public if it prohibited the necessary experiments. The GGI would therefore fail to meet the Complex Standard. One challenge of SRM, of course, is that if SRM experiments do turn out to be necessary, we might not recognize that fact until it is too late. Thus, we may not know that the GGI has been detrimental – and to that extent illegitimate – until after the fact.

4.1.2 International Atomic Development Agency

Before the Cold War set in, the United States envisioned a very different regime for managing nuclear weapons and nuclear technology generally. In June 1946, Bernard Baruch urged the UN to create a powerful international body – the International Atomic Development Agency (IADA) – that would effectively control all aspects of nuclear technology.[49] This so-called Baruch Plan largely followed an earlier report by the U.S. State Department, which had become known as the Acheson-Lilienthal Report. Under the Baruch Plan, the IADA would exercise close control over all phases of nuclear activity. Through ongoing surveys, it would identify all global deposits of

[49] Bernard M. Baruch, Statement of the United States Policy on Control of Atomic Energy as Presented by Bernard M. Baruch, Esq., to the United Nations Atomic Energy Commission (June 14, 1946), reprinted in U.S. DEP'T OF STATE, PUB. 2560, THE UNITED STATES AND THE UNITED NATIONS, REP. SERIES NO. 2, THE UNITED STATES ATOMIC ENERGY PROPOSALS (1946), *available at* http://www.atomicarchive.com/Docs/Deterrence/BaruchPlan.shtml [hereinafter Baruch Plan].

uranium and thorium, controlling the extraction of those minerals in an unspecified manner. It would "exercise complete managerial control" over plants producing fissile materials, and it would own and control the output of those plants.[50] The IADA would maintain a monopoly on research into nuclear explosives, although the manufacture of nuclear weapons would be prohibited, and it would become the world's leading authority on peaceful uses of nuclear energy. Through licensing and inspection arrangements, it would control any use of nuclear technology, providing materials for such activities "under lease or other arrangement."[51] This combination of expertise, ownership, management, and inspection would enable the IADA to understand, recognize, and detect misuses of nuclear technology while promoting its peaceful use in an equitable, secure fashion.[52]

Baruch insisted on swift sanctions against violators. Baruch specifically insisted that such sanctions be immune to veto by the permanent members of the UN Security Council.[53] By a bare majority vote in the Security Council, the UN would have been able to sanction states that the IADA ruled to be in violation of the international nuclear regime.

In part because of Baruch's insistence on veto-proof sanctions, his proposal ended in a diplomatic stalemate. The Soviet Union rejected the IADA out of concern that the United States would retain its nuclear arsenal and that the IADA would become an instrument of U.S. policy. In 1949, the Soviet Union detonated its first nuclear weapon. The arms race had begun, and the prospect of centralized global governance of nuclear weapons faded.

Given both its discretion in the development of nuclear fuel and certain kinds of nuclear research, as well as its power in sanctioning sovereign states, the IADA would have exercised considerable authority in making collective decisions about nuclear technology. Assuming that no state managed to evade the IADA long enough to develop nuclear weapons on its own, the IADA would have exercised a complete monopoly over a world-changing technology.

Thus, the IADA, as proposed by Baruch, constitutes a conceivable model for an SRM governance institution. It would have been an international organization for research into a sensitive, dangerous suite of technologies, about which it would have made important decisions on behalf of the international community – including decisions about experimental uses of the technology. If it had the will to do so, such an organization would be better positioned than any viable alternative to ensure that researchers behave ethically – both with respect to the political legitimacy of their

[50] *Id.* at 8.
[51] *Id.* at 9.
[52] *Id.* at 4–11.
[53] *Id.* at 5.

decisions and with respect to satisfying other requirements of ethical conduct, such as those in our proposed ethical framework.[54]

The degree to which such a program would actually ensure ethical conduct of SRM experiments, however, would depend heavily on the structure of its decision-making processes and on the degree to which its personnel meet the substantive conditions and exhibit the epistemic virtues required by the Complex Standard. One concern about such an organization is that the technocrats who run it may develop goals or preferences that diverge from the interests of the international community. Some staff members' enthusiasm for SRM might exceed that of the global public in dangerous ways. Some might be susceptible to pressure from particular states whose views differ from those of the international community, or might obscure information to protect or further their own careers at the expense of public transparency and accountability. Any of these factors could cause the organization to violate the second or third criteria of the Complex Standard. Thus, placing so much responsibility for SRM research in the hands of unelected technocrats might lead to politically illegitimate institutions or decisions.

The unhappy fate of the Baruch Plan, however, provides an instructive lesson for thinking about the conditions required for widespread international acceptance of an SRM GGI, as required by the Complex Standard. The Baruch Plan was infeasible because it concentrated too much power in an international organization. Some states may have bristled at ceding such power to an international body. Others, such as the Soviet Union, may have feared that the IADA would have been too beholden to the United States. Assuming that contemporary states would likewise reject any GGI that is either too powerful or too likely to be dominated by one or more great powers, the international community would need to design an SRM governance institution carefully in order to give it an appropriate amount of power and international accountability. Otherwise, the SRM GGI would be unlikely to secure the multilateral consent required for legitimacy. Still, if the international community decides to delegate limited authority for SRM experiments to an INGO, the IADA offers one possible conception for doing so.

4.2 *Institutions for Managing Global Commons*

A stable planetary climate represents a type of global commons – a global public good that no single country is capable of controlling.[55] SRM experiments involve a rapid, deliberate change in the climate – a change that could have negative conse-

[54] *See* Morrow et al., *supra* note 3, at 3–6.

[55] Marvin S. Soros, *Garret Hardin and Tragedies of Global Commons*, *in* HANDBOOK OF GLOBAL ENVIRONMENTAL POLITICS 35, 45 (Peter Dauvergne ed., 2006).

quences for some persons and ecosystems. Thus, GGIs designed to manage global commons provide another type of model for an SRM GGI.

Like a stable climate, Antarctica is viewed by many as a global commons. The Antarctic Treaty System (ATS), established in 1959 by the twelve countries active in Antarctica during the International Geophysical Year, sought to ensure the peaceful use of this commons for scientific exploration.[56] Today, the Treaty has forty-eight parties. Twenty-eight of these parties are active in Antarctica and therefore have decision-making authority as Consultative Members; the remaining twenty have observer status as Non-Consultative Members.[57]

The original Antarctic Treaty focused primarily on freezing territorial claims and establishing a legal framework for exploration. Environmental issues entered the ATS through later protocols, the most comprehensive of which is the 1991 Protocol on Environmental Protection (the Madrid Protocol). The Madrid Protocol, which entered into force in 1998, is perhaps most broadly known for establishing a fifty-year moratorium on exploiting mineral resources in the Antarctic; more relevant to our analysis, it also established a set of principles regarding environmental protection, an intergovernmental body of scientific experts to offer advice on environmental issues, a procedure for environmental impact assessment of activities in Antarctica, and a consultative process regarding these activities.[58]

Article 3 of the Protocol lays out a set of principles that gives primacy both to ethical concerns and scientific research. These principles require that activities in the Antarctic be planned and executed "so as to limit adverse impacts on the Antarctic."[59] The article also mandates monitoring of risky activities and requires that such activities be modified or stopped if monitoring reveals adverse impacts.[60] In principle, this article requires states parties to give significant weight to environmental, ethical, and even aesthetic values in regulating governmental and nongovernmental activities in the Antarctic. Among the ethical principles recognized are those akin to our Principles of Minimization and Respect.[61]

Article 11 establishes the Committee for Environmental Protection (CEP). The CEP consists of representatives from each Party to the Protocol, along with their advisors. Parties to the ATS who are not Parties to the Protocol, as well as relevant NGOs invited by the CEP, may attend meetings as observers. The Protocol instructs the CEP to provide technical advice on the implementation of the Protocol, including

[56] Antarctic Treaty, 1 Dec., 1959, 402 UNTS 71.

[57] Secretariat of the Antarctic Treaty System, "Parties" (2008), http://www.ats.aq/devAS/ats_parties.aspx.

[58] Madrid Protocol on Environmental Protection to the Antarctic Treaty, 4 Oct., 1991, 30 ILM 1455 [hereinafter Madrid Protocol].

[59] *Id.* at Art. 3, para. 2. (See the Appendix to this chapter for the complete text of Article 3.)

[60] *Id.* at Art. 3, para. 2(d)–(e), 4(b).

[61] *See* Morrow et al., *supra* note 3, at 3–6.

advice on the effectiveness of parties' efforts to comply with the Protocol.[62] Because the CEP must release reports on its sessions to states parties and to the public,[63] it could alert interested states and members of civil society to activities that run contrary to the Protocol. Ultimately, however, the CEP's role is strictly advisory; it has no power to affect decisions directly.

The states parties hold ultimate responsibility for assessing the environmental impact of their activities, although they must discuss their assessment of some activities with the other Parties and the CEP. As laid out in Article 8 and Annex I, the Protocol recognizes three tiers of activities in the Antarctic: those determined by national procedures to have "less than a minor or transitory impact," those "likely to have no more than a minor or transitory impact," and those likely to have "more than a minor or transitory impact."[64] Activities falling into the second category require an Initial Environmental Evaluation characterizing the activity, alternatives to the activity, and likely impacts.[65] Activities falling into the third category require a Comprehensive Environmental Evaluation (CEE), which describes the state of the environment prior to the activity; the activity and all relevant alternatives, including the alternative of not proceeding with the activity, along with the expected consequences of each alternative; the direct and indirect impacts of the proposed activity; the cumulative impact of the proposed activity, given existing and currently planned activities; the methodology and data used to forecast consequences; the measures that could be taken to monitor the effects of the activity and to minimize or mitigate them; a nontechnical summary of the above information; and the contact information for the author(s) of the CEE.[66]

The draft CEE must be circulated for review to the public, to the Antarctic Treaty parties, and to the CEP. In principle (although not always in practice), the activity cannot proceed until the draft CEE has been considered by the Antarctic Treaty Consultative Meeting on the advice of the Committee, and a final version of the CEE must respond to the comments raised in the review process. The draft and final CEE must be made publicly available.[67] Once the activity begins, its impacts must be monitored.[68]

Christopher Joyner highlights a number of potential weaknesses in the Madrid Protocol process.[69] The role of the Environmental Impact Assessment consultative

[62] Madrid Protocol, *supra* note 58, at Art. 11.

[63] *Id.* at Art. 11, para. 5.

[64] *Id.* at Art. 8, para. 1.

[65] Madrid Protocol, *supra* note 58, at Annex I, Art. 2.

[66] *Id.* at Annex I, Art. 3, para. (1)–(2).

[67] *Id.* at Annex I, Art. 3, para. (3)–(6).

[68] *Id.* at Art. 8.

[69] CHRISTOPHER C. JOYNER, GOVERNING THE FROZEN COMMONS: THE ANTARCTIC REGIME AND ENVIRONMENTAL PROTECTION 165–74 (1998).

process is fundamentally hortatory; although individual governments must respond to comments under the Protocol, they retain the final decision on whether to proceed with a specific activity. In addition, the boundaries between the different categories of activities are ill-defined, left to some combination of party judgment and the evolution of precedent. More broadly, the mechanism of enforcement of the Protocol in general is unclear: parties are to enforce it through laws and regulations, and shall exert "appropriate efforts, consistent with the Charter of the United Nations" to ensure that other parties do,[70] whereas an Arbitral Tribunal or the International Court of Justice is empowered to settle disputes, but again these are largely hortatory procedures.[71]

Despite these problems, the Madrid Protocol provides a GGI model that directly addresses elements of scientific research ethics and the Complex Standard. In particular, the consultative process for assessing proposed activities exemplifies the transparency and stakeholder engagement necessary for legitimacy. CEEs must contain nontechnical summaries, making them more easily digestible by states and civil society. Draft and final CEEs, along with reports on CEP sessions, are distributed to states parties and the public.[72] This increases the transparency of the international governance of Antarctic activity, as required by the Complex Standard.

The Madrid Protocol does not, however, provide an effective means for citizens of one state to hold another state or its citizens accountable for behavior that violates the Protocol. The hortatory nature of the EIA process would be even more problematic in the case of SRM, where the incentive to ignore the exhortations of other states might be much greater than in the Antarctic case. If an SRM GGI had no more power than the CEP does in Antarctica, then it could not deter even a moderately motivated state from conducting SRM experiments. Conversely, an SRM GGI that could, at its own discretion, prohibit certain experiments would be too powerful – too much like Baruch's proposed IADA – to be feasible, and a GGI that could prohibit experiments if and only if they violated constraints laid down in a treaty would be more like the CTBT than the CEP. Thus, replacing the hortatory model of the Madrid Protocol with something stronger brings us back to the nuclear weapons testing models.

Some elements of the Madrid Protocol could be readily adapted to the context of SRM research governance. Article 3 in particular would need just one major addition – impact on human populations – and a suite of minor contextual adaptations to address the global climate commons instead of the Antarctic "frozen commons." The conditions and processes for conducting environmental impact assessments

[70] Madrid Protocol, *supra* note 58, at Art. 13, para. 1–2.

[71] JOYNER, *supra* note 69, at 166.

[72] Madrid Protocol, *supra* note 58, at Annex I, Art. 3.

translate relatively easily to the SRM case as well. By requiring earlier involvement from other states and an SRM counterpart of the CEP, along with scientific peer review of SRM research proposals, an SRM GGI might be able to exert more influence on the shaping of proposals than the CEP exerts on proposed activities in the Antarctic. In any case, the Madrid Protocol provides a model for achieving the transparency required for legitimacy.

One challenging difference between the Antarctic context and the SRM context relates to the range of parties involved: the Antarctic Treaty engages in a consultative status with only the twenty-eight countries active in Antarctica, whereas a GGI focused on SRM research would need to engage not just the countries actively conducting research but the larger group of countries with populations potentially at risk. SRM experiments would also be likely to affect various states in more direct and more significant ways than Antarctic activities would, making disputes over SRM more heated than those over Antarctic activities. Broader and more contentious discussions over SRM experiments may increase the time it takes to complete an EIA for any proposed experiment, as compared to the time it takes to complete an EIA for proposed Antarctic activities. The broader constituency of an SRM GGI might also make it more difficult to craft an institution that enjoys sufficiently widespread acceptance to be legitimate.

As the preceding discussion shows, incorporating elements of the Madrid Protocol into an SRM GGI could help improve the chances that the GGI would retain its legitimacy under the Complex Standard, especially in terms of epistemic virtues and the delivery of positive net benefits to the global public. By providing a smaller but open forum for international deliberation about the decisions of individual states, it could also improve the GGI's ability to reach legitimate decisions about particular SRM experiments, without requiring unanimous consent from the international community about that experiment. As its processes are hortatory rather than coercive, however, an SRM GGI modeled on the Protocol would have little power to rein in states that decided to pursue SRM without international approval.

5. CONCLUSION

The international community is starting to consider SRM as a stopgap or emergency measure for coping with the possible inadequacy of medium-term mitigation efforts. As it would be foolish and unethical to deploy SRM without an adequate understanding of the technology, state or non-state actors may decide to pursue SRM research – potentially including large-scale experiments – in the near future.

Large-scale SRM experiments, such as those involving the injection of enough reflective aerosols into the stratosphere to produce detectable climatic changes at the regional or global scale, involve significant ethical challenges. One important challenge is the need for politically legitimate decisions about whether and how to conduct such experiments; this requirement parallels the requirement for informed consent in medical experiments. Given that an SRM experiment is a global one, the decision to undertake it must be made by a politically legitimate GGI.

The Complex Standard for the political legitimacy of GGIs involves three broad requirements. First, the GGI must enjoy the consent of most (democratic) states. Second, the GGI must meet certain substantive conditions, such as the avoidance of serious injustices, the delivery of positive net benefits, and the maintenance of institutional integrity. Third, the GGI must exhibit certain epistemic virtues, such as transparency and accountability.

Other GGIs hold lessons for the design of a legitimate SRM GGI. Our analysis of the international nuclear testing regime suggests that a legitimate SRM GGI might evolve through negotiations among a smaller, more manageable group of powers, as long as the GGI itself receives the approval of the UN. It also suggests that, in the absence of a GGI empowered to authorize particular experiments, an institution with the legitimate authority to prohibit certain classes of experiments could protect the world against the most ethically problematic ones. As our analysis of the Madrid Protocol to the Antarctic Treaty suggests, an institution requiring and facilitating international discussion of any proposed SRM experiment would fare well on the third, epistemic criterion of the Complex Standard. Spelling out a set of principles that SRM experiments must follow, as the Madrid Protocol does for Antarctic activities, might increase the likelihood that such an institution could deliver positive net benefits, as required by the second substantive criterion of the Complex Standard. Our analysis of the proposed IADA suggests that a more powerful GGI, which might have the power to authorize specific experiments, may find it more difficult to meet the Complex Standard; such a powerful GGI may never enjoy the widespread support required for legitimacy, it is more likely to depart from the wishes of its creators, and it may do so in ways that violate the substantive and epistemic criteria of the Complex Standard.

The lessons from these case studies are complementary. A single institution could prohibit certain classes of experiments, such as the TBTs, while facilitating international dialogue about the experiments proposed by states or non-state actors, like the Madrid Protocol. Such an institution would leave room for states to create a multilateral organization that combined research efforts without exercising the far-reaching powers and technological monopolies of the IADA. This is only a preliminary vision, of course, of an approach to managing SRM research, leaving many institutional issues open for further exploration.

The international community has never confronted a decision quite like that of conducting SRM experiments – much less to deploy SRM. That is why none of the institutions we consider provide perfect analogues for an SRM GGI. This is not the first time, however, that humanity has faced novel problems demanding unprecedented institutions. We believe that by learning from the successes and failures of the past, the international community can design an institution to manage decisions about SRM and SRM research in an ethically responsible way.

APPENDIX: ARTICLE 3 OF THE MADRID PROTOCOL

The complete text of Article 3 of the Madrid Protocol reads:

1 The protection of the Antarctic environment and dependent and associated ecosystems and the intrinsic value of Antarctica, including its wilderness and aesthetic values and its value as an area for the conduct of scientific research, in particular research essential to understanding the global environment, shall be fundamental considerations in the planning and conduct of all activities in the Antarctic Treaty area.

2 To this end:
   - (a) activities in the Antarctic Treaty area shall be planned and conducted so as to limit adverse impacts on the Antarctic environment and dependent and associated ecosystems;
   - (b) activities in the Antarctic Treaty area shall be planned and conducted so as to avoid:
     - (i) adverse effects on climate or weather patterns;
     - (ii) significant adverse effects on air or water quality;
     - (iii) significant changes in the atmospheric, terrestrial (including aquatic), glacial or marine environments;
     - (iv) detrimental changes in the distribution, abundance or productivity of species of populations of species of fauna and flora;
     - (v) further jeopardy to endangered or threatened species or populations of such species; or
     - (vi) degradation of, or substantial risk to, areas of biological, scientific, historic, aesthetic or wilderness significance;
   - (c) activities in the Antarctic Treaty area shall be planned and conducted on the basis of information sufficient to allow prior assessments of, and informed judgements about, their possible impacts on the Antarctic environment and dependent and associated ecosystems and on the value of Antarctica for the conduct of scientific research; such judgments shall take account of:
     - (i) the scope of the activity, including its area, duration and intensity;
     - (ii) the cumulative impacts of the activity, both by itself and in combination with other activities in the Antarctic Treaty area;

(iii) whether the activity will detrimentally affect any other activity in the Antarctic Treaty area;
(iv) whether technology and procedures are available to provide for environmentally safe operations;
(v) whether there exists the capacity to monitor key environmental parameters and ecosystem components so as to identify and provide early warning of any adverse effects of the activity and to provide for such modification of operating procedures as may be necessary in the light of the results of monitoring or increased knowledge of the Antarctic environment and dependent and associated ecosystems; and
(vi) whether there exists the capacity to respond promptly and effectively to accidents, particularly those with potential environmental effects;

(d) regular and effective monitoring shall take place to all assessment of the impacts of ongoing activities, including the verification of predicted impacts;
(e) regular and effective monitoring shall take place to facilitate early detection of the possible unforeseen effects of activities carried on both within and outside the Antarctic Treaty area on the Antarctic environment and dependent and associated ecosystems.

3 Activities shall be planned and conducted in the Antarctic Treaty area so as to accord priority to scientific research and to preserve the value of Antarctica as an area for the conduct of such research, including research essential to understanding the global environment.

4 Activities undertaken in the Antarctic Treaty area pursuant to scientific research programs, tourism and all other governmental and nongovernmental activities in the Antarctic Treaty area for which advance notice is required in accordance with Article VII (5) of the Antarctic Treaty, including associated logistic activities, shall:
(a) take place in a manner consistent with the principles in this Article; and
(b) be modified, suspended or cancelled if they result in or threaten to result in impacts upon the Antarctic environment or dependent or associated ecosystems inconsistent with those principles.[73]

ACKNOWLEDGMENTS

The authors wish to thank Ian D. Lloyd and Robert Keohane for helpful comments on drafts of this chapter.

[73] *Id.* at Art. 3.

7

Geoengineering and the Myth of Unilateralism

Pressures and Prospects for International Cooperation

Joshua B. Horton

1. INTRODUCTION

In recent years, discussions of geoengineering have intensified among scientists, policy makers, and other interested observers. The possibility that one state might unilaterally deploy geoengineering technology has become a fixture in these debates, and has cast a pall over substantive inquiry into climate intervention research and implementation. Speculation about "rogue" states pursuing geoengineering outside multilateral frameworks has given pause to calls for more robust experiments and field trials, and has contributed to the adoption of moratoria by the London Convention/London Protocol (LC/LP) and the Convention on Biological Diversity (CBD).[1] In sum, the fear of unilateralism has become an idée fixe in conversations about geoengineering, in effect putting the brakes on more ambitious research efforts and deliberations about governance issues.

In this chapter, I argue that this fear of unilateralism is largely misplaced, grounded more in unexamined policy assumptions than in reasoned analysis of the strategic situation faced by states. I will present this argument in five parts. First, I will document the widespread notion that unilateral geoengineering poses a genuine threat to the international order. Second, I will closely examine the interests and constraints that are likely to confront states contemplating intervention in the climate system. Third, I will demonstrate that international dynamics are more likely to create pressures leading to cooperation than to foster tendencies toward unilateralism. Fourth, I will consider different mechanisms for encouraging collaboration on climate intervention strategies. Finally, I will consider the implications of this argument for future discussions of geoengineering.

[1] In 2008, parties to the CBD agreed to prohibit ocean fertilization activities except in "coastal waters." *See* CBD Dec. IX/16, UNEP/CBD/COP/DEC/IX/16 (May 30, 2008). Later that year, parties to the LC/LP agreed to ban ocean fertilization activities except for "legitimate scientific research." *See* LC/LP Res. LC-LP.1, LC 30/16/Annex 6 (Oct. 31, 2008).

In what follows, I will focus specifically on stratospheric aerosol injections as the intervention strategy most likely to be selected to combat the effects of global climate change.[2] Solar Radiation Management (SRM) technologies are broadly regarded as primary candidates for geoengineering deployment, because of their relative simplicity, rapid effects, and low cost.[3] The injection of sulfate aerosols is regarded as particularly attractive given current knowledge and familiarity with natural analogs (i.e., volcanic eruptions).[4] Although the arguments outlined later on in this chapter are developed in reference to stratospheric aerosol injections, the general thesis that unilateralism is an exaggerated threat applies to geoengineering as a whole, with appropriate modifications for alternative intervention technologies.

2. UNILATERALISM AS CONVENTIONAL WISDOM

Conventional concerns about the threat of unilateral geoengineering typically run as follows: At some point in the decades to come, in a world of accelerating climate change and continuing failure to significantly reduce greenhouse gas (GHG) emissions, Country A decides that its interests would best be served by implementing a climate intervention strategy (in the present case, stratospheric aerosol injections). This decision may be driven by any number of developments: altered precipitation patterns, rising sea level, shifting disease vectors, disruptions to agriculture, desertification, etc. Unable to gain international support for the deployment of SRM, Country A deploys the technology on a unilateral basis, flouting the consensus among states, materially affecting the global climate system, and offending those who view climate intervention as "unnatural" – in other words, seizing control of the "global thermostat."[5] Such unilateral action inflames global public opinion, raises international tensions, and triggers a response in the form of sanctions, a trade war, or worse.

Variations of this scenario have been advanced most prominently by members of the policy community, especially foreign policy experts.[6] For example, Victor et al. write:

> [G]eoengineering is an option at the disposal of any reasonably advanced nation. A single country could deploy geoengineering systems from its own territory

[2] JASON J. BLACKSTOCK ET AL., CLIMATE ENGINEERING RESPONSES TO CLIMATE EMERGENCIES 4–16 (2009).

[3] Lee Lane & J. Eric Bickel, *Solar Radiation Management and Rethinking the Goals of COP-15*, *in* COPENHAGEN CONSENSUS ON CLIMATE: ADVICE FOR POLICYMAKERS 16–19 (2009).

[4] ROYAL SOCIETY, GEOENGINEERING THE CLIMATE: SCIENCE, GOVERNANCE AND UNCERTAINTY 29–32 (2009).

[5] Alan Robock, *20 Reasons Why Geoengineering May Be a Bad Idea*, 64 BULL. ATOMIC SCI. 14, 17 (2008).

[6] *See* Jason J. Blackstock & Jane C.S. Long, *The Politics of Geoengineering*, 327 SCI. 527 (2010); BLACKSTOCK ET AL., *supra* note 2, at 44; Lane & Bickel, *supra* note 3, at 19; John Virgoe, *International*

> without consulting the rest of the planet. Geoengineers keen to alter their own country's climate might not assess or even care about the dangers their actions could create for climates, ecosystems, and economies elsewhere. A unilateral geoengineering project could impose costs on other countries, such as changes in precipitation patterns and river flows or adverse impacts on agriculture, marine, fishing, and tourism.... At some point in the near future, it is conceivable that a nation that has not done enough to confront climate change will conclude that global warming has become so harmful to interests that it should unilaterally engage in geoengineering.... Unilateral action would create a crisis of legitimacy that could make it especially difficult to manage geoengineering schemes once they are under way.[7]

Policy makers and analysts worry about the effects of unilateral deployment on international peace and stability. Unilateral geoengineering is viewed as a challenge to the international order, in terms of both legitimacy and security, in a way reminiscent of nuclear proliferation.[8]

Climate scientists have also noted concerns about unilateral geoengineering.[9] For instance, Lawrence states:

> [I]t is easy to imagine a future scenario in which certain nations begin to undertake large-scale geoengineering efforts on their own. It is not uncommon for nations to act unilaterally in what they perceive as their own best interests, regardless of any international outcry about the consequences for the rest of the world.[10]

Lawrence continues:

> [W]ithout a good overview of potential geoengineering efforts which might eventually be undertaken, it would be difficult to monitor for the possibility of "covert" geoengineering.... a clear line will need to be drawn between allowed scientific experiments which are small-scale yet large enough to have statistically significant

Governance of a Possible Geoengineering Intervention to Combat Climate Change, 95 CLIMATIC CHANGE 103 (2009). These concerns were raised in 2009–2010 hearings on geoengineering held by the House Committee on Science and Technology, Subcommittee on Energy and Environment. *See* Memorandum from Richard Lattanzio & Emily Barbour, Cong. Research Serv., International Governance of Geoengineering (Mar. 11, 2010); and FRANK RUSCO, CLIMATE CHANGE: PRELIMINARY OBSERVATIONS ON GEOENGINEERING SCIENCE, FEDERAL EFFORTS, AND GOVERNANCE ISSUES 7 (2010).

7 David G. Victor et al., *The Geoengineering Option: A Last Resort against Global Warming?*, 88 FOREIGN AFF. 322, 333 (2009).

8 For an explicit comparison, *see* Clive Hamilton, The Return of Dr. Strangelove: The Politics of Climate Engineering as a Response to Global Warming (June 2010) (unpublished manuscript, *available at* http://www.clivehamilton.net.au/cms/media/documents/articles/dr_strangeloves_return.pdf).

9 *See* Michael C. MacCracken, *Geoengineering: Worthy of Cautious Evaluation?*, 77 CLIMATIC CHANGE 235, 238 (2006); Robock, *supra* note 5; ROYAL SOCIETY, *supra* note 4, at 40.

10 Mark G. Lawrence, *The Geoengineering Dilemma: To Speak or Not to Speak*, 77 CLIMATIC CHANGE 245, 246 (2006).

> signals, and what goes beyond this, so that "science" cannot be used as camouflage for unilateral attempts to undertake large-scale geoengineering efforts.[11]

Not only do members of the scientific community echo fears articulated by policy analysts, they express concern that unilateral climate interventions might undermine the integrity of the scientific process by parading as experimental in nature. The conduct of nominally "scientific" nuclear research by international pariah states stands as a cautionary tale in this regard.

Opponents of geoengineering have used the unilateral scenario to argue against research and deployment. ETC Group is arguably the most outspoken critic of geoengineering, and has identified "unilateral" as one of the "Best Reasons to Say No to Geoengineering."[12] The organization argues:

> It has been well established in the Stockholm Declaration (1972), the Rio Declaration (1992), the precedent-setting *Trail Smelter* case and in the UNFCCC [United Nations Framework Convention on Climate Change] itself that states are obliged to ensure that "activities within their jurisdiction or control do not cause damage to the environment of other states or of areas beyond the limits of national jurisdiction." The widely acknowledged potential for unilateral geoengineering deployment flies of the face of this principle.[13]

Undoubtedly, many of those scientists, policy analysts, and others who have raised the issue of unilateralism have done so not with the intent to inhibit geoengineering research and possible implementation, but rather to consider unilateralism as one potential complicating factor in any effort to deploy geoengineering technology. Nevertheless, speculation about unilateral intervention has injected an element of fear into the climate debate, and been warmly received by those opposed to geoengineering for primarily ideological reasons. Given the present stakes, it is important to show that the specter of unilateral deployment is largely illusory, and that multilateral cooperation is the outcome favored by events.

3. THE MYTH OF UNILATERAL DEPLOYMENT

Put simply, unilateral geoengineering deployment is unlikely to occur because the incentives faced by states do not support it. To illustrate, suppose Country B has both an interest in geoengineering deployment and the capacity to carry it out. In the present case, deployment would take the form of a stratospheric aerosol injection system. Country B would confront a multiplicity of practical constraints.

[11] *Id.*

[12] ETC GROUP, RETOOLING THE PLANET? – CLIMATE CHAOS IN THE GEOENGINEERING AGE 34 (2009).

[13] *Id.* at 39. The provision quoted by the ETC Group derives from Rio Declaration on Environment and Development para. 2, June 14, 1992, A/CONF.151/26 (Vol. I) Chapter I, Annex I.

First, the relative simplicity and affordability of stratospheric aerosol injection technology would make it widely available to members of the international community.[14] Other states or international bodies might inject stratospheric aerosols just as easily as Country B does. Given climate dynamics, multiple injections would interact in a variety of ways. Specifically, the results of aerosol injections by Country B would necessarily be mediated by the number of additional injection projects, the volumes of aerosols injected, the types of aerosols injected, the timing and phasing of other injections, and the location of injection sites.[15] The effects of such intervening variables are potentially large and would likely frustrate any deployment plan that failed to take them into account.

For example, one of the most common proposals for stratospheric aerosol injection involves regional deployment designed to stabilize the climate in the Arctic, including the Greenland ice sheet.[16] Preliminary models indicate that isolated, high-latitude aerosol dispersals would combine with the increased poleward water vapor transport characteristic of global warming to produce greater snowfall in the Arctic region.[17] This would result in enhanced regional albedo (reflectivity), reinforced snowpack, and moderated climate change in the north polar region. However, stratospheric injections carried out simultaneously at lower latitudes (by, say, a low-lying island state facing an existential threat from sea level rise) would have the effect of reducing water vapor transport, which in turn would reduce Arctic precipitation, reverse snowpack gains, and leave ice sheets in a deteriorating state. In the absence of international coordination, a regional Arctic rescue plan could come undone as a result of otherwise good intentions.

Aerosol chemistry provides another example. The aerosol most commonly suggested for stratospheric injection is sulfuric acid.[18] Plans call for delivering sulfate aerosols by dispersing gas-phase precursor materials. Precursor oxidation and aerosol formation involve complex processes with the potential to reduce the effectiveness of stratospheric insertion. For instance, coagulation could lead to excessively large sulfuric acid particles that sediment out of the stratosphere, lessening the effect of the initial dispersion.[19] Multiple independent injections would increase the

[14] BLACKSTOCK ET AL., *supra* note 2, at 46–51.

[15] *Id.* at 19–20; P.J. Crutzen, *Albedo Enhancement by Stratospheric Sulfur Injections: A Contribution to Resolve a Policy Dilemma?*, 77 CLIMATIC CHANGE 211 (2006); and K. Caldeira & L. Wood, *Global and Arctic Climate Engineering: Numerical Model Studies*, 366 PHILOSOPHICAL TRANSACTIONS ROYAL SOC'Y A – MATHEMATICAL, PHYSICAL, AND ENG. SCI. 3 4039 (2008).

[16] Workshop Report on Managing Solar Radiation (L. Lane et al. eds., Apr. 2007) (unpublished report, *available at* http://event.arc.nasa.gov/main/home/reports/SolarRadiationCP.pdf).

[17] Caldeira & Wood, *supra* note 15, at 4043–45.

[18] *See* Edward Teller, Roderick Hyde, & Lowell Wood, Active Climate Stabilization (Apr. 18, 2002) (unpublished manuscript, *available at* http://www.osti.gov/accomplishments/documents/fullText/ACC0233.pdf).

[19] BLACKSTOCK et al., *supra* note 2, at 22.

likelihood of such unintended consequences. Unsynchronized staging, scheduling, and delivery of sulfate aerosol injections would magnify the potential for perverse particulate interactions, and might jeopardize the success of geoengineering deployment. Lack of policy coordination may result in separate injection schemes and mutual suboptimality.

As these cases illustrate, stratospheric aerosol injections by Country B could not be kept separate from injections by other parties, and the consequent interactions could undermine the effectiveness of Country B deployment. Simultaneous injection schemes carried out by multiple countries may be unlikely, but the possibility is nontrivial, particularly given the technical simplicity and low cost of injection systems. The widespread availability of stratospheric injection technology, combined with its potential to hinder any unilateral deployment, provides strong incentives to coordinate implementation plans at the international level.

Second, other states may choose to pursue other types of geoengineering activities, including alternative SRM strategies and Carbon Dioxide Removal (CDR) techniques. As in the case described earlier in this section, parallel deployments of additional geoengineering technologies would have the potential to interfere with stratospheric injections by Country B. Consequently, Country B would also need to ensure that its stratospheric aerosol deployment was coordinated with any other geoengineering activities conducted by other international actors. The pursuit of strategies such as marine cloud whitening, ocean fertilization, artificial upwelling, etc. would affect the outcome of stratospheric aerosol deployment by modifying regional albedo, disrupting regional (and potentially global) circulation patterns, altering atmospheric chemistry, and affecting nutrient cycles, as well as in other less predictable ways.[20] To maximize the chances of success, Country B would need to account for these complicating factors in its aerosol injection plans, monitor concurrent climate interventions carried out by other actors, and adjust stratospheric aerosol deployment schedules and timetables on an ongoing basis.

Third, by undertaking a stratospheric aerosol injection project, Country B would be faced with the so-called termination problem: if emissions were not reduced simultaneously, project termination would result in rapid temperature increases and a destabilized climate system, so that in effect Country B would be committed to

[20] A. Oschlies et al., *Climate Engineering by Artificial Ocean Upwelling: Channelling the Sorcerer's Apprentice*, 37 GEOPHYSICAL RES. LETTERS 4701 (2010); Michael C. MacCracken, *On the Possible Use of Geoengineering to Moderate Specific Climate Change Impacts*, 4 ENVTL. RES. LETTERS 45107 (2009); Christine Bertram, *Ocean Iron Fertilization in the Context of the Kyoto Protocol and the Post-Kyoto Process*, 38 ENERGY POL'Y 1130 (2010); Raymond T. Pollard et al., *Southern Ocean Deep-Water Carbon Export Enhanced by Natural Iron Fertilization*, 457 NATURE 577 (2009); and Greg H. Rau, *Electrochemical CO_2 Capture and Storage with Hydrogen Generation*, 1 ENERGY PROCEDIA 823 (2009).

aerosol injections indefinitely.[21] This dilemma would create pressure for Country B to synchronize national deployment with international emissions mitigation efforts.[22] Unless its government was prepared to shoulder the burden of deployment on an essentially permanent basis, in effect providing a global public good, Country B would need to link deployment to a robust global emissions reduction policy and a more general decarbonization of the world economy.[23]

Fourth, the effects of deployment could be offset with a variety of countermeasures. Though rarely mentioned in the literature, states opposed to geoengineering have a number of tools at their disposal to counteract climate interventions.[24] In the case of stratospheric aerosol injections, for example, fluorocarbon gases could be deployed to offset cooling effects. Alternatively, the strategic use of black carbon could neutralize artificial albedo enhancement. The availability of effective countermeasures would serve as perhaps the most potent check on unilateral deployment of geoengineering technologies such as stratospheric aerosol injections.

Thus, the incentive structure faced by a state interested in implementing a climate intervention strategy would strongly discourage unilateral postures that dismissed the need for international agreement and coordination. Any country considering unilateral deployment would find itself tangled in a web of technical and political constraints and steered toward reaching some form of global consensus. Individual incentives may be inadequate to deter unilateralism on their own, but their collective weight is likely to tilt the playing field decisively in favor of multilateral cooperation. For instance, Country B may be sufficiently motivated to accept the costs associated with the termination problem and dispense with efforts to synchronize emissions mitigation policies. But once deployed, a large number of international actors would effectively exercise joint control over any injection system, frustrating any attempt by Country B to pursue a coherent climate intervention policy managed solely by its national government. Furthermore, any actor opposed to the project could easily (and anonymously) counter its effects using relatively simple means such as release of black carbon, thereby neutralizing the entire scheme.

[21] BLACKSTOCK et al., *supra* note 2, at 24.

[22] T.M. Lenton & N.E. Vaughan, *The Radiative Forcing Potential of Different Climate Geoengineering Options*, 9 ATMOSPHERIC CHEMISTRY & PHYSICS 5539 (2009); Rob Swart & Natasha Marinova, *Policy Options in a Worst Case Climate Change World*, 15 MITIGATION & ADAPTATION STRATEGIES FOR GLOBAL CHANGE 531 (2010); Mike MacCracken, *Beyond Mitigation: Potential Options for Counter-Balancing the Climatic and Environmental Consequences of the Rising Concentrations of Greenhouse Gases* (World Bank, Working Paper No. 4938, 2009); and H. Damon Matthews & Ken Caldeira, *Transient Climate–Carbon Simulations of Planetary Geoengineering*, 104 PROC. NAT'L ACAD. SCI. 9949 (2007).

[23] Although climate policy is sometimes presented as a choice between mitigation and intervention, the implications of the termination problem demonstrate that this is a false choice. Geoengineering cannot achieve the goal of climate stabilization without complementary carbon mitigation. Indeed, these two approaches are less mutually exclusive than mutually dependent.

[24] BLACKSTOCK et al., *supra* note 2, at 28.

This situation is ultimately attributable to the highly complex nature of the climate system.[25] Climate dynamics are multivariate and interdependent, determined by a range of factors including atmospheric and ocean chemistry, albedo, atmospheric circulation, the hydrologic cycle, ocean currents, vegetation coverage, biogeochemical cycles (carbon, nitrogen, etc.), and solar effects. Feedback mechanisms and nonlinearity are essential features of the climate system. Because climate complexity renders the outcome of any geoengineering effort contingent on the interplay of multiple natural (and anthropogenic) variables and processes, success hinges on coordinated international intervention.

In other words, geoengineering is ruled not by the threat of unilateral deployment, but rather by a "logic of multilateralism": in a world of multiple, competing nation-states, nature dictates that effective geoengineering and multilateral cooperation on climate intervention are identical pursuits. Any country that embarks on unilateral implementation will soon find its efforts frustrated by rivals and friends alike, whose actions across the entire policy spectrum are inextricably linked via the global climate system. The inherent complexity of the climate system makes the success of any climate intervention singularly dependent on the behaviors that other states do and do not engage in.[26] Recognizing the true collective action constraints associated with geoengineering removes the specter of unilateralism from discussions of climate engineering. Furthermore, recognizing the multilateral logic that underlies the geoengineering enterprise also points to a suite of diplomatic and institutional tools with proven capacity to promote international cooperation.

4. GETTING PROBLEMS RIGHT

To appreciate the divergent challenges posed by multilateralism as opposed to unilateralism, the concept of "problem structure" is helpful. Problem structure refers to the essential characteristics of interactive social problems. These characteristics shape attempts at problem resolution and strongly influence the likelihood of success. Arguably the most sophisticated and useful discussion of problem structure is provided by Arild Underdal, who treats it as a variable stretching along a continuum from purely benign to purely malign: "a perfectly benign problem would be one characterized by identical preferences. The further we get from that state of harmony, the more malign the problem becomes."[27] Malign problems are problems

[25] CLIMATE CHANGE 2007: SYNTHESIS REPORT (R.K. Pachauri & A. Reisinger eds., 2007); N. STERN, THE ECONOMICS OF CLIMATE CHANGE: THE STERN REVIEW (2007); and NAT'L RES. COUNCIL, AMERICA'S CLIMATE CHOICES: ADVANCING THE SCIENCE OF CLIMATE CHANGE (2010).

[26] For more on structural and behavioral consequences of system complexity, see ROBERT JERVIS, SYSTEM EFFECTS: COMPLEXITY IN POLITICAL AND SOCIAL LIFE (1997).

[27] Arild Underdal, *One Question, Two Answers*, *in* ENVIRONMENTAL REGIME EFFECTIVENESS: CONFRONTING THEORY WITH EVIDENCE 3, 15 (Edward L. Miles et al. eds., 2002).

of "incongruity" in which "the cost-benefit calculus of individual actors includes a nonproportional or biased sample (representation) of the actual *universe* of costs and benefits produced by his decisions and actions" (emphasis in original).[28] In other words, costs and benefits do not fall equally on all actors, so individual incentives differ from the overall group incentive, and suboptimal outcomes result.

In contrast, benign problems are problems of "coordination" in which "(1) the overall result depends on the compatibility of individual choices, (2) more than one route can lead to the collective optimum, and (3) the choice between or among these routes is not a trivial or obvious one, meaning that compatibility cannot be taken for granted even when actor interests are identical."[29] Because benign problems are easier to solve than malign ones, benign problem structures are more likely to result in successful outcomes than are malign structures. Conflicts at any level of analysis, from local to international, may be classified as comparatively benign or comparatively malign.[30]

Geoengineering, governed by a multilateral logic, represents a benign problem structure with relatively few obstacles to cooperation. Indeed, geoengineering features structural supports that make cooperation advantageous if not essential. Injecting aerosols into the stratosphere requires multistate coordination if it is to succeed in moderating global temperatures. By contrast, climate mitigation represents a malign problem structure.[31] Emission reductions confront states with the

[28] *Id.* at 17.

[29] *Id.* at 20.

[30] In the language of game theory, malign problems approximate "mixed-motive games," whereas benign problems approximate "coordination games." A game is defined by its payoff function or preference ordering, which represents an actor's preferences over possible outcomes. In a standard two-person game, each actor may choose one of two "strategies," cooperate, conventionally denoted as "C," or defect, "D." This choice situation gives rise to four possible outcomes: mutual cooperation, or "CC"; mutual defection or "DD"; the "sucker's payoff" or "CD," in which the first player cooperates and the second player defects; and the "temptation payoff" or "DC," in which the first player defects and the second player cooperates. Players rank these four outcomes from most to least preferred, depending on the particular characteristics of the situation, and this preference ordering defines the game. An equilibrium (or Nash equilibrium) is an outcome in which neither player has an incentive to change strategy if the other player does not, thereby rendering it stable. A Pareto optimal outcome is one in which it is impossible to improve one player's payoff without reducing the other player's payoff.

Two games are particularly significant. Stag hunt, also known as assurance, is defined by the payoff function CC>DC>DD>CD; although two equilibria, CC and DD, are possible, the fact that only CC is Pareto optimal means that mutual cooperation is the preferred outcome. Stag hunt belongs to a broad category of games called coordination games, which are characterized by the presence of multiple equilibria. In contrast, in a prisoners' dilemma, the payoff function DC>CC>DD>CD leads to the equilibrium DD. Because this is not Pareto optimal, the outcome of mutual defection is referred to as suboptimal. Prisoner's dilemma is the most prominent member of a class of games known as mixed-motive or collaboration games, which are distinguished by the fact that players are motivated both to cooperate and to defect simultaneously. For more on game theory, see MARTIN J. OSBORNE, AN INTRODUCTION TO GAME THEORY (2004).

[31] Scott Barrett, *The Incredible Economics of Geoengineering*, 39 ENVTL. & RES. ECON. 45 (2008).

following dilemma: although countries prefer collective carbon mitigation, they more strongly prefer that other countries reduce emissions while they pursue economic growth unburdened by an effective price on carbon.[32] The widespread temptation to "defect" from an emissions mitigation agreement has paralyzing effects. The credible threat of unilateral cheating results in global inaction and accelerating climate change.

The international nuclear arena offers a useful (and telling) parallel. Like climate mitigation, nuclear nonproliferation presents a malign problem structure in which states face powerful incentives to engage in unilateral violations of agreed nonproliferation rules. The consequence is broken rules and chronic mistrust, which currently threaten to undermine the nonproliferation regime. Arms control, on the other hand, stands as a comparatively benign problem in which bargains are struck with the help of coordination, monitoring, and verification mechanisms. The history of successful arms control pacts, from Strategic Arms Limitation Talks (SALT) I and II to Strategic Arms Reduction Treaties (START) I and II, testifies to the relative ease of resolving benign problem structures.[33]

Viewed in this way, the conventional tendency to hear nuclear echoes in contemporary debates about geoengineering is turned on its head. The governance challenges associated with geoengineering resemble the challenges associated with arms control much more than the difficulties plaguing nonproliferation efforts. It is not geoengineering but rather mitigation policy that is subject to the types of malign incentives typical of nuclear nonproliferation. Decades of unsuccessful attempts to reduce global carbon emissions underscore this similarity, and explain the need to consider geoengineering in the first place.

5. PROMOTING MULTILATERALISM

The benign problem structure underlying the question of geoengineering deployment can be addressed through the use of a portfolio of tactics known to political scientists as "international management theory."[34] Employing these instruments

[32] James K. Sebenius, *Towards a Winning Climate Coalition, in* NEGOTIATING CLIMATE CHANGE: THE INSIDE STORY OF THE RIO CONVENTION 277 (Irving Mintzer ed., 1994).

[33] *See generally* THOMAS GRAHAM, JR., DISARMAMENT SKETCHES: THREE DECADES OF ARMS CONTROL AND INTERNATIONAL LAW (2002).

[34] For some of the major works in the area, see generally MANAGING GLOBAL ISSUES: LESSONS LEARNED (P.J. Simmons & Chantal de Jonge Oudraat eds., 2001); Raimo Vayrynen, *Norms, Compliance, and Enforcement in Global Governance, in* GLOBALIZATION AND GLOBAL GOVERNANCE 25 (Raimo Vayrynen ed., 1999); THE IMPLEMENTATION AND EFFECTIVENESS OF INTERNATIONAL ENVIRONMENTAL COMMITMENTS: THEORY AND PRACTICE (David G. Victor et al. eds., 1998); Oran R. Young & George J. Demko, *Improving the Effectiveness of International Environmental Governance Systems, in* GLOBAL ENVIRONMENTAL CHANGE AND INTERNATIONAL GOVERNANCE 229 (Oran R. Young et al.

facilitates the achievement of multilateral coordination and effective international governance. By using these policy levers, interested parties are able to overcome precisely those kinds of obstacles that currently impede global cooperation on climate intervention. International management theorists cite three strategies as key to solving interdependence problems. Persuasion helps redefine national interests so that states favor international cooperation. Pressure brings national policies into alignment with harmonious interests. And provision of assistance enables states to pursue compliant policies. Together, these "three Ps" are the main instruments by which noncompliance is ameliorated and governance regimes are made successful.[35]

In the case of geoengineering, the strategy required to achieve cooperative, multilateral implementation of climate intervention technologies is persuasion. Persuasion consists of efforts to alter conceptions of interest so that actors are motivated to behave in ways that support successful outcomes. Learning may be central to the process of redefining interests. The objects of persuasion may include not only governments, but the private firms that are often the ultimate targets of international regulation, as well as larger publics whose beliefs and preferences can prove critical to regime performance. Implicit in arguments about the efficacy of persuasion as a problem-solving strategy is a conception of interests as constructed and subjective, and amenable to change.

In order to move toward deployment, major powers must regard geoengineering as serving their national interest. In the case of stratospheric aerosols, leading states must come to view injections as a necessary complement to carbon reductions. Without consensus on the need for intervention, there is little value in pressing for more robust action. And capacity does not present any serious obstacles. Indeed, an abundance of capacity in the form of available technologies such as aerosol injections represents a key structural imperative supporting global collective action.

Persuading governments to rethink their interests and recast their policies is, of course, a daunting challenge. National interests may be malleable, but they are

eds., 1996); ABRAM CHAYES & ANTONIA HANDLER CHAYES, THE NEW SOVEREIGNTY: COMPLIANCE WITH INTERNATIONAL REGULATORY AGREEMENTS (1995); Abram Chayes et al., *Managing Compliance: A Comparative Perspective*, *in* ENGAGING COUNTRIES: STRENGTHENING COMPLIANCE WITH INTERNATIONAL ENVIRONMENTAL ACCORDS 39 (Edith Brown Weiss & Harold K. Jacobson eds., 1998); INSTITUTIONS FOR THE EARTH: SOURCES OF EFFECTIVE INTERNATIONAL ENVIRONMENTAL PROTECTION (Peter M. Haas et al. eds., 1993); RONALD B. MITCHELL, INTENTIONAL OIL POLLUTION AT SEA: ENVIRONMENTAL POLICY AND TREATY COMPLIANCE (1994); and Oran R. Young, *The Effectiveness of International Institutions: Hard Cases and Critical Variables*, *in* GOVERNANCE WITHOUT GOVERNMENT: ORDER AND CHANGE IN WORLD POLITICS 160 (James N. Rosenau & Ernst-Otto Czempiel eds., 1992).

[35] Some analysts also cite the importance of clear rules and informational transparency in helping resolve international problems. However, empirical support for these propositions is mixed. *See* Joshua B. Horton, *Controlling the Wildlife Trade: CITES, Regime Effectiveness, and the Global Market for Wildlife Products* (Aug. 2006) (unpublished PhD dissertation, Johns Hopkins University, on file with Eisenhower Library, Johns Hopkins University).

typically entrenched and taken for granted. Although knowledge of geoengineering remains limited, and international opinion is inchoate, there exists unmistakable latent hostility to climate engineering in most national governments.[36] Opposition may become further entrenched as the variable distributional effects of climate change become clearer.[37] For example, Russia may prefer the improved agricultural productivity and greater accessibility linked to global warming to previously less hospitable conditions across the bulk of its landmass.[38]

However, such positions may be more amenable to change than commonly supposed. At an April 2009 geoengineering workshop in Lisbon held under the auspices of the International Risk Governance Council (IRGC), representatives from the United States, Canada, European Union (EU) Member States, Russia, China, and India all expressed some degree of openness to the goals of climate intervention.[39] Significantly, the world's first aerosol field trial was conducted in Russia in 2009, led by scientists with very close links to Russian government officials including Prime Minister Vladimir Putin.[40] China engaged in weather modification while hosting the 2008 Summer Olympics.[41]

In the United States, John Holdren, chief science advisor to President Barack Obama, famously noted in a 2009 interview on geoengineering, "It's got to be looked at. We don't have the luxury of taking any approach off the table." Holdren broached the subject again at a recent international conference on science and science policy.[42] Several official and semiofficial reports on geoengineering have been released in recent years, including reports from the Congressional Research Service (CRS), Government Accountability Office (GAO), RAND Corporation, Woodrow Wilson International Center for Scholars, and Bipartisan Policy Center (BPC).[43]

[36] Jason Blackstock, *The International Politics of Geoengineering Research*, in H.C. REP. NO. 221–5, at Ev 1 (2010).

[37] JOSHUA W. BUSBY, CLIMATE CHANGE AND NATIONAL SECURITY: AN AGENDA FOR ACTION (2007).

[38] Samuel Charap, *Russia's Lackluster Record on Climate Change*, 10 RUSSIAN ANALYTICAL DIG. 11 (2010).

[39] M. GRANGER MORGAN & KATHARINE RICKE, COOLING THE EARTH THROUGH SOLAR RADIATION MANAGEMENT: THE NEED FOR RESEARCH AND AN APPROACH TO ITS GOVERNANCE (forthcoming 2010).

[40] Yuri A. Izrael et al,. *Field Experiment on Studying Solar Radiation Passing through Aerosol Layers*, 34 RUSSIAN METEOROLOGY & HYDROLOGY 265 (2009).

[41] Andrew Jacobs, *China Hopes, and Tries, for Rain-Free Festivities*, N.Y. TIMES, Oct. 1, 2009.

[42] John P. Holdren, Assistant to the President for Science and Technology, and Director, Office of Science and Technology Policy, Executive Office of the President of the U.S., Climate-Change Science and Policy: What Do We Know? What Should We Do? (Sept. 6, 2010).

[43] CONG. RES. SERV., INTERNATIONAL GOVERNANCE OF GEOENGINEERING (Memorandum 2010); CONG. RES. SERV., GEOENGINEERING: GOVERNANCE AND TECHNOLOGY POLICY (Rep. 2010); U.S. GOV'T ACCOUNTABILITY OFF., GAO-10–903, A COORDINATED STRATEGY COULD FOCUS FEDERAL GEOENGINEERING RESEARCH AND INFORM GOVERNANCE EFFORTS (2010); U.S. GOV'T ACCOUNTABILITY OFF., GAO-11–71, CLIMATE ENGINEERING: TECHNICAL STATUS, FUTURE DIRECTIONS, AND POTENTIAL

Small geoengineering research programs have been established in the United Kingdom (UK) and Germany, the latter with EU funding support.[44]

Available public opinion data indicate that people are generally unaware of geoengineering, and their views on technologies such as SRM remain largely unformed.[45] The UK Natural Environment Research Council (NERC) recently sponsored preliminary research into public attitudes toward geoengineering.[46] Most participants in the study signaled openness to the idea of climate intervention so long as it is linked to ongoing mitigation efforts. A more thorough understanding of public opinions on geoengineering will shed light on the cultural contexts in which national decisions will be made.

6. IMPLICATIONS FOR GEOENGINEERING

This chapter has argued that the threat of unilateral deployment so often ascribed to geoengineering is unsubstantiated. Instead, geoengineering is characterized by a logic of multilateralism that renders problems associated with deployment manageable by means of familiar diplomatic tools. Several points follow from this conclusion.

First, the prospects for international cooperation on climate intervention are brighter than is generally believed. Although conventional wisdom is preoccupied with the notion of states "going rogue," a close examination of the incentive structure facing states interested in geoengineering reveals that the playing field is tilted in the direction of multilateral collaboration. In the case of stratospheric aerosol injections, for example, successful deployment depends on close coordination with other countries' climate mitigation and intervention policies. Such deployment exhibits a benign, rather than malign, problem structure, with a corresponding improvement in the likelihood of an effective global partnership to implement climate intervention technologies.

Second, subtle (and sometimes unsubtle) comparisons between geoengineering deployment and nuclear proliferation are inapt. Geoengineering presents a relatively

RESPONSES (2011); ROBERT J. LEMPERT & DON PROSNITZ, RAND CORP., GOVERNING GEOENGINEERING RESEARCH: A POLITICAL AND TECHNICAL VULNERABILITY ANALYSIS OF POTENTIAL NEAR-TERM OPTIONS (2011); ROBERT L. OLSON, WOODROW WILSON INTERNATIONAL CENTER FOR SCHOLARS, GEOENGINEERING FOR DECISION MAKERS (2011); and TASK FORCE ON CLIMATE REMEDIATION RES., BIPARTISAN POLICY CENTER, GEOENGINEERING: A NATIONAL STRATEGIC PLAN FOR RESEARCH ON THE POTENTIAL EFFECTIVENESS, FEASIBILITY, AND CONSEQUENCES OF CLIMATE REMEDIATION TECHNOLOGIES (2011).

[44] Eli Kintisch, *With Emissions Caps on Ice, Is Geoengineering Next Step in D.C. Climate Debate?*, SCIENCE, Sept. 27, 2010, http://news.sciencemag.org/scienceinsider/2010/09/with-emissions-caps-on-ice-is.html.

[45] ANTHONY LEISEROWITZ ET AL., CLIMATE CHANGE IN THE AMERICAN MIND: AMERICANS' CLIMATE CHANGE BELIEFS, ATTITUDES, POLICY PREFERENCES, AND ACTIONS (2010).

[46] IPSOS MORI, EXPERIMENT EARTH? REPORT ON A PUBLIC DIALOGUE ON GEOENGINEERING (2010).

tractable problem, whereas the issues associated with proliferation are more difficult to resolve given underlying incentives. Emissions reduction shares more in common with nuclear proliferation, as its record makes clear. In turn, geoengineering shares more in common with arms control, widely regarded as a model of international cooperation and regime performance. Put simply, nuclear proliferation is a poor analog for geoengineering, and attempts to make this connection do not spring from reasoned analysis. If anything, viewing geoengineering through the lens of nuclear security inspires confidence in our capacity to establish successful governance arrangements.

Third, recognizing climate intervention as relatively manageable brings into sharper focus a plausible strategy to promote geoengineering. Persuading countries to alter their interests is no simple task, yet the international record is replete with such instances.[47] The principal challenge to successful deployment is not institutional design, but persuading interested parties that climate intervention must be an essential complement to emissions reduction strategies. Recent developments suggest that global opinion is beginning to shift in this direction. A short time ago, geoengineering was regarded as a fringe, even taboo, subject, yet it has quickly gained attention, exposure, and credibility. If geoengineering attains legitimacy, its prospects as a critical tool in the fight against climate change will improve even more, and systemic attributes will favor a multilateral climate intervention solution.

Although there are many risks associated with geoengineering, and it is important they not be overlooked, it is equally important that research not be obstructed by unfounded fears. Unilateral deployment constitutes one such fear. The perceived threat of unilateral geoengineering has loomed large over discussions of climate interventions, inhibiting debate and discouraging legitimate scientific inquiry. This chapter has attempted to expose this threat as myth, and to demonstrate the multilateral bias inherent in the geoengineering enterprise. This logic of multilateralism in no way guarantees global collaboration on geoengineering, but it does suggest strategies for achieving consensus and cooperation. With success contingent on cooperation, it is time to dispense with fears that impede efforts to address climate change on a comprehensive, informed, and timely basis.

[47] For a particularly insightful example, see RICHARD ELLIOT BENEDICK, OZONE DIPLOMACY: NEW DIRECTIONS IN SAFEGUARDING THE PLANET (1991).

8

International Legal Regimes and Principles Relevant to Geoengineering

Albert C. Lin

1. INTRODUCTION

Geoengineering has only recently entered into serious scientific and policy discussions. Recognition of the need for geoengineering governance, whether formal or informal, is growing, but has yet to achieve a critical mass. At present, no international agreements directly address geoengineering. Nonetheless, various treaties as well as principles of international law are potentially relevant and will likely play a role in future geoengineering governance. These legal authorities fall into three categories: (1) treaties that may have applicability to geoengineering generally, regardless of the specific technique used; (2) treaties whose applicability may depend on the geoengineering method or the medium affected; and (3) non-treaty sources of law, including customary international law and other sources of legal norms. Given the arduous process of international treaty-making and the lack of specific treaty provisions that speak directly to geoengineering, the last of these categories could wind up playing the most significant role in international geoengineering governance; however this chapter will examine the potential role of all three of these categories

2. GENERAL TREATIES

This section considers the first category: international agreements whose ambit arguably extends to geoengineering projects and research in a general sense. Such agreements include: the United Nations Framework Convention on Climate Change,[1] the Convention on the Prohibition of Military or Any Other

[1] United Nations Framework Convention on Climate Change, May 9, 1992, S. TREATY DOC. NO. 102–38, 1771 U.N.T.S. 164 [hereinafter FCCC], *available at* http://untreaty.un.org/English/notpubl/unfccc_eng.pdf.

Hostile Use of Environmental Modification Techniques,[2] and the Convention on Biological Diversity.[3]

2.1 *United Nations Framework Convention on Climate Change*

The United Nations Framework Convention on Climate Change (UNFCCC) is a logical starting point for considering the potential locus of formal geoengineering governance because of its focus on addressing climate change. Addressing geoengineering through the UNFCCC is appealing because virtually all nations are parties to the UNFCCC, there are already well-established institutions for administering and implementing the treaty, and these institutions could coordinate any geoengineering efforts with mitigation and adaptation strategies to combat climate change.[4] Indeed, the UNFCCC is a framework convention that contemplates the formation of more-specific protocol agreements as further information develops and as support for international cooperation builds. As explained below, however, the commitments made in the UNFCCC are general in nature and create no clear obligations with respect to geoengineering.

The UNFCCC's objective, for example, is to "achieve ... stabilization of greenhouse gas concentrations in the atmosphere at a level that would prevent dangerous anthropogenic interference with the climate system."[5] As this statement suggests, the negotiations leading to the UNFCCC focused primarily on the reduction of greenhouse gas (GHG) emissions; geoengineering did not receive serious consideration as a means of dealing with the growing climate crisis.[6] Nonetheless, the UNFCCC's objective statement could serve as the basis for distinguishing between carbon dioxide removal (CDR) techniques, which can contribute to stabilizing GHG concentrations, and solar radiation management (SRM) techniques, which do not. Moreover, SRM techniques – and perhaps some CDR techniques – themselves may conflict with the UNFCCC's objective to the extent that they constitute dangerous anthropogenic interference with the climate system.[7]

[2] Convention on the Prohibition of Military or Any Other Hostile Use of Environmental Modification Techniques, May 18, 1977, 31 U.S.T. 333, T.I.A.S. 9614 [hereinafter ENMOD].

[3] Convention on Biological Diversity, Preamble, June 5, 1992, 1760 U.N.T.S. 143 [hereinafter CBD], *available at* http://www.cbd.int/convention/convention.shtml.

[4] *See* Albert C. Lin, *Geoengineering Governance*, ISSUES IN LEGAL SCHOLARSHIP, July 2009, at 15, 19.

[5] FCCC, *supra* note 1, art. 2.

[6] *See* Daniel Bodansky, *May We Engineer the Climate?*, 33 CLIMATIC CHANGE 309, 313 (1996).

[7] The use of stratospheric aerosols to block solar radiation, for example, could adversely modify the Asian and African summer monsoons. *See* Alan Robock et al., *Regional Climate Responses to Geoengineering with Tropical and Arctic SO_2 Injections*, 113 J. GEOPHYS. RES. D16101 (2008); It has also been suggested that ocean fertilization efforts could result in increased methane emissions that would undermine carbon removal efforts. *See* THE ROYAL SOCIETY, GEOENGINEERING THE CLIMATE: SCIENCE, GOVERNANCE AND UNCERTAINTY 18 (2009), *available at* http://royalsociety.org/Geoengineering-the-climate/.

Other provisions of the UNFCCC are also arguably relevant to geoengineering, but their precise application would be open to varying interpretations. For example, among the principles set out in Article 3 is the requirement that parties "protect the climate system for the benefit of present and future generations of humankind."[8] Whether SRM techniques such as stratospheric aerosols or cloud whitening "protect the climate system" is debatable. These techniques promise to ameliorate temperature increases, but would likely have their own adverse climatic effects. Given the UNFCCC's focus on emissions reductions, the better interpretation of "protect the climate system" is one that involves the maintenance of existing climate dynamics to the extent possible, and not just the partial replication of earlier climate conditions.

The precautionary principle, found in Article 3.3, is another provision of which application to geoengineering will likely be the subject of serious debate. Article 3.3 states that "lack of full scientific certainty should not be used as a reason for postponing" measures "to anticipate, prevent or minimize the causes of climate change and mitigate its adverse effects."[9] Because the precautionary principle generally is understood as an appeal for caution in the face of uncertainty, commentators tend to assume that given potential side effects, application of the principle would block the deployment of geoengineering projects.[10] Nonetheless, if the uncertainties and risks posed by climate change come to overshadow those posed by a particular geoengineering technique, the precautionary principle might actually support the deployment of techniques that mitigate climate change's adverse effects.[11] For example, if the climate system were to reach a tipping point beyond which there would be catastrophic effects such as the sudden melting of the West Antarctic ice sheet, the precautionary response might be to deploy an SRM technique.

Several of the commitments set out in Article 4 of the UNFCCC also may be relevant. Article 4.1(d) articulates the parties' obligation to "[p]romote and cooperate in the conservation and enhancement, as appropriate, of sinks and reservoirs of all greenhouse gases."[12] This language, it could be argued, may support the deployment of CDR geoengineering projects such as ocean fertilization. Similarly, Article 4.1(g) and (h) describe obligations to promote and cooperate in research

[8] FCCC, *supra* note 1, art. 3.1.

[9] *Id.*, art. 3.3.

[10] *See, e.g.*, Bodansky, *supra* note 6, at 319–20.; John Virgoe, *International Governance of a Possible Geoengineering Intervention to Combat Climate Change*, 95 CLIMATIC CHANGE **103**, 111 (2009); William Daniel Davis, Note, *What Does "Green" Mean?: Anthropogenic Climate Change, Geoengineering, and International Environmental Law*, 43 GA. L. REV. 901, 931–32 (2009).

[11] FCCC Article 3.3's declaration that the policies adopted should "cover all relevant sources, sinks and reservoirs of greenhouse gases" is arguably consistent with CDR geoengineering techniques.

[12] FCCC, *supra* note 1, art 4.1(d).

and information exchange relating to "the economic and social consequences of various response strategies," broad language that could be supportive of geoengineering research.[13]

In addition to the provisions already discussed, the UNFCCC regime may intersect with geoengineering in one other significant way: the issuance of carbon credits. Private interest in geoengineering, particularly ocean fertilization, has been sparked by the possibility that such projects could serve as a source of carbon credits, perhaps under the Clean Development Mechanism (CDM) of the Kyoto Protocol or in voluntary carbon markets.[14] However, satisfying CDM requirements that emissions reductions be "additional to any that would occur in the absence of the certified project activity" and that benefits be "real, measurable, and long-term"[15] is likely to be difficult.[16] Ascertaining the amount of carbon sequestered by ocean fertilization, for instance, requires not only accurate measurement of carbon flux over extended periods of time, but also complex modeling of the depletion of other nutrients that would no longer be available for phytoplankton growth.[17] Determining who should receive credits – the private entrepreneur undertaking a project, nations on whose territory the project is initiated, nations suffering adverse effects from the project, or the international community if a project takes place beyond national boundaries – is also likely to be a contentious issue.[18]

2.2 *ENMOD*

Another treaty with potential implications for geoengineering governance is the Convention on the Prohibition of Military or Any Other Hostile Use of Environmental Modification Techniques, or ENMOD.[19] Developed in response to American attempts to use weather modification techniques as a tool of warfare during the Vietnam War, ENMOD prohibits parties from "engag[ing] in military or any hostile use of environmental modification techniques having widespread, long-lasting or severe effects as the means of destruction, damage or injury to any

[13] *Id.*, art. 4.1(g), (h). Relatedly, Article 4.1(f) requires parties to conduct impact assessments of projects undertaken to mitigate climate change.

[14] *See* ELI KINTISCH, HACK THE PLANET: SCIENCE'S BEST HOPE – OR WORST NIGHTMARE – FOR AVERTING CLIMATE CATASTROPHE 132–36 (2010); Sallie W. Chisholm et al., *Dis-Crediting Ocean Fertilization*, 294 SCI. 309 (2001).

[15] Kyoto Protocol to the United Nations Framework Convention on Climate Change art. 12.5, Dec. 10, 1997, U.N. Doc. FCCC/CP/1997/L.7/ADD.1, 37 I.L.M. 32.

[16] *See* David Freestone & Rosemary Rayfuse, *Ocean Iron Fertilization and International Law*, 364 MARINE ECO. PROGRESS SERIES 227, 231 (2008) (noting that so far, almost none of projects approved under the CDM have involved carbon sinks of any type).

[17] *See* Chisholm et al., *supra* note 14, at 310.

[18] *See* ROYAL SOCIETY, *supra* note 7, at 41; Freestone & Rayfuse, *supra* note 16, at 231.

[19] *See* ENMOD, *supra* note 2.

other State Party."[20] There are thus three key elements required to trigger ENMOD: (1) environmental modification; (2) widespread, long-lasting or severe effects; and (3) military or hostile use. The treaty defines "environmental modification" as "any technique for changing – through the deliberate manipulation of natural processes – the dynamics, composition or structure of the Earth, including its biota, lithosphere, hydrosphere, and atmosphere or of outer space."[21] This broad definition encompasses virtually any geoengineering techniques that might be developed. Moreover, in order to be effective, geoengineering deployment would necessarily generate, or at least seek to generate, "widespread [or] long-lasting ... effects." The primary difficulty with applying ENMOD to geoengineering, however, is that the treaty is aimed specifically at the military or hostile use of environmental modification techniques. ENMOD explicitly provides that the use of such techniques for peaceful purposes is outside the treaty's scope.[22]

Under some circumstances, however, the deployment of geoengineering to counter climate change arguably would constitute hostile use. For example, suppose that a low-lying nation threatened by rising seas decided to implement a geoengineering project unilaterally, regardless of the adverse consequences on other countries. Although that nation might argue that its purpose was benign, adversely affected nations would certainly object, particularly if they were not warned or consulted. ENMOD's distinction between "military or any other hostile use" and "the use of environmental techniques for peaceful purposes" suggests that an actor's purpose is critical in determining the treaty's applicability. Nonetheless, objecting nations could reasonably argue that a party's failure to consult with affected nations or its knowledge of, recklessness, or even negligence with respect to the effects of a unilateral geoengineering project suffices to constitute "hostile use."[23]

Compared to other environmental treaties, ENMOD provides a relatively powerful mechanism for enforcement. Potential treaty violations are referred to the United Nations Security Council. If the Security Council's investigation determines that a violation has harmed or is likely to harm a party, other parties to ENMOD are to provide assistance to that party.[24]

[20] *Id.*, art. I. ENMOD neither addresses environmental modification undertaken by nonparties, including private actors nor does it govern the use of such techniques against nonparty states.

[21] *Id.*, art. II.

[22] *Id.*, art. III.

[23] The United States and the Soviet Union, which compiled the draft version of ENMOD, intended the term "hostile use" to be limited to hostile acts designed to cause destruction, damage, or injury to another state party. They did not intend to prohibit, for example, the use of environmental techniques during military training maneuvers or for scientific or economic purposes. *See* Susana Pimento & Edward Hammond, A Political Primer on the Environmental Modification Convention (ENMOD), CCD Negotiations: Article I (2002), http://www.sunshine-project.org/enmod/primer.html.

[24] *Id.*, art. V.3-V.5.

ENMOD is nevertheless subject to significant limitations. First, no party has ever been formally accused of violating ENMOD, and thus no referrals to the Security Council have taken place. Indeed, the treaty is rarely invoked and has been the subject of only two review conferences.[25] Moreover, less than half of the world's nations are parties to the agreement; nonparties include France, Indonesia, Saudi Arabia, and South Africa.[26] ENMOD's limited membership, combined with its narrow coverage, undermines its potential applicability.

The hypothetical situation described above, however, does underscore the importance of international norms of conduct, whether or not embodied in a formal treaty. Even relatively limited treaties such as ENMOD can serve as a foundation for the establishment of general norms. Thus, the principles underlying ENMOD would likely be invoked against nonparties who undertake hostile uses of geoengineering. Or, as other commentators have suggested, ENMOD might serve as a normative precedent against the use of geoengineering even for peaceful purposes.[27]

2.3 *Convention on Biological Diversity*

The Convention on Biological Diversity (CBD) is another treaty that could apply to geoengineering generally, and as described below, it has already come into play with respect to the governance of ocean fertilization geoengineering efforts. Signed in 1992, the CBD identifies the conservation of biodiversity as a "common concern" of humankind.[28] The CBD does not directly address geoengineering or climate change, but obviously is relevant to these topics insofar as they affect biodiversity. Rather than establishing comprehensive and binding international standards, however, the treaty relies primarily on national laws and policies to promote biodiversity.[29] Article 7(c), for instance, requires parties to "[i]dentify processes and categories of activities which have or are likely to have significant impacts on the conservation and sustainable use of biological diversity," but establishes no substantive duty to avoid or limit such impacts.

[25] *See* Susana Pimento Chamorro & Edward Hammond, *Addressing Environmental Modification in Post-Cold War Conflict* (2001), http://www.edmonds-institute.org/pimiento.html.

[26] *See* Status of Multilateral Arms Regulation and Disarmament Agreements: ENMOD, http://disarmament.un.org/TreatyStatus.nsf/ENMOD%20(in%20alphabetical%20order)?OpenView (listing 73 parties) (last visited Apr. 12, 2010).

[27] *See, e.g.*, Davis, *supra* note 10, at 936; William Pentland, *Is Geoengineering Legal?*, CLEANBETA, June 1, 2009, http://cleantechlawandbusiness.com/cleanbeta/index.php/2009/06/is-geoengineering-legal/.

[28] CBD, *supra* note 3, Preamble.

[29] *Id.*, arts. 8, 9, 10 (setting out obligations with respect to *in situ* conservation, *ex situ* conservation, and sustainable use of biological resources). The treaty does encourage parties to enter into agreements to notify and consult other states when activities carried out under a party's jurisdiction or control are likely to significantly and adversely affect biodiversity beyond that state. *Id.*, art. 14.1(c).

Growing interest in ocean fertilization prompted parties to the CBD to become involved in geoengineering governance. In May 2008, the Conference of the Parties to the CBD issued a decision "request[ing]" member states to ensure that ocean fertilization projects do not occur unless "there is an adequate scientific basis on which to justify such activities" and "a global, transparent and effective control and regulatory mechanism is in place for these activities."[30] The decision, however, does allow "small-scale scientific research studies within coastal waters" to proceed in the meantime, subject to several conditions.[31] Namely, such projects must be: (1) "justified by the need to gather specific scientific data"; (2) "subject to a thorough prior assessment of the potential impacts ... on the marine environment"; (3) "strictly controlled"; and (4) "not ... used for generating and selling carbon offsets or any other commercial purposes."[32]

One subsequent ocean fertilization experiment has raised serious questions about the administration of this exception and about potential conflicts between the CBD decision and treaties that address ocean pollution more specifically. In the February 2009 LOHAFEX experiment, German and Indian scientists released six tons of iron over a 300 km^2 section of the southern Atlantic Ocean.[33] Opponents contended that this release was neither small-scale nor within coastal waters. Proponents responded that the project was research-oriented and "coastal" because it was located in a region influenced by land.[34] Although the experiment ultimately went forward after the German government conducted further environmental review, the controversy highlighted criticisms of the scope of the CBD moratorium. Namely, critics have contended that limiting the research exception to small-scale studies in coastal waters is "arbitrary[] and counterproductive."[35] Small-scale studies in sensitive

[30] Ninth Meeting of the Conference of the Parties to Convention on Biological Diversity, Decision IX 16: Biodiversity and Climate Change, § C(4), UNEP/CBD/COP/DEC/IX/16 (Oct. 9, 2008), *available at* http://www.cbd.int/doc/decisions/cop-09/cop-09-dec-16-en.pdf. More recently, a scientific subcommittee of the CBD recommended that the Conference of the Parties adopt a similar position with respect to climate-related geoengineering generally. *See* Convention on Biological Diversity Subsidiary Body on Scientific, Technical and Technological Advice, *In-Depth Review of the Work on Biodiversity and Climate Change*, § A.8.(w), UNEP/CBD/SBSTTA/14/L.9 (May 14, 2010), *available at* http://www.cbd.int/sbstta14/meeting/in-session/?tab=2.

[31] *Id.*

[32] *Id.*

[33] *See* Richard Black, *Setback for Climate Technical Fix*, BBC NEWS, Mar. 23, 2009, http://news.bbc.co.uk/2/hi/7959570.stm. The experiment's outcome cast doubt on the efficacy of iron fertilization as a means of sequestering carbon because much of the resultant phytoplankton growth entered the food chain rather than sinking to the bottom of the oceans.

[34] *See* Editorial, *The Law of the Sea*, 2 GEOSCI. 153 (2009).

[35] Intergovernmental Oceanographic Commission (of UNESCO), *Report on the IMO London Convention Scientific Group Meeting on Ocean Fertilization*, at 4, IOC.INF-1247 (June 15, 2008), http://www.ioc-unesco.org/index.php?option=com_oe&task=viewDocumentRecord&docID=2002; *See also* Editorial, *supra* note 34, at 153. The United Nations General Assembly, in contrast, issued a

coastal areas could be more environmentally damaging than large-scale studies in less-sensitive areas, and useful information may not be available without performing large-scale experiments.[36] Furthermore, the CBD moratorium is arguably inconsistent with restrictions issued under the London Convention and Protocol, which are discussed below.

3. MEDIA-SPECIFIC TREATIES

In contrast to treaties of general applicability such as the UNFCCC, other treaties may apply only to particular types of geoengineering projects, depending on the nature of the projects or their potential environmental impacts. Specialized treaty regimes can offer potential advantages in terms of expertise and contextual considerations, but exclusive reliance on these regimes may result in gaps or inadequacies in oversight.[37] And as with the general treaties already discussed, these specialized (or media-specific) treaties were developed in response to other circumstances and may ultimately represent a poor fit for geoengineering regulation.[38]

3.1 *Ocean Fertilization*

For geoengineering projects involving ocean fertilization, the London Convention and London Protocol,[39] United Nations Convention on the Law of the Sea,[40] and various regional agreements may be relevant.[41]

resolution "welcom[ing]" the CBD decision. Oceans and the Law of the Sea, G.A. Res. 63/11, ¶ 116, U.N. Doc. A/RES/63/111 (Dec. 5, 2008), *available at* http://daccess-dds-ny.un.org/doc/UNDOC/GEN/N08/477/45/PDF/N0847745.pdf?OpenElement.

[36] Intergovernmental Oceanographic Commission, *supra* note 35, at 2.

[37] , *See, e.g.*, ROBERT V. PERCIVAL ET AL., *Criticisms of Media-Specific Statutes in U.S. Environmental Regulation*, *in* ENVIRONMENTAL REGULATION: LAW, SCIENCE, & POLICY 96 (5th ed., 2006), may apply similarly to patchwork global regulation of geoengineering.

[38] *Cf.* Bodansky, *supra* note 6, at 316 (urging caution "about drawing conclusions from existing legal rules, for the simple reason that these rules were not developed with climate engineering in mind").

[39] Convention on the Prevention of Marine Pollution by Dumping of Wastes and Other Matter, Dec. 29, 1972, 1046 U.N.T.S. 120 [hereinafter London Convention]; 1996 Protocol to the Convention on the Prevention of Marine Pollution by Dumping of Wastes and Other Matter, 1972, Nov. 7, 1996, 36 I.L.M. 1 [hereinafter London Protocol].

[40] United Nations Convention on the Law of the Sea, Dec. 10, 1982, 1833 U.N.T.S. 397 [hereinafter UNCLOS], *available at* http://www.un.org/Depts/los/convention_agreements/texts/unclos/closindx.htm.

[41] The discussion here does not consider the storage of CO_2 in or under the seabed, which is generally categorized as a form of carbon capture and storage rather than as a type of geoengineering. For analyses of regulatory issues with respect to those techniques, see Ray Purdy, *The Legal Implications of Carbon Capture and Storage under the Sea*, 7 SUSTAINABLE DEV. L. & POL'Y 22 (2006); INTERGOVERNMENTAL PANEL ON CLIMATE CHANGE, IPCC SPECIAL REPORT ON CARBON DIOXIDE CAPTURE AND STORAGE 254–55, 308–09 (2005).

3.1.1 London Convention/London Protocol

The London Convention and London Protocol (LC/LP) seek to control sources of marine pollution by regulating the dumping of waste into the sea. The 1972 Convention, which eighty-six nations have ratified or acceded to,[42] prohibits, or requires a permit for, the dumping of specifically listed items at sea.[43] The more stringent 1996 Protocol, which is intended to replace the 1972 Convention and currently has been ratified by thirty-eight nations,[44] bans ocean dumping in general except for explicitly listed items.[45] Under each treaty, dumping is defined to include "any deliberate disposal into the sea of wastes or other matter from vessels, aircraft, platforms or other man-made structures at sea."[46] "Placement of matter for a purpose other than the mere disposal thereof," in contrast to dumping, is allowed as long as such placement is not contrary to the purposes of the treaty.[47]

Recent ocean fertilization proposals, including proposals to generate carbon offsets through commercial projects, have attracted the attention of the LC/LP parties.[48] In 2007, the meeting of the parties agreed that ocean fertilization falls within the jurisdiction of the LC/LP and that "given the present state of knowledge regarding ocean fertilization ... large-scale operations [are] currently not justified."[49] In 2008, the meeting of the parties adopted a resolution distinguishing between "legitimate scientific research" – which would be regarded as "placement of matter for a purpose other than the mere disposal thereof" – and other ocean fertilization activities, which "should not be allowed."[50] Drawing the line between legitimate

[42] *See* International Maritime Organization, Status of Conventions Summary (Feb. 28, 2011), http://www.imo.org/About/Conventions/StatusOfConventions/Pages/Default.aspx.

[43] London Convention, *supra* note 39, art. IV.1.

[44] *See* Conventions Summary, *supra* note 42,

[45] London Protocol, *supra* note 39, art. 4.1. The dumping of items listed in Annex 1 to the Protocol is subject to a permitting process. *Id.*, art. 4.1.2. One category of materials listed in Annex 1 is "inert, inorganic geological material," which arguably includes the iron dust that would be used in fertilization efforts. *See* Jennie Dean, *Iron Fertilization: A Scientific Review with International Policy Recommendations*, 32 ENVIRONS 321, 336 (2009).

[46] London Protocol, *supra* note 39, art. 1.4.1; London Convention, *supra* note 39, art. III.1(a).

[47] London Protocol, *supra* note 39, art. 1.4.2.2; London Convention, *supra* note 39, art. III.1(b)(ii).

[48] *See* Chris Vivian, *Towards Regulation of Ocean Fertilisation by the London Convention and London Protocol – The Story So Far*, GEOENGINEERING QUARTERLY (Mar. 20, 2010).

[49] International Maritime Organization [IMO], *Report of the Twenty-Ninth Consultative Meeting and the Second Meeting of Contracting Parties*, ¶ 4.23, LC 29/17 (Dec. 14, 2007), *available at* http://www.imo.org/includes/blastDataOnly.asp/data_id%3D20797/17.pdf. The meeting also endorsed a "Statement of Concern" prepared by scientific working groups declaring that "knowledge about the effectiveness and potential environmental impacts of ocean iron fertilization currently was insufficient to justify large-scale operations and that this could have negative impacts on the marine environment and human health." *Id.* ¶¶ 4.14, 4.23 (quoting LC/SG 30/14, ¶¶ 2.23 to 2.25).

[50] IMO, *Thirtieth Meeting of the Contracting Parties to the London Convention and the Third Meeting of the Contracting Parties to the London Protocol, Resolution LC-LP.1 (2008) on the Regulation of Ocean*

research and non-research activities, of course, will pose a challenging task. As for legitimate research, parties are to evaluate such proposals using "utmost caution and the best available guidance" pending the development of an assessment framework by scientific advisory groups to the LC/LP.[51] As of this writing, working groups continue to develop that framework and to analyze options for further regulating ocean fertilization under the LC/LP.[52] Significant questions to be resolved include: whether research activities should be subject to a permit; whether treaty amendments specifically addressing ocean fertilization are necessary; and how potential commercial benefits (e.g., from generating and selling carbon credits) should be addressed.[53] The governing bodies for the LC/LP are expected to consider adoption of the assessment framework in October 2010.[54]

3.1.2 Law of the Sea

The UN Convention on the Law of the Sea (UNCLOS) is a general regime for ocean governance that largely codifies existing customary international law. The duties set out in the treaty may relate to geoengineering in two fundamental ways. First, states have a general obligation to "protect and preserve the marine environment,"[55] including the obligation to "take ... all measures ... necessary to prevent, reduce and control pollution of the marine environment from any source."[56] To the extent that ocean fertilization projects generate marine pollution or harm the marine environment, those projects may be regulated or even prohibited. Second, the requirement to protect the marine environment arguably creates an affirmative obligation to adopt measures to combat ocean acidification and other adverse effects of higher GHG concentrations, including ocean fertilization and other carbon removal techniques.

With respect to the first theory, UNCLOS broadly defines "pollution of the marine environment" as "the introduction by man, directly or indirectly, of substances or energy into the marine environment, ... which results or is likely to result in such deleterious effects as harm to living resources and marine life [and] hazards

Iron Fertilization, Res. LC-LP.1 (2008) (Oct. 31, 2008), http://www.imo.org/includes/blastDataOnly.asp/data_id%3D24337/LC-LP1%2830%29.pdf.

51 *Id.*

52 *See* IMO, *Report of the Thirty-First Consultative Meeting and the Fourth Meeting of Contracting Parties*, ¶¶ 4.14–4.39, LC 31/15 (Nov. 30, 2009), *available at* http://www.imo.org/includes/blastDataOnly.asp/data_id%3D27809/15.pdf. For a draft version of the assessment framework, see Draft "Assessment Framework for Scientific Research Involving Ocean Fertilization," LC/SG 32/15 Annex 2 (June 29, 2009), *available at* http://www.imo.org/includes/blastDataOnly.asp/data_id%3D26427/15.pdf.

53 *See* LC 31/15, *supra* note 52, ¶¶ 4.33–4.39.

54 *See* Vivian, *supra* note 48.

55 UNCLOS, *supra* note 40, art. 192.

56 *Id.*, art. 194.1.

to human health...."[57] Ocean fertilization undoubtedly would change the composition of the phytoplankton community and as a result would alter food webs and biogeochemical cycles and decrease oxygen levels in the oceans.[58] The duty to control pollution set out in UNCLOS includes the duty to ensure that pollution from activities under a state's control does not cause damage to other states and the duty not to transform one type of pollution into another.[59] Ocean fertilization projects will run afoul of this latter duty to the extent that they transform atmospheric pollution into marine pollution. Moreover, Article 210 of UNCLOS specifically requires states to adopt measures governing pollution of the marine environment by dumping, and such measures are to be no less stringent than global rules and standards[60] – that is, the standards set out under the LC/LP.[61] Although UNCLOS, like the LC/LP, distinguishes between dumping and placement for purposes other than disposal, ocean fertilization efforts are more accurately characterized as dumping rather than placement. Granted, iron fertilization would not be undertaken for the purpose of disposing iron. However, as David Freestone and Rosemary Rayfuse have argued, iron fertilization would serve as the means of placing excess carbon dioxide in the ocean for purposes of disposal.[62]

As under the LC/LP, one might distinguish between ocean fertilization research and ocean fertilization deployment under UNCLOS. UNCLOS explicitly protects the right to conduct marine scientific research.[63] Nonetheless, researchers hoping to conduct ocean fertilization experiments do not have free rein because this right is subject to the treaty's provisions for protecting the marine environment.[64] In other words, ocean fertilization research that may harm living resources and marine life would be subject to regulation under UNCLOS.

Although the standards applied under UNCLOS largely reiterate those established by the LC/LP, UNCLOS does offer some advantages in terms of enforcement. The LC/LP lacks an enforcement mechanism,[65] whereas UNCLOS prescribes compulsory dispute resolution procedures.[66] In addition, because of its near-universal membership,[67] UNCLOS also offers potentially broader coverage than the LC/

57 *Id.*, art. 1.1(4).
58 *See* Chisholm et al., *supra* note 14, at 310.
59 *Id.*, arts. 194.2, 195.
60 *Id.*, art 210.6.
61 *See* Freestone & Rayfuse, *supra* note 16, at 229.
62 *Id.*
63 UNCLOS, *supra* note 40, art. 238.
64 *Id.*, art. 240.
65 *See* DAVID HUNTER ET AL., INTERNATIONAL ENVIRONMENTAL LAW & POLICY 819 (3d ed. 2007).
66 UNCLOS, *supra* note 40, arts. 279–99.
67 As of January 1, 2010, 160 nations have ratified or acceded to UNCLOS. *See* Table Recapitulating the Status of the Convention and of the Related Agreements, http://www.un.org/Depts/los/reference_files/status2010.pdf.

LP. However, both regimes rely on member states to implement standards adopted under the respective treaties, which may lead to uneven application and enforcement. As ocean fertilization projects will most likely occur on the high seas, flag states will have primary responsibility for enforcing applicable standards.[68] As such, project sponsors will have incentives to arrange for their projects to occur under the flag of states with weak or nonexistent enforcement regimes.[69]

As to the second theory, UNCLOS's duty to protect the marine environment could provide authority that would support ocean fertilization and other carbon sequestration projects. UNCLOS neither prescribes specific measures states must take in carrying out this duty nor does it specifically address climate change. Nevertheless, given that ocean acidification is associated with higher atmospheric carbon dioxide levels and the resultant adverse consequences on coral reefs and other marine life,[70] some CDR techniques could be defended as consistent with this duty. In particular, UNCLOS parties have an obligation to adopt laws to reduce and control pollution of the marine environment from or through the atmosphere.[71]

3.1.3 Regional Treaties

Regional agreements may also be relevant to particular ocean fertilization projects, depending on their location. For example, a number of experiments have focused on the "Southern Ocean," a region where iron fertilization might be effective because of the relatively large quantities of surface macronutrients returning to the deep ocean in that area.[72] For projects south of 60° south latitude, the Antarctic Treaty System would come into play.[73] The 1959 Antarctic Treaty addresses environmental matters only in passing; it recognizes the preservation and conservation of living resources in Antarctica as a "matter[] of common interest."[74] The 1991 Protocol on Environmental Protection, however, requires that activities in the Antarctic Treaty area be planned and conducted so as to avoid or limit adverse impacts on the environment.[75] The Protocol also requires the preparation of prior environmental

[68] UNCLOS, *supra* note 40, art. 216. Coastal states have enforcement authority over dumping occurring in their territorial waters and exclusive economic zones, and states where material to be dumped is loaded also have enforcement authority. *Id.*

[69] *See* Freestone & Rayfuse, *supra* note 16, at 230.

[70] *See, e.g.*, James C. Orr et al., *Anthropogenic Ocean Acidification over the Twenty-First Century and Its Impact on Calcifying Organisms*, 437 NATURE 681 (2005).

[71] UNCLOS, *supra* note 40, art. 212.

[72] *See* Ken O. Buesseler & Philip W. Boyd, *Will Ocean Fertilization Work?*, 300 Sci. 67 (2003).

[73] Antarctic Treaty art. VI, Dec. 1, 1959, 19 I.L.M. 860 (defining geographical scope of treaty provisions).

[74] *Id.*, art. IX.1(f).

[75] Protocol on Environmental Protection to the Antarctic Treaty art. 3, Oct. 4, 1991, 30 I.L.M. 1455.

assessments.[76] None of these provisions specifically addresses ocean fertilization. Nevertheless, as Daniel Bodansky has contended, any ocean fertilization projects in the Antarctic Treaty area would almost certainly be reviewed by the treaty parties, who have established a fairly well-developed and manageable system of international governance that includes a mandatory dispute settlement procedure.[77]

Another example of a potentially relevant regional agreement is the 1992 Convention for the Protection of the Marine Environment of the North-East Atlantic (OSPAR), whose members include fifteen European countries bordering the Atlantic Ocean.[78] In prohibiting ocean dumping, OSPAR essentially tracks the London Convention and Protocol regime. By providing a regional governance mechanism, however, OSPAR does offer an additional and potentially more credible enforcement option.[79]

3.2 *Atmosphere-Based Geoengineering*

Proposals in this category include the release of sulfur aerosols into the stratosphere to block the sun's radiation,[80] as well as the seeding of clouds with seawater particles to increase their reflectivity.[81] The discussion here focuses on proposals involving stratospheric aerosols, which have received much attention because of their apparent advantages in cost and flexibility of deployment.[82] In contrast to the UNCLOS governance regime for the oceans, no global instrument governs the atmosphere.[83] Rather, states have sovereignty over the air space above their territories, subject to international norms regarding transboundary harm.[84] Regional air pollution agreements, however, as well as the Montreal Protocol,[85] may come into play.

[76] *Id.*, art. 8.

[77] *Id.*, arts. 18–20; *see* Bodansky, *supra* note 6, at 315.

[78] Convention for the Protection of the Marine Environment of the North-East Atlantic, Sept. 22, 1992, 32 I.L.M. 1069, http://www.ospar.org/html_documents/ospar/html/OSPAR_Convention_e_updated_text_2007.pdf [hereinafter OSPAR]; *see also* OSPAR Commission, About OSPAR, http://www.ospar.org/content/content.asp?menu=00010100000000_000000_000000 (last visited Apr. 2, 2010).

[79] OSPAR Annex II; *see* HUNTER ET AL., *supra* note 65, at 824.

[80] *See* Paul Crutzen, *Albedo Enhancement by Stratospheric Sulfur Injections: A Contribution to Resolve a Policy Dilemma?*, 77 CLIMATIC CHANGE 211, 211–12 (2006).

[81] *See* John Latham, *Control of Global Warming?*, 347 NATURE 339 (1990); John Latham et al., *Global Temperature Stabilization via Controlled Albedo Enhancement of Low-Level Maritime Clouds*, 366 PHIL. TRANS. ROY. SOC'Y A 3969 (2008), http://rsta.royalsocietypublishing.org/content/366/1882/3969.full.pdf.

[82] *See* David G. Victor et al., *The Geoengineering Option*, FOREIGN AFF., Mar./Apr. 2009, at 64, 69; Graeme Wood, *Moving Heaven and Earth*, ATLANTIC, July/Aug. 2009, at 70, 72.

[83] ROYAL SOCIETY, *supra* note 7, at 40.

[84] *Id.*

[85] Montreal Protocol on Substances That Deplete the Ozone Layer, Sept. 16, 1987, S. Treaty Doc. No. 100–10, 1522 U.N.T.S. 29 [hereinafter Montreal Protocol].

3.2.1 LRTAP

The Convention on Long-Range Transboundary Air Pollution (LRTAP)[86] is a regional framework agreement that obligates parties "to limit and, as far as possible, gradually reduce and prevent air pollution including long-range transboundary air pollution."[87] LRTAP's coverage is fairly broad, encompassing fifty-one nations in North America, Europe, and the former Soviet Union.[88]

LRTAP defines air pollution broadly as "the introduction by man ... of substances or energy into the air resulting in deleterious effects of such a nature as to endanger human health, harm living resources and ecosystems and material property."[89] Although LRTAP itself contains relatively "soft" requirements, eight subsequent protocols to the agreement do set out binding obligations governing specific classes of pollutants.[90] Two of those protocols address sulfate emissions: the 1985 Protocol requires parties to reduce such emissions by 30 percent, and the 1994 Protocol mandates further reductions.[91] More specifically, the 1994 Protocol requires parties to "control and reduce their sulphur emissions in order to protect human health and the environment from adverse effects, in particular acidifying effects."[92] These protocols, which were intended to reduce acid precipitation, at first glance might appear to be a potentially significant hurdle to the implementation of sulfate aerosol geoengineering. The ultimate effect of these protocols, however, would depend largely on the amount of sulfur injected into the stratosphere in any geoengineering effort. Sizeable uncertainty surrounds the amount of sulfur that ultimately would be needed, given the complexities of atmospheric processes and unresolved details regarding how sulfur would be released and the aerosol particle sizes that would result.[93] Nonetheless, one study concluded that "the additional sulfate deposition that would result from geoengineering will not be sufficient to negatively impact

[86] Convention on Long-Range Transboundary Air Pollution, Nov. 13, 1979, 18 I.L.M. 1442 [hereinafter LRTAP].

[87] *Id.*, art. 2.

[88] *See* U.N. Economic Commission for Europe, Status of Ratification of the 1979 Geneva Convention on Long-Range Transboundary Air Pollution as of 1 March 2011, http://www.unece.org/env/lrtap/status/lrtap_st.htm (last visited June 10, 2010).

[89] LRTAP, *supra* note 86, art. 1(a).

[90] *See* PHILIPPE SANDS & PAOLO GALIZZI, DOCUMENTS IN INTERNATIONAL ENVIRONMENTAL LAW 33 (2d ed. 2004).

[91] Protocol to the 1979 Convention on Long-Range Transboundary Air Pollution on the Reduction of Sulphur Emissions or Their Transboundary Fluxes by at Least 30 Per Cent, July 8, 1985, 27 I.L.M. 707, 1480 U.N.T.S. 217; Protocol to the 1979 Convention on Long-Range Transboundary Air Pollution on Further Reduction of Sulphur Emissions, June 14, 1994, 33 I.L.M. 1542 [hereinafter 1994 Protocol]. The United States is a party to LRTAP, but not to either of the Sulphur Protocols.

[92] 1994 Protocol, *supra* note 91, art. 2.1.

[93] *See* Philip J. Rasch et al., *Exploring the Geoengineering of Climate Using Stratospheric Sulfate Aerosols: The Role of Particle Size*, 35 GEOPHYS. RES. LETT. L02809 (2008).

most ecosystems"[94] in terms of the direct effects of acid precipitation. In other words, the LRTAP Protocols would most likely not be an insuperable barrier to the use of sulfate aerosols, unless their scope is more expansively understood to include adverse environmental effects other than acid precipitation.

3.2.2 Montreal Protocol

The Montreal Protocol to the Vienna Convention for the Protection of the Ozone Layer restricts the consumption and production of ozone-depleting substances.[95] Sulfate aerosols themselves do not destroy ozone directly. If injected into the stratosphere, however, they provide a surface for the activation of ozone-destroying chlorine gases already present, thereby intensifying their ozone-depleting effect and delaying recovery of the ozone layer.[96] The Montreal Protocol regime does not presently regulate sulfates that could wind up in the stratosphere.[97] However, given the potential for stratospheric aerosols to undermine the fundamental objective of the Protocol, the parties to the Protocol would likely take action to address geoengineering projects involving the release of stratospheric aerosols.[98] The Protocol requires the parties to assess and review its control measures at least every four years, and it authorizes the adoption of new control measures as needed.[99]

3.2.3 Space-Based Geoengineering

Serious geoengineering discussions to date have focused primarily on ocean fertilization and on land-based or atmosphere-based proposals. Nonetheless, there are also proposals to deploy shields or other means of blocking solar radiation in outer space.[100] Compared to the use of stratospheric aerosols, such an approach would be far more costly and would face more complicated barriers to implementation.[101]

[94] Ben Kravitz et al., *Sulfuric Acid Deposition from Stratospheric Geoengineering with Sulfate Aerosols*, 114 J. GEOPHYS. RES. D14109 (2009).

[95] Montreal Protocol, *supra* note 85.

[96] *See* Simone Tilmes et al., *The Sensitivity of Polar Ozone Depletion to Proposed Geoengineering Schemes*, 320 SCI. 1201 (2008).

[97] Montreal Protocol (as amended), *supra* note 85, art. 2A-2I, Annex A-Annex C.

[98] *Cf.* Virgoe, *supra* note 10, at 111.

[99] Montreal Protocol (as amended), *supra* note 85, arts. 2.10, 6. The parties to the Protocol have made frequent use of the Protocol's adjustment and amendment processes. *See* HUNTER ET AL., *supra* note 65, at 589–94.

[100] *See, e.g.*, Roger Angel, *Feasibility of Cooling the Earth with a Cloud of Small Spacecraft Near the Inner Lagrange Point*, 103 PROC. NAT'L ACAD. SCI. 17,184 (2006).

[101] Under one proposal, approximately 16 trillion discs would need to be manufactured and placed in orbit, at an estimated cost of $5 trillion. *See* Oliver Morton, *Is This What It Takes to Save the World?*, 447 NATURE 132, 136 (2007).

With respect to space-based geoengineering, the most pertinent international agreement is the 1967 Outer Space Treaty.[102] Established to prevent a race to militarize or colonize outer space, this treaty declares outer space to be the "province of all mankind," "free for exploration and use by all States."[103] The treaty further provides that parties are to conduct research or activities in outer space "with due regard to the corresponding interests" of other parties "so as to avoid their harmful contamination and also adverse changes in the environment of the Earth resulting from the introduction of extraterrestrial matter."[104] Although one can imagine arguments that a space-based geoengineering project would be contrary to the interests of a party and adverse to the environment,[105] the language of these provisions is hardly definitive or dispositive, particularly because it focuses on harms "resulting from the introduction of extraterrestrial matter." In addition, the treaty lacks a dispute settlement mechanism that might address objections to a geoengineering project.[106] Nonetheless, like ENMOD, the treaty might serve as a source of norms regarding international consultation and cooperation with respect to geoengineering.

4. NORMS

Although a sizeable number of existing multilateral agreements could apply to geoengineering, none of them provides a complete or direct response to the challenges raised. Given the significant gaps left by existing treaties, customary international law and general principles will likely play a critical, if not predominant, role in geoengineering governance. Several international environmental norms reflected in various treaties and other international documents are likely to be invoked. These norms include principles regarding transboundary harm, the precautionary principle, and the principle of intergenerational equity.[107]

With respect to transboundary harms, there are several relevant norms. First, a nation that carries out an activity resulting in transboundary harm has an obligation

[102] Treaty on Principles Governing the Activities of States in the Exploration and Use of Outer Space, Including the Moon and Other Celestial Bodies, Jan. 27, 1967, 18 U.S.T. 2410, 610 U.N.T.S. 205, *reprinted in* 6 I.L.M. 386 (1967).

[103] *Id.*, Preamble, art. I.

[104] *Id.*, art. IX.

[105] For example, a country that benefits from more moderate temperatures and increased rainfall as a result of climate change might object that it would be harmed by geoengineering efforts.

[106] *See* Bodansky, *supra* note 6, at 314.

[107] *See* ROYAL SOCIETY, *supra* note 7, at 40. The U.N. Environment Programme's Weather Modification Guidelines, although concerned with modification of weather rather than climate, articulate several of these norms. U.N. Environment Programme, *Provisions for Cooperation between States in Weather Modification*, Dec. 8/7/A of the Governing Council (Apr. 29, 1980), http://www.unep.org/Law/PDF/UNEPEnv-LawGuide&PrincNo3.pdf.

to notify and consult with potentially affected states.[108] Given the potential for intended and unintended consequences of geoengineering efforts to affect many nations, notification, consultation, and transboundary environmental impact assessment would almost certainly be required.[109] Second, a nation has an obligation not to cause environmental harm to others, or at least to take practicable steps to control such harm.[110] Although this norm is well-established,[111] its applicability may depend on the amount of harm resulting from a geoengineering project and the degree of care taken by the responsible state. Third, to the extent that harm does occur, a nation is responsible for the costs of mitigating or compensating such harm.[112] This norm may well require the establishment of a compensation fund and a procedure for making and resolving compensation claims prior to the execution of a geoengineering project.

The roles and effects of other potentially relevant norms are less clear. As discussed above, the precautionary principle would likely counsel caution in the deployment of geoengineering,[113] but the principle is a subject of some controversy and arguably has not attained the status of customary international law.[114] The principle of intergenerational equity, which counsels that present generations not leave future generations in a worse position with respect to options and resources,[115] is more widely accepted. But its precise application with respect to geoengineering efforts is unclear; one of the attractions of at least some types of geoengineering is their apparent cost advantage over emission reduction efforts. Assuming that such a cost advantage exists, some might contend that geoengineering would not run afoul of the principle so long as it leaves future generations with greater financial resources,

[108] *See* Rio Declaration on Environment and Development Principle 19, June 14, 1992, U.N. Doc. A/CONF. 151/5, 31 I.L.M. 8744 (1992) [hereinafter Rio Declaration].

[109] *See* HUNTER ET AL., *supra* note 65, at 532 ("In the transboundary context, many commentators believe that the duty to conduct an EIA is probably now a requirement of customary law."); NEIL CRAIK, THE INTERNATIONAL LAW OF ENVIRONMENTAL IMPACT ASSESSMENT: PROCESS, SUBSTANCE AND INTEGRATION 15 (2008) (noting "existence of a large number of treaty-based EIA commitments").

[110] *See* Stockholm Declaration of the United Nations Conference on the Human Environment Principle 21, June 16, 1972, U.N. Doc. A/CONF.48/14 (1972), *reprinted in* 11 I.L.M. 1416 (1972) [hereinafter Stockholm Declaration]; Rio Declaration, *supra* note 108, Principle 2.

[111] *See, e.g.*, HUNTER ET AL., *supra* note 65, at 502 (describing the obligation not to cause environmental harm as "[a] central principle of international environmental law" and "a part of customary international law").

[112] *See* Stockholm Declaration, *supra* note 110, Principle 22; Rio Declaration, *supra* note 108, Principle 2.

[113] *See supra* text accompanying notes 9–11.

[114] *See* Rio Declaration, *supra* note 108, Principle 15; Jutta Brunnee, *The Stockholm Declaration and the Structure and Processes of International Environmental Law*, *in* THE STOCKHOLM DECLARATION AND LAW OF THE MARINE ENVIRONMENT 67, 77 (Myron H. Nordquist et al. eds., 2003).

[115] *See* Stockholm Declaration, *supra* note 110, Principles 1, 2; Rio Declaration, *supra* note 108, Principle 3.

even if they would be living in a world with lesser natural bounty. Conversely, many geoengineering schemes, once deployed, would potentially tie the hands of future generations by requiring them to continue geoengineering efforts for many years in order to avoid a rebound effect from their sudden cessation.

5. CONCLUSION

International cooperation on deciding how to proceed with geoengineering, if at all, is hardly assured. Global governance of geoengineering could occur through existing treaties, new treaty instruments, or ad hoc responses to individual geoengineering proposals or projects. Whatever governance does occur is likely to be driven by international norms such as those regarding transboundary harm and equity, rather than by the formal requirements of existing treaty regimes. Developing a governance structure to address geoengineering research and deployment at an early stage, rather than relying on ad hoc responses to later crises, would be preferable for a number of reasons: it can help assure that research is carried out with the blessing of the international community and with proper safeguards, establish oversight of geoengineering efforts, and minimize the influence that vested interests might have on governance structures and decisions.[116] But even in the absence of specifically applicable treaty provisions, geoengineering will be too important a subject, with effects too universal, for the international community to ignore.

[116] *See* Lin, *supra* note 4, at 19–20.

9

Climate Geoengineering

Solar Radiation Management and Its Implications for Intergenerational Equity

Wil C. G. Burns

1. INTRODUCTION

As David Victor recently observed, climate geoengineering, broadly defined as "the deliberate large-scale manipulation of the planetary environment to counteract anthropogenic climate change,"[1] was once viewed as "a freak show in otherwise serious discussions of climate science and policy."[2] However, in the past few years, the feckless response of the world community to burgeoning greenhouse gas emissions[3]

[1] The Royal Society, *Geoengineering the Climate: Science, Governance and Uncertainty* (Sept. 2009), at 11, http://royalsociety.org/Geoengineering-the-climate/ (last visited Mar. 28, 2011).

[2] David G. Victor, *On the Regulation of Geoengineering*, 24(2) OXFORD REV. ECON. POL'Y 322, 323 (2008).

[3] Since the Kyoto Protocol was signed in the 1990s, the annual rate of greenhouse gas emissions has actually accelerated from 1.3 percent in the 1990s to 3.3 percent from 2000 to 2006, although that rate has slowed during the current economic downturn. Stefan Folster & Johan Nystrom, *Climate Policy to Defeat the Green Paradox*, 39 AMBIO 223, 223 (2010); A.J. Dolman et al., *A Carbon Cycle Science Update since IPCC AR-4*, 39 AMBIO 402, 403 (2010). As a consequence, even limiting projected temperature increases to below 4 degrees Celsius above preindustrial levels may require a "radical reframing of both the climate change agenda, and the economic characterization of contemporary society." Kevin Anderson & Alice Bows, *Reframing the Climate Change Challenge in Light of Post-2000*, PHIL. TRANSACTIONS ROYAL SOC'Y A, Aug. 29, 2008, at 18. *See also* Joeri Rogelj et al., *Analysis of the Copenhagen Accord Pledges and Its Global Climatic Impacts – A Snapshot of Dissonant Ambitions*, 5 ENVTL. RES. LETTERS 034013 (2010), at 7 (Pledges made by the Parties in the Copenhagen Accord at the 15th Conference of the Parties may result in a temperature increase of 2.5 to 4.2 degrees Celsius by 2100, with temperatures continuing to increase after this point); International Energy Agency, *World Energy Outlook 2010*, Executive Summary (2010), at 11, http://www.worldenergyoutlook.org/docs/weo2010/WEO2010_ES_English.pdf, (last visited Nov. 11, 2010) (Copenhagen Accord pledges put us on track for more than a 3.5 degrees Celsius increase in temperatures). This is an extremely foreboding development, as most scientists and policy makers now believe that even a 2 degrees Celsius increase from preindustrial levels will result in serious impacts on human institutions and ecosystems. German Advisory Council for Global Change, *New Impetus for Climate Policy: Making the Most of Germany's Dual Presidency*, German Advisory Council on Global Change, WBGU Policy Paper 5 (2007); Commission of European Communities, *Communication from the Commission to the Council, the European Parliament, the European Economic and Social Committee and the Committee of the Regions, Limiting Global Climate Change to 2°C the Way Ahead for 2020*

has led to increasingly serious consideration of geoengineering as a potential means to avoid a "climate emergency,"[4] such as rapid melting of the Greenland and West Antarctic ice sheets,[5] or as a stopgap measure to buy time for effective emissions mitigation responses.[6] Indeed, a number of recent studies indicate that geoengineering

and Beyond (2007); James Hansen et al., *Dangerous Human-Made Interference with Climate: A GISS Model Study*, 7 ATMOSPHERIC CHEMISTRY & PHYSICS 2287–12 (2007), *available at* http://pubs.giss.nasa.gov/docs/2007/2007_Hansen_etal_1.pdf (last visited Oct. 21, 2008).

4 Ken Caldeira & David W. Keith, *The Need for Climate Engineering Research*, ISSUES SCI. & TECH. 57, 57 (Fall 2010).

5 Jason J. Blackstock et al., *Climate Engineering Responses to Climate Emergencies*, Novim (July 29, 2009), at 1–2; Peter J. Irvine et al., *The Fate of the Greenland Ice Sheet in a Geoengineered, High CO_2 World*, 4 ENVTL. RES. LETTERS, 045109 (2009), at 2. A complete melting of the Greenland ice sheet could occur with temperature increases of 2–3 degrees Celsius. Stephen Schneider, *The Worst-Case Scenario*, 458 NATURE 1104, 1104 (2009). This could raise global sea level by approximately 7 meters and trigger a slowdown or collapse of the ocean thermohaline circulation, which could result in significant cooling over much of the northern hemisphere. Jason A. Lowe et al., *The Role of Sea-Level Rise and the Greenland Ice Sheet in Dangerous Climate Change: Implications for the Stabilisation of Climate*, *in* AVOIDING DANGEROUS CLIMATE CHANGE 30 (Hans Joachim Schellnhuber ed., 2006); Julian A. Dowdeswell, *The Greenland Ice Sheet and Global Sea-Level Rise*, 311 SCI. 963, 963 (2004). Global average temperature increases of 1–4 degrees Celsius relative to 1990–2000 could result in a sea level rise of 4–6 meters. Intergovernmental Panel on Climate Change, Working Group II Contribution, *Climate Change 2007: Climate Change Impacts, Adaptation and Vulnerability* 17 (2007). Even a 5-meter rise in sea level could affect 5 percent of the world's population and threaten $2 trillion of Gross Domestic Product. United Nations Framework Convention on Climate Change, *Mechanisms to Manage Financial Risks from Direct Impacts of Climate Change in Developing Countries*, FCCC/TP/2008/9, Nov. 21, 2008, at 35. Some proponents of geoengineering also cite concern about temperatures reaching a critical "tipping point," or a "regime shift," triggering "non-linear self-reinforcing further warming or other dangerous environmental effects beyond those resulting immediately from the temperature rise itself." Alan Carlin, *Why a Different Approach Is Required if Global Climate Change Is to Be Controlled Efficiently or even at All*, 32 WM. & MARY ENVTL. L. REV. 685, 706–07 (2008). *See also* Rob Swart & Natasha Marinova, *Policy Options in a Worst Case Climate Change World*, 15 MITIGATION & ADAPTATION STRATEGIES FOR GLOBAL CHANGE 531, 532–33 (2010). Potential regime shifts could include the complete disappearance of Arctic sea ice in summer, leading to drastic changes in ocean circulation and climate patterns across the whole Northern Hemisphere; acceleration of ice loss from the Greenland and Antarctic ice sheets; ocean acidification from carbon dioxide absorption, potentially wreaking havoc on ocean ecosystems, massive dieback of forests, and shutdown of the Atlantic Thermohaline Circulation system. Alan Hastings & Derin B. Wysham, *Regime Shifts in Ecological Systems Can Occur with No Warning*, 13 ECO. LETTERS 464–72 (2010); *Runaway Tipping Points of No Return*, RealClimate, July 5, 2006, http://www.realclimate.org/index.php/archives/2006/07/runaway-tipping-points-of-no-return/ (last visited Feb. 13, 2011). Moreover, a temperature increase of this magnitude could lead to a release of greenhouse gases double that produced by humans to date, triggering a "runaway greenhouse effect." Carlin, *supra*, at 691. *See also* Kevin Schaefer et al., *Amount and Timing of Permafrost Carbon Release in Response to Climate Warming*, Tellus, Feb. 15, 2011 (Early View), http://onlinelibrary.wiley.com/doi/10.1111/j.1600–0889.2011.00527.x/pdf, (last visited Feb. 26, 2011); Fred Pearce, *Climate Warming as Siberia Melts*, NEW SCI., Aug. 11, 2005, at 12.

6 Martin Bunzl, *Research Geoengineering: Should Not or Could Not?*, 4 ENVTL. RES. LETTERS 045104 (Oct.–Dec. 2009), http://iopscience.iop.org/1748–9326/4/4/045104/fulltext (last visited Sept. 27, 2010); Christopher Mims, *"Albedo Yachts" and Marine Clouds: A Cure for Climate Change?*, SCI. AM., Oct. 21, 2009, at 3.

schemes could potentially mitigate the climatic impacts associated with a doubling of atmospheric carbon dioxide levels from preindustrial levels.[7]

However, many policy makers and commentators, even including some who have signaled tentative support for geoengineering options, have expressed serious reservations. Most of the focus of these concerns has been on *intragenerational* considerations associated with the two major categories of geoengineering, solar radiation management (SRM)[8] and carbon dioxide removal (CDR)[9] schemes. For example, several recent studies have concluded that stratospheric sulfate aerosol injection,[10] perhaps the most widely discussed SRM scheme,[11] could lead to a substantial

[7] B. Govindasamy, K. Caldeira & P.B. Duffy, *Geoengineering Earth's Radiation Balance to Mitigate Climate Change from a Quadrupling of CO_2*, 37 GLOBAL & PLANETARY CHANGE 157, 158 (2003); Ken Caldeira & Lowell Wood, *Global and Arctic Climate Engineering: Numerical Model Studies*, 366 PHIL. TRANSACTIONS ROYAL SOC'Y A 4039, 4044 (2008).

[8] Climate geoengineering schemes seek to reduce net radiative forcing by balancing the forcing associated with greenhouse gases with negative forcing by reducing the amount and characteristics of solar radiation. Carlin, *supra* note 5, at 688. It has been calculated that solar irradiance would have to be reduced by 1.8 percent to offset the radiative forcing associated with a doubling of carbon dioxide concentrations from preindustrial levels. The Royal Society, *supra* note 1, at 23. Solar radiation management methods seek to reduce net incoming shortwave solar radiation by deflecting sunlight or increasing the reflectivity of the atmosphere, clouds, or the earth's surface.

[9] Carbon dioxide removal (CDR) schemes remove carbon dioxide from the atmosphere after they have been released, facilitating the escape of more long-wave heat radiation. The Royal Society, *supra* note 1, at 1, http://royalsociety.org/geoengineering-the-climate/, (last visited Sept. 28, 2010). The most prominent CDR approaches include ocean iron fertilization, *see infra* note 17; air capture of carbon dioxide, David Biello, *Pulling CO2 from the Air: Promising Idea, Big Price Tag*, Yale Environment 360, http://www.360.yale.edu/content/print.msp?id=2197 (last visited Oct. 12, 2010); and mineral sequestration of carbon dioxide by combining it with suitable rocks such as olivine or serpentine, or injection into the ground to react with local mineral rock, Klaus S. Lackner, *Air Capture and Mineral Sequestration*, U.S. House of Representatives, Hearings Before the House Science/Technology Subcommittee, Feb. 4, 2010, http://democrats.science.house.gov/Media/file/Commdocs/hearings/2010/Energy/4feb/Lackner_Testimony.pdf (last visited Oct. 12, 2010), at 7; and biochar and biomass methods, The Royal Society, *supra* note 1, at 11.

[10] Stratospheric sulfate aerosol injection is a geoengineering scheme that involves the release of large quantities of sulfur into the stratosphere, or a precursor gas that oxidizes in the stratosphere, for the purpose of scattering incoming solar radiation. Sulfur injection would be facilitated by utilizing a delivery system such as a high flying jet, artillery shells or balloons. Philip J. Rasch et al., *An Overview of Geoengineering of Climate Using Stratospheric Sulphate Aerosols*, 366 PHIL. TRANSACTIONS ROYAL SOC'Y A 4007, 4013–14 (2009). Although sulfur dioxide is the most widely discussed candidate for atmospheric injection, other candidates include hydrogen sulfide (H_2S), carbonyl sulfide, ammonium sulfide and engineering nanoscale particles. *See, e.g.*, Ben Kravitz et al., *Sulfuric Acid Deposition from Stratospheric Geoengineering with Sulfate Aerosols*, 114 J. GEOPHYSICAL RES., D14109 (2009), at 2; David W. Keith, *Photophoretic Levitation of Engineered Aerosols for Geoengineering*, 108(38) PNAS 16428–16431.

[11] Eric Bickel & Lee Lane, *An Analysis of Climate Engineering as a Response to Climate Change*, Copenhagen Consensus Center (2009), at 17, http://fixtheclimate.com/fileadmin/templates/page/scripts/downloadpdf.php?file=/uploads/tx_templavoila/AP_Climate_Engineering_Bickel_Lane_v.5.0.pdf (last visited Nov. 19, 2010). One commentator has identified sulfur injection as "probably the most seriously discussed geoengineering proposal." Albert C. Lin, *Balancing the Risks:*

reduction in precipitation in monsoon regions in East and South-East Asia and Africa. This could result in a severe reduction in monsoonal intensity, potentially undermining the food security of 2 billion people in the region.[12] Diebacks of tropics forests could also be triggered by substantial precipitation declines in the Amazon and Congo valleys.[13] Additionally, sulfate aerosol loading of the atmosphere could accelerate the hydroxyl-catalyzed ozone depstructure cycles, resulting in significant depletion of the stratospheric ozone layer.[14]. For example, a recent study concluded that sulfate aerosol loading could result in an annual 4.5 percent decrease in stratospheric ozone levels, more than the annual mean global total loss due to the emission of anthropogenic ozone depleting substances in recent years.[15] Several studies have also indicated that ocean iron fertilization,[16] a CDR approach, could undermine

Managing Technology and Dangerous Climate Change, 8(3) ISSUES IN LEGAL SCHOLARSHIP 1, 4 (2009).

[12] Victor Brovkin et al., *Geoengineering Climate by Stratospheric Sulfur Injections: Earth System Vulnerability to Technological Failure*, 92 CLIMATIC CHANGE 243, 252 (2009); Alan Robock, Luke Oman & Georgiy L. Stenchikov, *Regional Climate Responses to Geoengineering with Tropical and Arctic SO_2 Injections*, 113 J. GEOPHYSICAL RES. D16101 (2008), at 13.

[13] Alexey V. Eliseev et al., *Global Warming Mitigation by Sulphur Loading in the Stratosphere: Dependence of Required Emissions on Allowable Residual Warming Rate*, 101 THEORETICAL APPLIED CLIMATOLOGY 67, 79 (2010).

[14] P. Heckendorn et al., *The Impact of Geoengineering Aerosols on Stratospheric Temperature and Ozone*, 4 ENVTL. RES. LETTERS 1, 7 (2009).

[15] *Id. See also* Alan Robock, *Whither Geoengineering?*, 320 SCI. 1166, 1166 (2008). Of course this could also have intergenerational implications. Recent research has indicated that sulfate injection schemes could delay recovery of the stratospheric ozone layer by as much as seventy years, thus impacting future generations. The Royal Society, *supra* note 1, at 31. Some researchers have suggested that injection of engineered nanoparticles could substantially reduce the potential threat to the stratospheric ozone layer by facilitating the lofting aerosols out of the stratosphere. David W. Keith, *Photophoretic Levitation of Engineered Aerosols for Geoengineering*, 107(38) PROC. NAT'L ACAD. SCI. 1628, 1630 (2010).

[16] Ocean iron fertilization (OIF) techniques seek to stimulate the production of phytoplankton through the addition of iron to ocean regions that are allegedly deficient in this micronutrient. Christine Bertram, *Ocean Iron Fertilization in the Context of the Kyoto Protocol and the Post-Kyoto Process*, 38(2) ENERGY POL'Y 1130, 1131 (2010); Philip Boyd, *Ironing Out Algal Issues in the Southern Ocean*, 305 SCI. 396–97 (2004). Phytoplankton take up carbon dioxide from seawater to carry out photosynthesis and to build up particulate organic carbon (POC). *Id.* Ultimately, part of the POC sinks to the deep ocean where it can be stored for a century or more. *Id.* The potential effectiveness of OIF remains contested. Although some proponents claim that OIF could result in a substantial drawdown of atmospheric carbon dioxide concentrations, O. Aumont & L. Bopp, *Globalizing Results from Ocean in situ Iron Fertilization Studies*, 20 GLOBAL BIOGEOCHEMICAL CYCLES 1, 1 (2006) (massive OIF could reduce atmospheric concentrations of carbon dioxide by between 50–107 parts per million after 100 years of fertilization), more recent research, informed by a number of recent field experiments, are not nearly as sanguine, *see* M.J.C. Crabbe, *Modelling Effects of Geoengineering Options in Response to Climate Change and Global Warming: Implications for Coral Reefs*, 33 COMPUTATIONAL BIO. & CHEMISTRY 415, 418 (2009) (OIF of 20 percent of world's oceans would only reduce atmospheric carbon dioxide by less or equal to 15 ppm at expected levels of 700 ppm in 2100 for business-as-usual scenarios of greenhouse gas emission); R.S. Lampitt, *Ocean Fertilization: A Potential Means of Geoengineering?*, 366 PHIL.

biological productivity in non-fertilized regions,[17] cause widespread eutrophication and anoxia,[18] and stimulate toxic algal blooms.[19]

This chapter will advance the argument that one category of geoengineering approaches, SRM schemes, could also severely circumscribe the options of *future generations* in the context of climate change policy, as well as potentially visit catastrophic negative climatic impacts. As such, this approach would, under all but the most stringent protocols, violate the tenets of an important principle of international law, intergenerational equity. Considerations of intergenerational equity, as such, are critical in the context of the pursuit of climate justice, defined as "special problems of obligation and participation posed by climate impacts and policies for their management."[20]

In developing this argument, I will: (1) present an overview of the principle of intergenerational equity, (2) discuss the application of intergenerational equity obligations in the context of SRM climate geoengineering, and (3). Discuss the implications of intergenerational equity for CDR geoengineering options.

2. SRM GEOENGINEERING AND INTERGENERATIONAL EQUITY CONSIDERATIONS

2.1 *Overview of Intergenerational Equity as a Principle of International Law*

Intergenerational equity is a "principle of distributive justice"[21] that calls for "fairness in the utilization of resources between human generations past, present and future."[22] It is ultimately grounded in the premise that human survival is a salutary goal, and the correlated moral obligations to support human continuity by sound stewardship

TRANSACTIONS ROYAL SOC'Y A 3919, 3928 (2008) (OIF could only draw down atmospheric levels of carbon dioxide by 10 ppm).

[17] Karen N. Scott, *The Day after Tomorrow: Ocean CO_2 Sequestration and the Future of Climate Change*, 18 GEO. INT'L ENVTL. L. REV. 57, 95 (2005).

[18] Lampitt, *supra* note 16, at 3930.

[19] Charles G. Trick, et al. *Iron Enrichment Stimulates Toxic Diatom Production in High-Nitrate, Low-Chlorophyll Areas*, PNAS EARLY EDITION, Mar. 10, 2010, http://www.pnas.org/cgi/doi/10.1073/pnas.0910579107, at 5887 (last visited Oct. 12, 2010).

[20] Ludvig Beckman & Edward A. Page, *Perspectives on Justice, Democracy and Global Climate Change*, 17(4) ENVTL. POL. 527, 527 (2008).

[21] Brett M. Frischmann, *Some Thoughts on Shortsightedness and Intergenerational Equity*, 36 LOY. U. CHI. L.J. 457, 460 (2005). "Distributive justice is concerned with sharing the benefits and burdens of social cooperation." Lawrence B. Solum, *To Our Children's Children's Children: The Problems of Intergenerational Ethics*, 35 LOY. L.A. L. REV. 163, 175 (2001).

[22] G.F. Maggio, *Inter/intragenerational Equity: Current Applications under International Law for Promoting the Sustainable Development of Natural Resources*, 4 BUFF. ENVT'L. L.J. 161, 163 (1997).

of the resources essential for life, as well as to ensure the dignity and well-being of earth's inhabitants.[23] As such, it "demands that present generations should not create benefits for themselves in exchange for burdens on future generations."[24]

Our obligation to conform our behavior to the principle of intergenerational equity derives from several different rationales. From social contract theory, all generations can be viewed as partners in an open-ended social contract that defines their rights, duties, and obligations. As Edmund Burke contended, because society's objectives cannot be achieved in a single generation, it is imperative that each generation protect the interests of those to come.[25]

Another basis for imposing intergenerational obligations is grounded in the equitable notions that underpin the "original position" theory formulated by John Rawls. As Edith Brown Weiss contends:

> In order to define what intergenerational equity then means, it is useful to view the human community as a partnership encompassing all generations, the purpose of which is to realize and protect the well-being of every generation and to conserve the planet for the use of all generations. Although all generations are members of this partnership, no generation knows in advance when it will be living, how many members it will have, nor even how many generations there will be.
>
> It is appropriate to adopt the perspective of a generation which is placed somewhere on the spectrum of time, but does not know in advance where … Such a generation would want to receive the planet in at least as good condition as every other generation receives it and to be able to use it for its own benefit. This requires that each generation pass on the planet in no worse condition than received and have equitable access to its resources.[26]

The notion of unjust enrichment is another rationale advanced as a basis for the existence of duties toward future generations. Our generation is indebted to past generations for conserving the resources that ensure our well-being. In turn, it can be argued that we hold these resources in trust and have a responsibility to pass them

[23] Dinah Shelton, *Intergenerational Equity*, in SOLIDARITY: A STRUCTURAL PRINCIPLE OF INTERNATIONAL LAW 131 (Rüdiger Wolfrum & Chie Kojima eds., 2010); Edith Brown Weiss, *Climate Change, Intergenerational Equity and International Law: An Introductory Note*, 15 CLIMATIC CHANGE 327, 330 (1989) ("Each generation is both a trustee and a beneficiary, or a custodian and user, of the planet").

[24] Marlos Goes, Klaus Keller & Nancy Tuana, *The Economics (or Lack Thereof) of Aerosol Engineering*, at 1 http://citeseerx.ist.psu.edu/viewdoc/summary?doi=10.1.1.144.446 (last visited Mar. 27, 2011).

[25] "[A]s the ends of such a partnership cannot be obtained in many generations, it becomes a partnership not only between those who are living but between those who are living, those who are dead, and those who are to be born." Edmund Burke, *Reflections on the Revolution in France* (1790), *in* 2 WORKS OF EDMUND BURKE 130–40 (1854). *See also* Robin Attfield, *Environmental Ethics and Intergenerational Equity*, 41(2) INQUIRY 207, 219 (1998).

[26] Edith Brown Weiss, *Climate Change, Intergenerational Equity and International Law: An Introductory Note*, 15 CLIMATIC CHANGE 330, 335 (1989).

on in no worse condition than we received them. To not do so would constitute a form of unjust enrichment.[27] Finally, intergenerational equity can be viewed as an extension of the public trust doctrine, mandating that the state protect the interests of both this generation and future generations in terms of conservation of options, quality and access to the earth's resources.[28]

The equitable considerations that support the principle of intergenerational equity mandate that "later generations [should] not be worse off than previous generations."[29] In the context of environmental resources, this includes both the form of resource stocks and the shape of environmental problems that current generations bestow on future generations.[30] More broadly, intergenerational equity also requires that future generations are accorded freedom of choice as to their political, economic, and social systems.[31]

Brown Weiss outlines three basic obligations of intergenerational equity:

1. **Conservation of options**. "[E]ach generation should be required to conserve the diversity of the natural and cultural base, so that it does not unduly restrict the options available to future generations in solving their problems and satisfying their own values ...";
2. **Conservation of quality**. "[E]ach generation should be required to maintain the quality of the planet so that it is passed on in no worse condition than that in which it was received ...";
3. **Conservation of access**. "[E]ach generation should provide its members with equitable rights of access to the legacy of past generations and should conserve this access for future generations."[32]

These three categories of "Planetary Obligations" are further disarticulated into five duties of use: (1) the duty to conserve resources; (2) the duty to ensure equitable use;

[27] Shelton, *supra* note 23, at 132.

[28] E.B. Weiss, *Intergenerational Equity: A Legal Framework for Global Environmental Change, in* E.B. WEISS, ENVIRONMENTAL CHANGE AND INTERNATIONAL LAW 395 (1992); Donna R. Christie, *Marine Reserves, The Public Trust Doctrine and Intergenerational Equity*, 19 J. LAND USE 427, 434 (2004), http://www.law.fsu.edu/journals/landuse/vol19_2/1achristie.pdf (last visited Feb. 12, 2011).

[29] Edith Brown Weiss, *What Obligation Does Our Generation Owe to the Next? An Approach to Global Environmental Responsibility: Our Rights and Obligations to Future Generations for the Environment*, 84 AM. J. INT'L L. 198, 200 (1990).

[30] Lars Osberg, *Meaning and Measurement in Intergenerational Equity* (1997), at 4, http://myweb.dal.ca/osberg/classification/book%20chapters/Meaning%20and%20Measurement%20in%20Intergenerational%20Equity/Meaning%20and%20Measurement%20in%20Intergenerational%20Equity.pdf (last visited Oct. 2, 2010).

[31] UNESCO, *Declaration on the Responsibilities of the Present Generations Towards Future Generations* (1997), at art. 3, http://portal.unesco.org/en/ev.php-URL_ID=13178&URL_DO=DO_PRINTPAGE&URL_SECTION=201.html (last visited Oct. 3, 2010).

[32] Weiss, *supra* note 29, at 201–02.

(3) the duty to avoid adverse impacts; (4) the duty to prevent disasters, minimize damage, and provide emergency assistance; and (5) the duty to compensate for environmental harm.[33]

Intergenerational equity is a binding principle of international law with broad application.[34] Most pertinent in the context of climate change policy making, the United Nations Framework Convention on Climate Change,[35] which has 194 parties,[36] incorporates the principle in Article 3(1), providing that "The Parties should protect the climate system for the benefit of present and future generations of humankind, on the basis of equity ..."[37] It can also be argued that intergenerational equity is a binding principle of customary international environmental law given its incorporation in a wide array of treaties,[38] domestic and international case law,[39] domestic legislation,[40] and soft law

[33] EDITH BROWN WEISS, IN FAIRNESS TO FUTURE GENERATIONS: INTERNATIONAL LAW, COMMON PATRIMONY, AND INTERGENERATIONAL EQUITY 51–60 (1989).

[34] Maggio, *supra* note 22, at 161.

[35] United Nations Framework Convention on Climate Change, May 9, 1992, 31 I.L.M. 849 (hereinafter UNFCCC).

[36] UNFCCC Secretariat, *Status of Ratification of the Convention*, http://unfccc.int/essential_background/convention/status_of_ratification/items/2631.php (last visited Oct. 3, 2010).

[37] UNFCCC, *supra* note 35, at art. 3(1).

[38] Convention on International Trade in Endangered Species of Wild Fauna and Flora, 12 ILM 1086 (1973), Preamble; Amazonian Co-Operation Treaty, 17 ILM 1045 (1978), Preamble; Convention on Biological Diversity, 31 ILM 818 (1992), Preamble, para. 23; Convention on the Conservation of Migratory Species of Wild Animals, 19 ILM 15 (1980), Preamble; Convention on the Conservation of European Wildlife and Natural Habitats, UKTS no. 56 (1982), Cmnd. 8738, Preamble; Convention on Access to Information, Public Participation in Decision-Making, and Access to Justice in Environmental Matters (Aarhus, June 25, 1998), Preamble. *See also* North–East Atlantic Fisheries Commission, *Information on the Protection of Biodiversity and Mitigating Impact of Fisheries in the North East Atlantic* (2010), at 2 ("Fishing communities and societies have the right to pursue their legitimate business of establishing economic development that meets the needs of the present generation without compromising the ability of future generations to meet their needs"). It should be noted, however, that the UNFCCC is the only treaty that includes intergenerational equity considerations in non-preambular provisions. Intergenerational equity principles are also incorporated into the first paragraph of the United Nations Charter ("We the peoples of the United Nations, determined to save future generations from the scourge of war), United Nations, The Charter of the United Nations (1945), Preamble.

[39] *See, e.g.*, Denmark v. Norway, 1993 ICJ 38 (Separate Opinion of Judge Weeramantry), at 274; Minors Oposa v. Secretary of the Dept. of Environment and Natural Resources, 33 ILM 173, 185 (1994); Legality of the Threat or Use of Nuclear Weapons Case, Advisory Opinion, [1996] I.C.J. Rep. 226, at 243–44; State of Himachal Pradesh and others (Appellants) *v.* Ganesh Wood Products and others (Respondents), AIR 1996 Supreme Court 149, 158 (1995), http://www.ecolex.org/server2.php/libcat/docs/COU/Full/En/COU-143787E.pdf (last visited Oct. 28, 2010).

[40] "The domestic Constitutions of twenty-two countries explicitly recognize the environmental interests of future generations." Lynda M. Collins, *Revisiting the Doctrine of Intergenerational Equity in Global Environmental Governance*, 30 DALHOUSIE L.J. 79, 98 (2007). *See* France ("in order to secure sustainable development, the choices made to meet the needs of the present should not jeopardize the ability of future generations and other peoples to meet their own needs," Article 2 Constitutional Amendment

instruments.[41] Moreover, the principle has been characterized as "a fundamental principle of sustainable development,"[42] a concept that many believe has now in its own right emerged as a principle of customary law.[43]

In the next section of this article, I will assess the applicability of the principle of intergenerational equity to potential climate geoengineering options.

2.2 *Intergenerational Equity and SRM Geoengineering*

As indicated above, SRM geoengineering schemes seek to ameliorate potential increases in temperature associated with the buildup of greenhouse gases in the atmosphere by deflecting incoming solar radiation, or increasing the reflectivity of the atmosphere, clouds, or earth's surfaces.[44] In addition to sulfate aerosol injection,[45] the other primary SRM schemes that have been proposed are seeding marine

on the Environment Charter); Sweden ("public institutions shall promote sustainable development leading to a good environment for present and future generations," Article 2 Constitution of Sweden); Switzerland (Swiss people and cantons "conscious of their common achievements and their responsibility towards future generations," Preamble Constitution of Switzerland); Ukraine (Parliament "aware of our responsibility before God, our own conscience, past, present and future generations," Preamble to Constitution of Ukraine); Poland ("bequeath to future generations all that is valuable from our [...] heritage," Preamble to Constitution of Poland); South Africa (right of South African citizens "to have the environment protected, for the benefits of future and present generations," Article 24 Constitution of South Africa). *See also* J. C. Tremmel, *Establishing Intergenerational Justice in National Constitutions, in* HANDBOOK OF INTERGENERATIONAL JUSTICE 187–214 (J.C. Tremmel ed., 2006).

[41] *See* European Parliament, *The Charter of Fundamental Rights of the EU* (2000/C 364/01, 7 Dec. 2000), para. 6, http://www.ec.europa.eu/justice_home/unit/charter/index_en.html, (last visited Oct. 28, 2010); IUCN Draft International Covenant on Environment and Development, Environmental Policy & Law Paper No. 31 Rev. 2 (2004), at art. 5, http://www.i-c-e-l.org/english/EPLP31EN_rev2.pdf (last visited Oct. 4, 2010); UNEP *Proposal for a Basic Law on Environmental Protection and the Promotion of Sustainable Development*, Document Series on Environmental Law No. 1, UNEP Regional Office for Latin America and the Caribbean, Mexico D.F. (1993); *Goa Guidelines on Intergenerational Equity Adopted by the Advisory Committee to the United Nations University Project on International Law, Common Patrimony and Intergenerational Equity*, Feb. 15, 1988. *See also* United Nations Commission on Sustainable Development, *Report of the Expert Group Meeting on Identification of Principles of International Law for Sustainable Development, Geneva, Switzerland, 26–28 September 1995*, at para. 38.

[42] OECD, *National Strategies for Sustainable Development: Good Practices in OECD Countries*, SG/SD(2005)6 at para. 16, reviewed in UNDSD, *Expert Group Meeting on Reviewing National Sustainable Development Strategies New York, 10–11 October 2005* UNDSD/EGM/NSDS/2005/CRP. 9, http://www.un.org/esa/sustdev/natlinfo/nsds/egm/crp_9.pdf (last visited Oct. 13, 2010). *See also* United Nations Conference on Environment and Development, Rio de Janeiro, Brazil, June 13, 1992, *Report of the United Nations Conference on Environment*, U.N. Doc. A/CONF.151/26 (vol. 1) (Aug. 12, 1992); U.N. Environment Programme, *Final Report of the Expert Group Workshop on International Environmental Law Aiming at Sustainable Development*, UNEP/IEL/WS/3/2 (1996) 13–14, para. 30, 44–45.

[43] PHILIPPE SANDS, PRINCIPLES OF INTERNATIONAL ENVIRONMENTAL LAW 254–55 (2003); Hari M. Osofsky, *Defining Sustainable Development after Earth Summit 2002*, 26 LOYOLA L.A. INT'L & COMP. L. REV. 111, 112 (2003).

[44] *See supra* note 8.

[45] *See supra* note 10 and accompanying text.

stratiform clouds with sulfur aerosols to increase reflectivity,[46] and the deployment of space-based "sunshades" to reduce incoming solar radiation inflows,[47] or the injection of huge amounts of dust particles in the equatorial plane between altitudes of 2000 and 4000 kilometers to reflect and scatter solar radiation.[48]

Proponents of SRM approaches tout their potential for offsetting the projected warming associated with a doubling or more of atmospheric levels of greenhouse gases from preindustrial times.[49] However, the potential effectiveness of SRM schemes also sows the seeds of major peril for future generations. Imagine a scenario in which a single nation,[50] or group of nations, deploys an SRM scheme, and it proves successful in abating temperature increases and other phenomena associated with climate change. Many analysts believe successful deployment of geoengineering technologies would severely undermine development of effective mitigation responses to climate change. As The Royal Society concluded:

> The very discussion of geoengineering is controversial in some quarters because of a concern that it may weaken conventional mitigation efforts, or be seen as a "get out of jail free" card by policy makers ... This is referred to as the "moral hazard" argument, a term derived from insurance, and arises where a newly-insured party is more inclined to take risky behavior than previously because compensation is available. In the context of geoengineering, the risk is that major efforts in geoengineering may lead to a reduction of effort in mitigation and/or adaptation because of a premature conviction that geoengineering has provided "insurance" against climate change.[51]

[46] Stephen Salter, Graham Sortino & John Latham, *Sea-Going Hardware for the Cloud Albedo Method of Reversing Global Warming*, 366 PHIL. TRANSACTIONS ROYAL SOC'Y A 3989, 3989 (2008). Cloud albedo enhancement geoengineering would seek to increase the number of cloud-condensation nuclei in low-level marine clouds. Large numbers of small cloud micro-droplets scatter and reflect more incoming solar radiation than larger droplets of the same total mass. The Royal Society, *supra* note 1, at 27; T.M. Lenton & N.E. Vaughan, *The Radiative Forcing Potential of Different Climate Geoengineering Options*, 9 ATMOSPHERE, CHEMISTRY & PHYSICS 5539, 5543 (2009).

[47] Takanobu Kosugi, *Role of Sunshades in Space as a Climate Control System*, 67 ACTA ASTRONAUTICA 241–53 (2010); Roger Angel, *Feasibility of Cooling the Earth with a Cloud of Small Spacecraft Near the Inner Lagrange Point (L1)*, 103 PROC. NAT'L ACAD. SCI. 17184–189 (Nov. 14, 2006). There are two major options for deployment of sunshades, boosting them into orbit around the earth, or placing them at an optimal point between the sun and earth. Kosugi, *supra*, at 242.

[48] The Royal Society, *supra* note 1, at 32.

[49] Kosugi, *supra* note 47, at 242; Oliver Morton, *Great White Hope*, 458 NATURE 1097, 1098–99 (2009); Michael C. MacCracken, *On the Possible Use of Geoengineering to Moderate Specific Climate Impacts*, 4 ENVTL. RES. LETTERS 1, 4 (2009).

[50] The cost of many geoengineering options might be "well within the budget of almost all nations," as well as a handful of wealthy individuals. Katharine Ricke et al., *Unilateral Geoengineering*, Non-Technical Briefing Notes for a Workshop at the Council on Foreign Relations, May 5, 2008, http://d1027732.mydomainwebhost.com/articles/articles/cfr_geoengineering.pdf (last visited Oct. 25, 2010); Lin, *supra* note 11, at 16.

[51] The Royal Society, *supra* note 1, at 37. *See also* David W. Keith, *Geoengineering the Climate: History and Prospect*, 25 ANN. REV. ENERGY ENV'T 245, 276 (2000).

Beyond empirical evidence of moral hazards in the context of insurance,[52] there is ample cause for concern that deployment of geoengineering technology could seriously undermine society's commitment to reducing greenhouse gas emissions and ultimately decarbonizing the world's economy. This is true for several reasons. First, although accurate cost assessments of geoengineering technologies are difficult at this protean stage, several studies have indicated that some SRM options could cost as little as 1 percent or less of the cost of dramatically reducing emissions,[53] exerting a potentially very powerful pull away from mitigation initiatives. Moreover, because geoengineering options "leave … powerful actors and their interests relatively intact,"[54] they are likely to be backed by influential constituencies going forward. Indeed, there are growing advocacy initiatives for geoengineering by think tanks funded by fossil fuel interests,[55] as well as support by powerful politicians such as former Speaker of the House, Newt Gingrich.[56] Finally, there would likely be substantial public support for geoengineering options because they would not require fundamental changes in lifestyles.[57]

Unfortunately, although a commitment to SRM geoengineering approaches in lieu of effective mitigation responses might prove effective and politically palatable

[52] Dianne Dumanoski, *Resisting the Dangerous Allure of Global Warming Technofixes*, Yale Environment 360, http://e360.yale.edu/feature/the_dangerous_allure_of_global_warming_technofixes/2224/ (last visited Sept. 30, 2010); H. Kunreuther, *Disaster Mitigation and Insurance: Learning from Katrina*, 605 ANNALS AM. ACAD. POL. & SOC. SCI. 208–27 (2006).

[53] Edward Parson & M. Granger Morgan, *Research on Global Sun Block Needed Now*, 463 NATURE 426, 426 (2010); Graeme Wood, *Re-Engineering the Earth*, ATLANTIC, July/Aug. 2009, at 1, http://www.theatlantic.com/magazine/archive/2009/07/re-engineering-the-earth/7552/ (last visited Sept. 30, 2010); David G. Victor, et al., *The Geoengineering Option: A Last Resort against Global Warming?*, 88(2) FOREIGN AFF. 64, 69 (2009). However, it should be emphasized that the costs of monitoring systems would likely substantially increase the cost of deploying such systems. Caldeira & Keith, *supra* note 4, at 60.

[54] Jay Michaelson, *Geoengineering: A Climate Change Manhattan Project*, 17 STAN. ENVTL. L.J. 73, 113 (1998).

[55] Alan Robock, *Geoengineering Shouldn't Distract from Investing in Emissions Reduction*, Bull. Atom. Scientists, May 29, 2008, http://www.thebulletin.org/web-edition/roundtables/has-the-time-come-geoengineering (last visited Oct. 7, 2010); Bjønar Egede-Nissen & Henry David Venema, *Desperate Times, Desperate Measures: Advancing the Geoengineering Debate at the Arctic Council*, International Institute for Sustainable Development, Aug. 2009, at 9, http://www.iisd.org/publications/pub.aspx?id=1162 (last visited Oct. 10, 2010); American Enterprise Institute for Public Policy Research Events, *Geoengineering: A Revolutionary Approach to Climate Change*, June 3, 2008, http://www.aei.org/event/1728>, site visited on Jan. 20, 2011.

[56] Newt Gingrich, *Can Geoengineering Address Concerns about Global Warming?*, http://www.newt.org/newt-direct/stop-green-pig-defeat-boxer-warner-lieberman-green-pork-bill-capping-american-jobs-and-t (last visited Jan. 20, 2011).

[57] J. Eric Bickel & Lee Lane, *An Analysis of Climate Engineering as a Response to Climate Change*, Copenhagen Consensus Center (2009), http://fixtheclimate.com/component-1/the-solutions-new-research/climate-engineering/ (last visited Oct. 10, 2010); John Virgoe, *International Governance of a Possible Geoengineering Intervention to Combat Climate Change*, 95 CLIMATIC CHANGE 103, 105 (2009).

for our generation, future generations may not feel the same way because of the threat posed by the "termination" effect.[58] The termination effect refers to the potential for a huge multi-decadal pulse of warming should the use of an SRM scheme be terminated abruptly in the future due to technological failure or a decision by future policy makers. This would be a consequence of the buildup of carbon dioxide that had accrued in the atmosphere in the interim, with its suppressed warming effect, as well as the temporary suppression of climate-carbon feedbacks.[59]

The ramifications of the termination effect could be "catastrophic."[60] As one study recently concluded:

> [S]hould the engineered system later fail for technical or policy reasons, the downside is dramatic ... The climate suppression has only been temporary, and the now CO_2-loaded atmosphere quickly bites back, leading to severe and rapid climate change with rates up to 20 times the current rate of warming of ≈0.2°C per decade ...[61]

As a consequence, temperatures could increase 6–10 degrees Celsius in the winter in the Arctic region *within 30 years of termination of the use of SRM technology*, with northern landmasses seeing increases of 6 degrees Celsius in the summer.[62] Moreover, temperatures could jump 7 degrees Celsius in the tropics in thirty years.[63] Projected temperature increases after termination would occur more rapidly than during one of the most extreme and abrupt global warming events in history, the Paleocene-Eocene Thermal Maximum.[64] It is beyond contention that climatic changes of this magnitude "could trigger unimaginable ecological effects."[65] To put

58 United Kingdom, House of Commons, Science and Technology Committee, *The Regulation of Geoengineering*, Fifth Report Session 2009–10, Mar. 10, 2010, http://www.publications.parliament.uk/pa/cm200910/cmselect/cmsctech/221/221.pdf (last visited Oct. 11, 2010).

59 H. Damon Matthews & Ken Caldeira, *Transient Climate-Carbon* 104(2) PROC. NAT'L ACAD. SCI. 9951 (2007).

60 B. Govindasamy et al., *Impact of Geoengineering Schemes on the Terrestrial Biosphere*, 29(22) GEOPHYSICAL RES. LETTERS 18–1, 18–3 (2002).

61 Peter G. Brewer, *Evaluating a Technological Fix for Climate*, 104(24) PROC. NAT'L ACAD. SCI. 9915, 9915 (2007). *See also* J.C. Moore, S. Jevrejeva & A. Grinstad, *Efficacy of Geoengineering to Limit 21st Century Sea-Level Rise*, PNAS Early Edition (2010), http://www.pnas.org/cgi/doi/10.1073/pnas.1008153107 (last visited Oct. 10, 2010).

62 Victor Brovkin et al., *Geoengineering Climate by Stratospheric Sulfur Injections: Earth System Vulnerability to Technological Failure*, 92 CLIMATIC CHANGE 243, 254 (2009).

63 Eli Kintisch, *Scientists Say Continued Warming Warrants Closer Look at Drastic Fixes*, 318 SCI. 1054, 1055 (2007).

64 *Id.* Additionally, a sudden surge of warming would also cause atmospheric releases of carbon stored in land and in the ocean, contributing to further warming. G. Bala, *Counteracting Climate Change Via Solar Radiation Management*, 101(11) CURRENT SCI. 1418, 1419 (2010).

65 Kintisch, *supra note* 63, at 1055. *See also* Andrew Ross & H. Damon Matthews, *Climate Engineering and the Risk of Rapid Climate Change*, 4 ENVTL. RES. LETTERS 045103 (Oct.–Dec. 2009), http://iopscience.iop.org/1748–9326/4/4/045103 (last visited Oct. 11, 2010 ("It seems likely that two decades of

this rate of temperature increase in perspective, a recent study concluded that even a warming rate of greater than 0.1 degrees Celsius per decade could threaten most major ecosystems and decrease their ability to adapt.[66] Should temperatures increase at a rate of 0.3 degrees Celsius per decade, only 30 percent of all impacted ecosystems and only 17 percent of all impacted forests would be able to adapt.[67] Moreover, temperature increases of this magnitude and rapidity would imperil many human institutions.[68]

As indicated above, a future generation would face the grave implications of the termination effect if an SRM scheme failed. This would contravene the second obligation of intergenerational equity outlined by Brown Weiss, conservation of quality, because the failure of our generation to substantially reduce its greenhouse gas emissions would result in greatly degraded planetary conditions for future generations under such a scenario.

Alternatively, even if a future generation was not compelled to forego or terminate deployment of an SRM scheme, it might deem it judicious to do so on policy or ethical grounds. For example, as indicated earlier in this article, atmospheric sulfur dioxide injection might result in adverse regional impacts on precipitation, undermining the interests of inhabitants in Asia and Africa.[69] Also, although another SRM scheme, marine cloud seeding, might substantially reduce incoming solar radiation, it could also result in sharp declines in precipitation in South America, including particularly serious impacts on the Amazon rain forest.[70]

Although our generation might deem such "collateral effects" acceptable, a future generation might not, especially if regional impacts were exacerbated by other factors, such as rising populations or declines in food production attributable to other causes, or if affected states threatened war.[71] However, leaders might feel that their

very high rates of warming would be sufficient to severely stress the adaptive capacity of many species and ecosystems, especially if preceded by some period of engineered climate stability").

[66] A. Vliet & R. Leemans, *Rapid Species' Response to Changes in Climate Require Stringent Climate Protection Targets*, AVOIDING DANGEROUS CLIMATE CHANGE 135–41 (2006).

[67] R. Leemans & B. Eickhout, *Another Reason for Concern: Regional and Global Impacts on Ecosystems for Different Levels of Climate Change*, 14 GLOBAL ENVTL. CHANGE-HUMAN POLICY DIMENSIONS 219–28 (2004).

[68] Dumanoski, *supra* note 52; Brewer, *supra* note 61, at 9915. Kintisch, *supra* note 63, at 1055.

[69] \ *Supra* notes 10 and 12, and accompanying text.

[70] G. Bala et al., *Albedo Enhancement of Marine Clouds to Counteract Global Warming: Impacts on the Hydrological Cycle*, 6 CLIMATE DYNAMICS, DOI 10.1007/s00382–010–0868–1 (2010), at 2; Andy Jones, Jim Haywood & Olivier Boucher, *Climate Impacts of Geoengineering Marine Stratocumulus Clouds*, 114 J. GEOPHYSICAL RES. D10106 (2009), at 5.

[71] Kevin Bullis, *The Geoengineering Gambit*, Tech. Rev., Jan./Feb. 2010, http://www.technologyreview.com/energy/24157/ (last visited Oct. 13, 2010); Testimony of Alan Robock, U.S. House Committee on Science and Technology Hearing, *Geoengineering: Assessing the Implications of Large-Scale Climate Intervention*, Nov. 5, 2009, at 11, http://democrats.science.house.gov/Media/file/Commdocs/hearings/2009/Full/5nov/Robock_Testimony.pdf (last visited Oct. 13, 2010); One plausible scenario

hands were tied given the potentially catastrophic global implications of suspending the use of SRM technologies. Indeed, some of the proponents of geoengineering strategies even tout the threat of the rebound effect as a way to ensure "policy continuity" in the future.[72] Placing future generations on the horns of such a dilemma would violate the first obligation of intergenerational equity outlined by Brown Weiss, conservation of options, because it would severely circumscribe the ability of those future generations to choose climate change policies that reflect their values.

It should also be emphasized that SRM technologies would have to be deployed for five hundred to a thousand years unless we can find a way to remove carbon dioxide from the atmosphere.[73] As a consequence, the intergenerational implications of SRM geoengineering would extend for a breathtaking period of time, threatening the interests of tens of billions of future inhabitants of this planet.

2.3 *Could SRM Schemes Be Deployed in a Way That Comports with Principles of Intergenerational Equity?*

Proponents of SRM geoengineering might contend that a geoengineering governance regime could condition deployment of an SRM scheme on a scheduled reduction in greenhouse gas emissions of sufficient magnitude to ensure that future generations would not face the threat of the termination effect. Unfortunately, this approach could prove problematic for several reasons. First, it is by no means clear that any current international regime would have jurisdiction over SRM schemes. For example, the United Nations Framework Convention on Climate Change (UNFCCC),[74] the most logical locus for international regulation of geoengineering, likely could not currently assert jurisdiction over SRM deployment. Although the treaty generally mandates that the parties "protect the climate system,"[75] it is made clear in Article 2 that this objective is to be effectuated by stabilizing the concentration of greenhouse gases at a level that prevents "dangerous anthropogenic interference."[76] Thus, the objective of the regime is to control atmospheric levels of

that would compel termination of an SRM scheme could be a threat of war by nations that might be potentially negatively impacted. David Roberts, *What Could Possibly Go Wrong: Blotting out the Sun*, POPSCI, Feb. 3, 2011, http://www.popsci.com/science/article/2011–01/what-could-possibly-go-wrong-blotting-out-sun (last visited Feb. 27, 2011).

[72] Bickel & Lane, *supra* note 57, at 27.

[73] Dumanoski, *supra* note 52; Brovkin et al., *supra* note 12, at 255. "[A] significant fraction of the effect will need to be maintained for >1,000 years, because approximately 20% of the CO_2 added to the atmosphere is only removed by natural sedimentation and weathering processes on timescales of 10,000 to 1,000,000 years." Naomi E. Vaughan & Timothy M. Lenton, *A Review of Climate Geoengineering Proposals*, CLIMATIC CHANGE, Online First, Mar. 22, 2011.

[74] UNFCCC, *supra* note 35.

[75] *Id.* at art. 3(1).

[76] *Id.* at art. 2(1).

greenhouse gases, whereas SRM approaches focus on reducing the amount of solar radiation incident on the surface of the earth. This conclusion is reinforced by the Commitments provisions of Article 4, which include the following:

(a) Develop, periodically update, publish and make available to the Conference of the Parties, in accordance with Article 12, *national inventories of anthropogenic emissions by sources and removals by sinks of all greenhouse gases not controlled by the Montreal Protocol*, using comparable methodologies to be agreed upon by the Conference of the Parties;
(b) *Formulate, implement, publish and regularly update national and, where appropriate, regional programmes containing measures to mitigate climate change by addressing anthropogenic emissions by sources and removals by sinks of all greenhouse gases* not controlled by the Montreal Protocol, and measures to facilitate adequate adaptation to climate change;
(c) Promote and cooperate in the development, application and diffusion, including transfer, of technologies, practices and processes *that control, reduce or prevent anthropogenic emissions of greenhouse gases not controlled by the Montreal Protocol in all relevant sectors*, including the energy, transport, industry, agriculture, forestry and waste management sectors;
(d) *Promote sustainable management, and promote and cooperate in the conservation and enhancement, as appropriate, of sinks and reservoirs of all greenhouse gases not controlled by the Montreal Protocol*, including biomass, forests and oceans as well as other terrestrial, coastal and marine ecosystems;

2. The developed country Parties and other Parties included in Annex I commit themselves specifically as provided for in the following:
   (a) Each of these Parties *shall adopt national policies and take corresponding measures on the mitigation of climate change, by limiting its anthropogenic emissions of greenhouse gases and protecting and enhancing its greenhouse gas sinks and reservoirs*. [emphasis added]

Again, the scope of obligations herein are restricted to reducing greenhouse emissions and enhancing sinks. The parties to the UNFCCC arguably could assert jurisdiction over CDR schemes under Article 4(1)(d) or 4(2)(a), as their deployment could enhance carbon dioxide sinks.[77] However, SRM schemes would fall outside the ambit of Article 4 because these technologies would neither enhance sinks nor contribute to reduction of greenhouse gas emissions.[78] Although the UNFCCC

[77] Under the UNFCCC a "sink" "means any process, activity or mechanism which removes a greenhouse gas, an aerosol or a precursor of a greenhouse gas from the atmosphere." *Id.* at art. 1(8).

[78] *See also* Virgoe, *supra* note 57, at 110. The Royal Society contends that any geoengineering scheme would be subject to UNFCCC jurisdiction. The Royal Society, *supra* note 1, at 41. In support of

could potentially be amended to assert jurisdiction over SRM deployment, it is difficult to be sanguine about the prospects given the very high bar for passage of amendments to the treaty,[79] as well as the resistance of many of the states that would most likely develop geoengineering systems to accept binding international mandates to address climate change.[80]

Second, even if there was authority under the UNFCCC to condition deployment of SRM technology on a commitment to reduce greenhouse gas emissions, it is far from clear that the political will exists to operationalize such a mandate. As indicated in the introduction to this chapter, the very impetus for geoengineering has been the abject failure of the world's major greenhouse-gas–emitting states to curb their emissions.[81] This is despite the fact that there is nearly universal recognition by states of the serious impacts that climate change will visit upon nations throughout the world.[82] Despite this fact, the latest "International Energy Outlook" assessment by the U.S. Energy Information Administration projects that energy-related carbon dioxide emissions may rise 43 percent by 2035 from 2007 levels.[83] If the world community has not been willing to make a meaningful commitment to reduce emissions in the face of a looming threat of extremely serious climatic impacts, why would it do so merely because the threat of those impacts could be reduced by deployment of geoengineering technologies?[84]

this proposition it cites a provision of the treaty that requires the parties to minimize "adverse effects on the economy, on public health and on the quality of the environment, of projects or measures undertaken by them to mitigate or adapt to climate change." UNFCCC, *supra* note 35, at art. 4(1)(f). However, as indicated before, SRM technologies could not be construed as measures to "mitigate" climate change under the UNFCCC because Article 4 restricts such measures to those that address sources or sinks. Moreover, most commentators and policy makers draw a distinction between geoengineering responses and adaptation responses, American Meteorological Society, *Proposals to Geoengineer Climate Require More Research, Cautious Consideration, and Appropriate Restrictions*, AMS News, July 21, 2009, http://www.ametsoc.org/policy/2009geoengineeringclimate_amsstatement.html (last visited Oct. 14, 2010); William R. Travis, Discussion Paper, *Geo-Engineering the Climate: An Emerging Technology Assessment* (2009), at 1, http://www.colorado.edu/ibs/pubs/eb/es2008–0002.pdf (last visited Oct. 14, 2010), again rendering 4(1)(f) non-applicable.

79 "The Parties shall make every effort to reach agreement on any proposed amendment to the Convention by consensus. If all efforts at consensus have been exhausted, and no agreement reached, the amendment shall as a last resort be adopted by a three-fourths majority vote of the Parties present and voting at the meeting ... " UNFCCC, *supra* note 35, at art. 15(3).

80 For example, China, the United States, and India have neither ratified the Kyoto Protocol nor committed themselves to binding long-term commitments.

81 *Supra* note 3 and accompanying text.

82 *See* UNFCCC, *Copenhagen Accord*, Draft Decision CP.15, FCCC/CP/2009/L.7, Dec. 18, 2009, at para. 1 ("[C]limate change is one of the greatest challenges of our time").

83 U.S. Energy Information Administration, *International Energy Outlook 2010 – Highlights*, http://www.eia.doe.gov/oiaf/ieo/highlights.html (last visited Oct. 14, 2010).

84 *See also* Chuck Greene, Bruce Monger & Mark Huntley, *Geoengineering: The Inescapable Truth of Getting to 350*, Solutions Journal, http://www.thesolutionsjournal.com/node/771 (last visited Oct. 24, 2010):"First, given a rapidly growing global population and the desire of most developing nations to

There are other regimes that might assert jurisdiction over SRM schemes; however, it is hard to be sanguine about their prospects to protect the interests of future generations. The Convention on Prohibition of Military or any Other Hostile Use of Environmental Modification Techniques (ENMOD)[85] prohibits states from engaging "in military or any other hostile use of environmental modification techniques having widespread, long-lasting or severe effects as the means of destruction, damage or injury to any other State Party."[86] The Convention's scope would encompass deployment of SRM technologies, as definition of the term "environmental modification techniques" under the treaty includes any technique "for changing – through the deliberate manipulation of natural processes – the dynamics, composition or structure of the Earth, including its ... atmosphere, or of outer space."[87] However, Article III exempts environmental modification techniques designed for "peaceful purposes" from the Convention's purview. Thus, a strong case could be made that geoengineering schemes would not be proscribed given the purposes for which they would be deployed.[88] Further, ENMOD is a treaty of limited subscription, with only seventy-three parties, and it does not govern attacks by a party state against a nonparty state.[89]

The Convention on Biological Diversity (CBD)[90] could also be germane given the clear link between SRM deployment and potential threats to biodiversity. This could include the potential impacts of SRM schemes on the stratospheric ozone layer[91] as

achieve an improved standard of living, society currently lacks the sense of urgency and political willpower necessary to alter its energy consumption habits in the short amount of time available."

85 May 18, 1977, 31 U.S.T. 333, 1108 U.N.T.S. 152, *reprinted in* 16 I.L.M. 88.

86 *Id.* at art. I(1).

87 *Id.* at art. II.

88 Daniel Bodansky, *May We Engineer the Climate?*, 33 CLIMATIC CHANGE 309, 311 (1996). However, some commentators have argued that to the extent that geoengineering schemes might result in serious negative implications, its use could be construed as "hostile," prohibiting their deployment under ENMOD. Bidisha Banerjee, *ENMOD Squad*, Slate, Sept. 23, 2010, http://www.slate.com/id/2268123/pagenum/all/ (last visited Mar. 27, 2011); Alan Robock, *20 Reasons Why Geoengineering May Be a Bad Idea*, 64(2) BULL. ATOM. SCIENTISTS 14, 17 (2008), http://cmapspublic3.ihmc.us/rid=1226664705437_1636398002_9066/Robock_2008_20%20reasons%20against%20geoengineering.pdf (last visited Mar. 27, 2011); William Daniel Davis, *What Does "Green" Mean?: Anthropogenic Climate Change, Geoengineering, and International Environmental Law*, 43 GA. L. REV. 901, 935 (2009).

89 Albert C. Lin, *Geoengineering Governance*, 8(3) ISSUES IN LEGAL SCHOLARSHIP 1, 20 (2009), http://www.bepress.com/ils/vol8/iss3/art2 (last visited Mar. 27, 2011).

90 Convention on Biological Diversity, June 5, 1992, 1760 U.N.T.S. 79.

91 Ultraviolet radiation can inhibit photosynthesis in phytoplankton. United Nations Environmental Programme, World Conservation Monitoring Centre, *Changing Oceans: Effects on Biodiversity*, http://www.unep-wcmc.org/climate/oceans/biodiv.aspx (last visited on Mar. 27, 2011); Sara Chesiuk, *Ozone Layer 101*, Canadian Wildlife Federation, http://www.cwf-fcf.org/en/what-we-do/wildlife/featured-species/flora/ozone-layer-101.html (last visited Mar. 27, 2011). This, in turn, threatens a large number of species that depend on phytoplankton as their primary food source, including seals, whales, fish, and more than fifty species of birds in the Antarctic. Science Network Western Australia, *Is the*

well as the threat that rising temperatures could pose to many species.[92] Indeed, the CBD has already engaged on geoengineering issues in the context of ocean iron fertilization experiments. Its states parties have passed two resolutions calling on its members to limit such experiments to small-scale scientific research and to conduct them only after undertaking a stringent environmental impact assessment.[93] However, resolutions of the CBD are not legally binding on its parties.[94] Moreover the parties to the CBD have suggested that other regimes might be more appropriate for regulation of geoengineering activities.[95]

2.4 *Intergenerational Equity and CDR Geoengineering*

A strong argument could be made that deployment of CDR geoengineering schemes[96] would not present the same kind of intergenerational threats posed by SRM approaches. As the Science and Technology Committee of the House of Commons in the United Kingdom observed, whereas SRM technologies would treat only the "symptom" (i.e., global warming), CDR schemes would address the "root issue" (i.e., rising levels of carbon dioxide).[97] As a consequence, the specter of

Ozone Hole Threatening Antarctic Wildlife?, http://www.sciencewa.net.au/topics/environment/869-is-the-ozone-hole-threatening-antarctic-wildlife.html (last visited Mar. 27, 2011). Increased ultraviolet radiation associated with ozone depletion may also adversely threaten other species, including amphibians, Christina Lydick, *Evaluating Amphibian Abnormalities on Wildlife Refuges*, BNET, http://findarticles.com/p/articles/mi_qa4444/is_1_25/ai_n52942966/ (last visited Mar. 27, 2011); Andrew Blaustein et al., *Ambient Ultraviolet Radiation Causes Mortality in Salamander Eggs*, 5(3) ECO. APPLICATIONS 740–43 (1995); and coral reef species, D.F. Gleason & G.M. Wellington, *Ultraviolet Radiation and Coral Bleaching*, 365 NATURE 836–38 (1993); M.P. Lesser et al., *Bleaching of Coral Reef Anthozoans: Effects of Irradiance, Ultraviolet Radiation and Temperature on the Activities of Protective Enzymes against Active Oxygen*, 8 CORAL REEFS 225–32 (1990).

[92] Carolyn Kousky et al., *Responding to Threats of Climate Change Mega-Catastrophes*, Policy Research Working Paper, WPS5127 (2009), at 5; Chris D. Thomas et al., *Extinction Risk from Climate Change*, 427 NATURE 145, 145 (2004).

[93] Convention on Biological Diversity, 10th Meeting of the Conference of the Parties, *Biodiversity and Climate Change*, Decision X/33 (2010), at para. 8(w), http://www.cbd.int/doc/decisions/cop-10/cop-10-dec-33-en.pdf (last visited Mar. 27, 2011); Convention on Biological Diversity, 9th Meeting of the Conference of the Parties, *Biodiversity and Climate Change*, Decision IX/16 (2008), http://www.cbd.int/decision/cop/?id=11659 (last site Mar. 27, 2011). For an analysis of the decision at the 10th Conference of the Parties to the CBD, see Masahiro Sugiyama & Taishi Sugiyama, *Interpretation of CBD COP10 Decision on Geoengineering*, SECR Decision Paper 10013 (2010), http://criepi.denken.or.jp/en/serc/research_re/download/10013dp.pdf (last visited Mar. 27, 2011).

[94] Alexander Proelss, *Legal Opinion on the Legality of the LOHAFEX Marine Research Experiment under International Law*, http://criepi.denken.or.jp/en/serc/research_re/download/10013dp.pdf (last visited Mar. 27, 2011).

[95] Sugiyama & Sugiyama, *supra* note 93, at 13.

[96] *See* note 9.

[97] United Kingdom, House of Commons, Science and Technology Committee, *supra* note 58, at 14.

abrupt climatic changes associated with a massive carbon dioxide pulse would not exist should use of such technologies be terminated by a future generation.

Conversely, deployment of CDR technologies could still pose a moral hazard problem by reducing the current generation's commitment to decarbonizing the economy. This would pass the responsibility on to future generations to address this issue, while again compelling them to continue to deploy CDR technologies in the interim. At the very least, this would deny them the full panoply of options that the principle of intergenerational equity demands. However, given the far less serious implications of deploying technologies that do not pose the threat of a termination effect, the ethical questions associated with deploying CDR technologies, at least from an intergenerational perspective, would be far less pressing.

3. IS GEOENGINEERING ACTUALLY A MEANS TO ACHIEVE INTERGENERATIONAL EQUITY?

Some proponents of geoengineering have suggested to me that research and/or deployment of geoengineering schemes would actually comport with principles of intergenerational equity. The contention is that geoengineering schemes could shield future generations from the potentially very serious impacts of climate change that may ensue in this century and beyond, and thus fulfill this generation's obligations to our successors. However, I believe that this presents a false dichotomy of potential policy options between failing to take meaningful measures to reduce our greenhouse gas emissions, or using geoengineering as a bandage to cover the wound that such a failure to act would inflict on our successors on this planet. As Attfield observes, intergenerational equity is only effectuated if future generations "receive an intact and renewable environmental and cultural heritage ... not preempted by the squandering of resources or the bequeathing of injustices by the generations now alive."[98] This necessarily dictates that our generation maximize its efforts to mitigate greenhouse gas emissions so that our children and grandchildren are not compelled to live with a Sword of Damocles over their head.

The convenient truth is that climate change can be addressed effectively, and in a manner that will not undermine the welfare of the current generation, through an aggressive program of mitigation. One study concluded that effectuating reductions in emissions by 75–90 percent by 2100 would cost about 3–6 percent of cumulative GNP during this century, certainly by no means an insubstantial sum.[99] However, to

[98] Attfield, *supra* note 25, at 210.

[99] Christian Azar & Stephen H. Schneider, *Are the Economic Costs of Stabilizing the Atmosphere Prohibitive?*, 42(1–2) ECO. ECON. 73, 76 (2002).

put this commitment in perspective, even a 6 percent reduction in GNP would result in the world community being ten times richer in 2102 instead of in 2100. In accord, McKinsey & Associates and the Vattenfall Institute have identified twenty-seven gigatons of annual potential carbon dioxide equivalent abatement consistent with stabilizing atmospheric concentrations at 450–550 ppmv, at a cost of less than $40 per ton.[100] The mechanisms to achieve these goals include a massive commitment to renewable energy sources, enhanced energy efficiency, and deployment of carbon capture and storage technologies.[101]

Moreover, although many supporters of geoengineering cite the threat of passing critical climatic thresholds in the shorter term,[102] there are also alternatives that can help us buy time as we make a transition to a decarbonized world economy. For example, a study by the United Nations Environment Program and the World Meteorological Organization concluded that implementation of a full set of measures to reduce black carbon[103] and ozone emissions by 2030 could reduce the potential increase in global temperature projected for 2050 by 50 percent.[104] This would translate into a reduction of temperatures by 0.5 degrees Celsius.[105] Moreover, it would yield substantial co-benefits, including the avoidance of more than 2 million

[100] Eric Beinhocker et al., *The Carbon Productivity Challenge: Curbing Climate Change and Sustaining Economic Growth*, McKinsey Global Institute (June 2008), at 8, http://www.mckinsey.com/mgi/publications/Carbon_Productivity/index.asp (last s visited Apr. 1, 2011).

[101] Shruti Mittal, *Tapping the Untapped: Renewing the Nation: Focus on Renewable Sources Especially Solar Energy*, CUTS Centre for International Trade, Economics & Environment Discussion Paper (2010), at 1–33; LESTER B. BROWN, PLAN B 4.0: MOBILIZING TO SAVE CIVILIZATION 109–42 (2009); David Hodas, *Imagining the Unimaginable: Reducing U.S. Greenhouse Gas Emissions by Forty Percent*, 26 VA. ENVTL. L.J. 271–90 (2008); BBC News, *Green Energy "Revolution" Needed*, June 6, 2008, http://news.bbc.co.uk/2/hi/business/7439338.stm (last visited Apr. 1, 2011).

[102] *See supra* notes 4 & 5 and accompanying text.

[103] Black carbon is a constituent element of the combustion product known as soot. Indoor sources are primarily due to cooking with biofuels, including dung, wood and crop residue. The primary outdoor source is attributable to fossil fuel combustion (diesel and coal), open biomass burning, and cooking with biofuels. V. Ramanathan & G. Carmichael, *Global and Regional Climate Changes due to Black Carbon*, 1 NATURE GEOSCI. 221, 221 (2008). Recent studies indicate that black carbon emissions are the second-largest contributor to global warming, as much as 55 percent of the forcing associated with carbon dioxide. *Reducing Black Carbon, or Soot, Is the Fastest Strategy for Slowing Climate Change*, IGSD Briefing Note, Mar. 27, 2008, http://www.igsd.org/docs/BC%20Briefing%20Note%2027Mar08.pdf (last visited Apr. 2, 2011); Jonathan Lash, *Black Carbon an Easy Target for Climate Change, Innovations*, Carnegie Council (Feb. 9, 2009), http://www.policyinnovations.org/ideas/innovations/data/000084 (last visited Apr. 2, 2011).

[104] United Nations Environment Program, Twenty-Sixth Session of the Governing Council/Global Ministerial Environment Forum, *Summary for Decision Makers of the Integrated Assessment of Black Carbon and Tropospheric Ozone*, UNEP/GC.26/INF/20 (Feb. 2011), at 3, http://www.unep.org/gc/gc26/download.asp?ID=2197 (last visited Apr. 1, 2011). *See also* Almut Arneth et al., *Clean the Air, Heat the Planet?*, 326 SCI. 672, 672 (2009).

[105] United Nations Environment Program, *supra* note 104, at 3.

premature deaths and the annual loss of 1–4 percent of global production of maize, rice, soybeans, and wheat.[106]

4. CONCLUSION

As Frischmann concludes, "the present generation has mastered the art of pushing the costs of shortsighted decisions onto future generations."[107] Deployment of SRM geoengineering technologies in the future could constitute the quintessential act of generational selfishness, compelling untold future generations to "stick with the program" or face catastrophic impacts. The potential intergenerational consequences of climate geoengineering counsel strongly in favor of doubling our resolve to address an issue for which this generation is profoundly responsible.

[106] *Id.*

[107] Frischmann, *supra* note 22, at 459.

10

Ocean Iron Fertilization

Science, Law, and Uncertainty

Randall S. Abate

1. INTRODUCTION

The international community's efforts to reduce greenhouse gas emissions through traditional regulatory strategies under the Kyoto Protocol[1] have reached a critical turning point. The Kyoto Protocol expired in 2012. Much-anticipated international climate change negotiations were conducted in December 2009 in Copenhagen,[2] in December 2010 in Cancun,[3] and in December 2011 in Durban,[4] to determine what type of international agreement would succeed the Kyoto Protocol. In the wake of these Conferences of the Parties (COPs) to the United Nations Framework

[1] Kyoto Protocol to the United Nations Framework Convention on Climate Change, *opened for signature* Dec. 11, 1997, 2303 U.N.T.S. 162 (entered into force Feb. 16, 2005) [hereinafter Kyoto Protocol]. The term "traditional regulatory strategies" refers to the mandatory emissions reductions commitments that developed countries had to meet within a prescribed time frame. To help ensure that parties meet these commitments, the Kyoto Protocol allows parties to rely on the "flexibility mechanisms" available under the agreement, such as emissions trading and the Clean Development Mechanism. However, climate geoengineering techniques are not available to parties as a method of compliance with their commitments under the Kyoto Protocol.

[2] The negotiations in Copenhagen yielded only the Copenhagen Accord, a nonbinding agreement that was widely regarded as a disappointing outcome. U.N. Framework Convention on Climate Change Conference of the Parties, Copenhagen, Denmark, Dec. 7–19, 2009, *Copenhagen Accord*, U.N. Doc. FCCC/CP/2009/L.7 (Dec. 18, 2009), *available at* http://unfccc.int/resource/docs/2009/cop15/eng/11a01.pdf#page=4.pdf.

[3] The outcome of the negotiations in Cancun offered only slightly more hope in producing an international agreement on reducing emissions of deforestation and degradation (REDD). *See* David Biello, *Cancun Talks Yield Climate Compromise*, SCI. AM., Dec. 11, 2010, http://www.scientificamerican.com/article.cfm?id=cancun-talks-yield-climate.

[4] The Durban negotiations succeeded in renewing the Kyoto Protocol and establishing a climate adaptation fund for developing countries; however, the outcome did little more than merely keep the international climate negotiation process viable for potentially more meaningful gains at future sessions. *See* John M. Broder, *Climate Talks in Durban Yield Limited Agreement*, N.Y. TIMES, Dec. 11, 2011, http://www.nytimes.com/2011/12/12/science/earth/countries-at-un-conference-agree-to-draft-new-emissions-treaty.html.

Convention on Climate Change, there appears to be little hope for a binding and effective successor agreement to Kyoto that would include the full participation of the major industrial nations and major emerging economies necessary to ensure its success.[5]

This uncertain future for traditional international climate change regulation has prompted a corresponding interest in deliberate interventions in the earth's climate system to address global climate change; these are commonly known as climate geoengineering techniques. Climate geoengineering has come a long way within the past few years from being considered "fringe science"[6] to finding a place in the mainstream of international climate change in the Cancun negotiations.[7]

One of these geoengineering strategies, ocean iron fertilization (OIF) has been the subject of considerable attention recently in domestic and international climate change regulation contexts. The purpose of OIF is to sequester carbon from the atmosphere through deliberate alteration of ocean iron composition. Covering approximately 70 percent of the earth's surface, the oceans may offer potentially effective opportunities for carbon dioxide removal as part of the international community's response to climate change.

These opportunities notwithstanding, there are several obstacles in seeking to enhance the oceans' carbon absorption capacity in the fight against global climate change. OIF is hampered by opposition grounded in the "moral hazard"[8] of

[5] Although the United States, China, and India agreed in Durban to a binding greenhouse gas emissions reduction treaty to succeed the Kyoto Protocol, experts deemed the agreement to be inadequate because of its insufficiently ambitious proposed reductions and because it would not take effect until 2020. *See* Louise Gray, *Durban Climate Change Conference: Big Three of US, China, and India Agree to Cut Carbon Emissions*, TELEGRAPH, Dec. 11, 2011, http://www.telegraph.co.uk/earth/environment/climatechange/8949317/Durban-climate-change-conference-Big-three-of-US-China-and-India-agree-to-cut-carbon-emissions.html; Damian Carrington, *Climate Deal: A Guarantee Our Children Will Be Worse Off than Us*, GUARDIAN, Dec. 11, 2011, http://www.guardian.co.uk/environment/damian-carrington-blog/2011/dec/11/durban-climate-change-conference-2011-climate-change

[6] *See* David G. Victor, *On the Regulation of Geoengineering*, 24 OXFORD REV. ECON. POL'Y 322, 328 (2008) (noting that climate geoengineering was once viewed as "a freak show in otherwise serious discussions of climate science and policy.")

[7] *See* Charles J. Hanley, *Geoengineering Talks Surface as UN Climate Change Talks in Cancun Falter*, HUFFINGTON POST, Dec. 4, 2010, http://www.huffingtonpost.com/2010/12/06/geoengineering-debate-sur_n_792409.html.

[8] "Moral hazard" refers to the notion that if people are insured against some risk, they will be more likely to behave in ways that increase the likelihood of the negative outcome occurring because they have no incentive to mitigate the risk. Originally coined in the insurance context, this term also has been applied to climate geoengineering measures, including OIF, in that some fear that these techniques would be perceived to be a substitute for and retreat from pursuing aggressive targets and timetables for greenhouse gas emission reductions for all nations. *See* Randall S. Abate, *A Tale of Two Carbon Sinks: Can Forest Carbon Management Serve as a Framework to Implement Ocean Iron Fertilization as a Climate Change Treaty Compliance Mechanism?*, 1 SEATTLE J. ENVTL. L. 1, 9 (2011).

employing it and the potential adverse environmental consequences associated with this climate change mitigation strategy. Other concerns involve whether OIF can work as effectively as its proponents suggest and whether carbon sequestered by this technique can be reliably monitored and verified. OIF also faces several daunting challenges regarding how it should be regulated within the international and domestic law frameworks.

2. THE SCIENCE OF OIF

This section of the chapter describes the scientific processes underlying OIF and the obstacles that this strategy faces in securing legitimacy as an accepted form of climate change mitigation. Like many other controversial international environmental regulatory challenges, including climate change regulation itself, one of the most difficult hurdles OIF faces is acceptance of the science underlying its mechanisms as legitimate and reliable.

2.1 *Basic Mechanisms of OIF*

The underlying premise of OIF is elegant in its simplicity. A pioneer of this method, the late John Martin, notoriously proclaimed, "Give me half a tanker of iron, and I'll give you an ice age."[9] Although this statement overstates the simplicity of the process, OIF involves the release of iron dust particles into ocean areas where iron exists in low concentrations such that its absence limits phytoplankton growth.[10] Oceans absorb carbon dioxide through the photosynthetic activity of planktonic algae.[11] Consequently, OIF is designed to "stimulate the rapid growth of phytoplankton whose photosynthetic activity could potentially absorb heat trapping carbon."[12] The ultimate objective of OIF is to absorb CO_2 and store it in the ocean interior for an adequate duration and in a sufficient quantity so as "to make a significant reduction in the increase of atmospheric CO_2 in a verifiable manner, without deleterious unintended side effects."[13]

[9] Hugh Powell, *Fertilizing the Ocean with Iron: Is This a Viable Way to Help Reduce Carbon Dioxide Levels in the Atmosphere?*, 46 OCEANUS 4 (2008), *available at* http://www.whoi.edu/cms/files/OceanusIron_Fertilizing_30749.pdf.

[10] Randall S. Abate & Andrew B. Greenlee, *Sowing Seeds Uncertain: Ocean Iron Fertilization, Climate Change, and the International Environmental Law Framework*. 27 PACE ENVTL. L. REV. 555, 561 (2010).

[11] *Bacteria and Climate Change, Invisible Carbon Pumps*, ECONOMIST, Sept. 9, 2010, at 1, *available at* http://www.economist.com/node/16990766?story_id=16990766.

[12] Abate & Greenlee, *supra* note 10, at 560–61.

[13] U.N. Educ. Scientific & Cultural Org. (UNESCO), Intergovernmental Oceanographic Commission, *Ocean Fertilization: A Scientific Summary for Policy Maker*, Part 2: *Why Fertilize the Ocean?*, Draft Version for Member States Comments, 3 (2010) [hereinafter UNESCO].

The photosynthetic activity of planktonic algae is the driving mechanism of OIF. "Plankton take up carbon in surface waters during photosynthesis, creating a bloom that others feed upon."[14] Carbon from the plankton is integrated into the waste products from these organisms and settles to the sea floor as "sea snow" in a process called "the biological pump."[15] Iron added to the ocean surface increases phytoplankton, so in theory fertilizing the ocean with iron would mean that more carbon would be removed from the surface water to the deep ocean.[16] Once in the deep ocean, the carbon would be "sequestered" or isolated in deep waters for centuries.[17] Enhancing the ocean sink could in theory help control atmospheric carbon dioxide levels and ultimately regulate climate.[18]

OIF experiments have supported the viability of OIF as a carbon mitigation technique.[19] Twelve open ocean experiments that began in the early 1990s have shown that adding iron to iron-limited regions of the ocean triggers large blooms of phytoplankton and that some carbon is sequestered.[20] These experiments attempted to duplicate the effect of large dust storms that annually deposit tens of millions of tons of iron in the ocean.[21] Minimal amounts of iron were applied in the experiments to stimulate large phytoplankton blooms.[22]

2.2 *Limitations of OIF Science*

Despite OIF's vast potential for effective climate change mitigation, it has been plagued by controversy and skepticism. Criticisms have focused on three primary concerns: (1) OIF may not work as effectively as its proponents suggest; (2) OIF

[14] News Release, Woods Hole Oceanogrpahic Institute, Effects of Ocean Iron Fertilization with Iron to Remove Carbon Dioxide from Atmosphere Reported (Apr. 16, 2004), *available at* http://www.whoi.edu/page.do?cid=886&ct=162&pid=9779&tid=282.

[15] *Id.* Martin was also the first to suggest that OIF could be used as a technique to sequester large amounts of carbon from the atmosphere by "stimulating the biological pump with iron." Dr. Margaret Leinen et al., *Why Ocean Iron Fertilization?* 9 (2009), http://www.climos.com/pubs/2009/Climos_Why_OIF-2009-03-12.pdf.

[16] Dawicki, *supra* note 13.

[17] *Id.*

[18] *Id.*

[19] Leinen et al., *supra* note 15, at 10.

[20] Philip W. Boyd et al., *Mesoscale Iron Enrichment Experiments 1993–2005: Synthesis and Future Directions*, 315 SCI. J. 612, 612–17 (2007).

[21] T.D. Jickells et al., *Global Iron Connections between Desert Dust, Ocean Biogeochemistry, and Climate*, 308 SCI. J. 67, 67–71 (2005).

[22] Kenneth H. Coale et al., *A Massive Phytoplankton Bloom Induced by an Ecosystem-Scale Iron Fertilization Experiment in the Equatorial Pacific Ocean*, 383 NATURE 495, 495–501 (1996). For a detailed description of OIF proponents' claims regarding the potential impact of full-scale OIF deployment and its impact on atmospheric concentrations of carbon dioxide, see Leinen et al., *supra* note 15, at 13–14.

could cause a wide range of adverse environmental consequences; and (3) the effectiveness of OIF is difficult to monitor and verify.[23]

2.2.1 Effectiveness of OIF's Carbon Sequestration

The effectiveness of OIF's carbon sequestration can be compromised in two principal ways. First, the location of OIF experiments is critically important.[24] Second, the proportion of carbon that sinks from the surface into deeper waters, where it is less easily returned to the atmosphere, needs to be maximized. This process is called "export efficiency."[25]

OIF's algal blooms are location-sensitive in two respects. First, the appropriate ocean region is vitally important, and second, the particular area of water within that region that has been selected to receive the iron slurry is also highly relevant.[26] Ocean waters known as high-nutrient, low-chlorophyll (HNLC) regions have been the focus of OIF experiments.[27] Although iron-deficient, these areas have high levels of other nutrients that plankton need to grow, including nitrate, phosphate, and silicic acid.[28]

HNLC waters occur in the northern and equatorial Pacific and in the Southern Ocean.[29] The Southern Ocean is regarded as the best location for OIF experiments.[30] It has a much larger area with much higher nutrient levels, which should increase the total size of blooms that could be stimulated. It has such an abundance of nutrients that they actually sink before they can be utilized – unless more iron is supplied."[31]

In addition to the location of OIF experiments, carbon export efficiency also is a concern. If the carbon drawn down from the atmosphere does not make it to the deep ocean, then it has not been effectively or "permanently" sequestered.[32] Carbon

[23] Abate & Greenlee, *supra* note 10, at 561.

[24] *Id.* at 563.

[25] Hugh Powell, Part 2: *Will Ocean Iron Fertilization Work? Getting Carbon into the Ocean Is One Thing. Keeping It There Is Another*, OCEANUS (2008), http://www.whoi.edu/cms/files/OceanusIron_Will_It_Work_30747.pdf

[26] *Id.*

[27] Abate & Greenlee, *supra* note 10, at 563–64.

[28] *Id.*

[29] Powell, *supra* note 25.

[30] *Id.* ("Logistically, equatorial waters would be easiest to work in: It's warm and sunny all year, the seas tend to be fairly calm, and the warm waters encourage rapid growth. But there is already enough plankton growth in equatorial waters to eventually use up their nutrient supply anyway; adding iron there just creates a faster, concentrated bloom in a specific location, but the net effect on atmospheric carbon dioxide levels is arguably negligible.").

[31] *Id.*

[32] Jennie Dean, *Iron Fertilization: A Scientific Review with International Policy Recommendations*, 32 ENVIRONS ENVTL. L. & POL'Y J. 321, 329 (2009).

export efficiency is compromised when carbon is re-released into the atmosphere after it has initially been sequestered beneath the ocean's surface. Once phytoplankton have initially sequestered the carbon, it will be effectively "transported" if it sinks. However, if the phytoplankton are merely consumed by zooplankton, then a significant percentage of the carbon will be re-released through the metabolic processes of the zooplankton, and only a small amount of the carbon will be transported to the deep ocean through fecal pellets.[33]

Environmental factors, such as vertical mixing and the currents found in the HNLC regions, also limit the drawdown of the carbon to the deep ocean.[34] Given that the HNLC regions, especially the Southern Ocean, are so variable in their physical and biological characteristics, the effects of OIF on carbon sequestration will be difficult to predict.[35] For example, one model projected that only 2–44 percent of the initial carbon sequestered using OIF techniques would be removed from the atmosphere during a one-hundred-year period.[36]

OIF experiments have failed to meet their projected outcomes. The first of these failures is the ratio of iron incorporated versus the amount added to the ocean. OIF experiments involve the application of significant quantities of iron dust particles to ocean waters. Unless conditions are ideal, a large percentage of iron is lost due to clumping and sinking before phytoplankton can process it.[37] Second, "projections from bottle experiments overestimated the drawdown potential of iron fertilization."[38] The initial large projections of what iron fertilization was capable of were based upon the observed rates of bottled experiments. In general, the bottle experiments consisted of adding particulate iron to 1-liter containers of seawater collected from the HNLC regions and observing the phytoplankton growth and change in CO_2 concentration. Additionally, with a bottle, there is no escaping through sinking, so the iron can be more fully utilized than it would be in the real ocean.[39]

These shortfalls are minor compared to a larger concern regarding the effectiveness of OIF experiments. Even if OIF worked perfectly, it would nonetheless fall

[33] *Id.*

[34] *Id.* at 332.

[35] *Id.* at 329 ("Difficulties with drawdown to the deep ocean are only expected to worsen in the coming years as global warming progresses. As the ocean's temperature rises, its ability to absorb carbon dioxide through the solubility pump will be reduced because gases are less soluble in warmer waters. Furthermore, it has been suggested that the warmer waters will result in a shutdown of many of the planet's currents, both across ocean surfaces and between different ocean layers. The shutdown of currents will have obvious negative consequences for nutrient recycling, biological distributions, and water temperatures.").

[36] *Id.* Model simulations for large-scale OIF typically last for ten- to one hundred-year periods. UNESCO, *supra* note 13, at 8.

[37] Dean, *supra* 32, at 331.

[38] *Id.* at 332.

[39] *Id.* at 331.

short of making a significant impact in sequestering CO_2 because it would require continual fertilization of the HNLC regions to remove a sufficient quantity of CO_2 to have an impact in combating climate change.[40] Even when employing the highest estimates for both carbon export ratios and atmospheric uptake efficiencies, there is still very limited potential for OIF to remove CO_2 from the atmosphere in sufficient quantities to mitigate climate change.[41]

2.2.2 Adverse Environmental Consequences

The potential adverse environmental consequences that may flow from OIF projects have been a significant focus of critics' attacks on this climate change mitigation method. Human intervention in any ecological system can prompt both foreseeable and unforeseeable consequences, the results of which are complex and difficult to monitor. The OIF context is no different. Therefore, although OIF could become a potent tool in combating climate change, critics have expressed concerns about the potentially devastating ecological and geophysical impacts of OIF.[42]

OIF poses the threat of several possible adverse environmental consequences that have received considerable attention. The most prominent of these concerns has been the potential OIF poses to produce toxic algae blooms.[43] Phytoplankton growth from OIF can produce domoic acid, a potent neurotoxin.[44] This neurotoxin can move up through the food chain as other animals eat the phytoplankton, harming birds, fish, marine mammals, and, ultimately, humans who consume the toxin-contaminated seafood.[45]

Concerns regarding the potential connection between OIF activities and harmful algal blooms may be overblown. Harmful algal blooms develop predominantly in coastal areas.[46] Such blooms rarely occur in the open ocean.[47] Unlike algal blooms in coastal areas, which release domoic acid as they proliferate in those nutrient-rich

[40] *Id.* at 332 (for example, one study indicates that fertilization of the entire Southern Ocean for the next fifty years would reduce atmospheric carbon by only 6 ppm, which is 1/14 of what experts believe we need to reduce to stabilize atmospheric concentrations).

[41] UNESCO, *supra* note 13, at 8.

[42] Abate & Greenlee, *supra* note 10, at 566.

[43] *Id.* at 566–67. *See generally* Charles G. Trick et al., *Iron Enrichment Stimulates Toxic Diatom Production in High-Nitrate, Low Chlorophyll Areas*, 107 PROC. NATL. ACAD. SCI., Mar. 15, 2010, *available at* http://www.pnas.org/content/early/2010/02/24/0910579107.full.pdf.

[44] Jessica Marshall, *Ocean Geoengineering Scheme May Prove Lethal: Seeding the Oceans with Iron Could Result in the Production of Potential Neurotoxin, Putting the Lives of Birds, Fish and even Humans at Risk*, DISCOVERY NEWS, Mar. 15, 2010, http://news.discovery.com/earth/geoengineering-carbon-sequestration-phytoplankton.html.

[45] *Id.*

[46] Leinen et al., *supra* note 15, at 19–20.

[47] *Id.*

environments, "open-ocean algal blooms have been determined to be nontoxic."[48] Some studies have concluded that the high urea concentrations derived from coastal pollution are necessary to stimulate domoic acid production.[49]

The potential harm to marine wildlife from algae bloom neurotoxins also has been the subject of considerable concern in the environmental community. These neurotoxins "can ascend the food chain and contaminate food webs on which marine life feed, which can lead to illness and mortality of thousands of marine mammals and birds along the coast of North America."[50] As a result, "[h]uman mortality also may ensue from consuming seafood that contains the toxin."[51]

Regardless of how much of a concern harmful algae blooms may present, there are several other potential adverse consequences from OIF. For example, OIF may "influence food web dynamics because phytoplankton is at the bottom of the food chain."[52] In addition, anoxia in the subsurface ocean could occur due to large-scale ocean iron fertilization.[53] Other adverse consequences of OIF could include nutrient depletion and lower primary productivity downstream of the fertilization site,[54] enhanced production of the nitrous oxide and methane,[55] and increased ocean acidification.[56] These documented environmental concerns are more than enough by themselves to render OIF a potentially undesirable policy option, yet OIF also faces several other challenging technological and legal hurdles.

[48] Sid Perkins, *Iron Fertilization in Ocean Nourishes Toxic Algae. Carbon Sequestration Efforts Could Trigger Harmful Algal Blooms*, SCI. NEWS, Mar. 15, 2010, http://www.sciencenews.org/view/generic/id/57318/title/Iron_fertilization_in_ocean_nourishes_toxic_algae,%20p.2.%20.

[49] Leinen et al., *supra* note 15, at 19–20.

[50] Abate, *supra* note 8, at 10–11.

[51] *Id.* at 11.

[52] Christine Bertram, Kiel Policy Brief No. 3, *Ocean Iron Fertilization: An Option for Mitigating Climate Change?* 3 (2009), http://www.ifw-kiel.de/wirtschaftspolitik/politikberatung/kiel-policy-brief/kiel_policy_brief_3.pdf [hereinafter Bertram].

[53] *Id.* (such anoxic conditions can cause significant die-offs of marine life, including fish, shellfish, and invertebrates, like one that occurred in 2006 in the productive coastal region off Oregon. Years of large-scale fertilization of the ocean could produce similar "low-oxygen events").

[54] Dean, *supra* note 32, at 329–30 ("[T]he goal of the iron addition is to maximize the utilization of nutrients in the HNLC regions. This is beneficial to the life in those HNLC regions but has the secondary effect of producing a nutrient deficit in the waters by the time they reach other areas of the world, such as the tropics. Lack of nutrients in these areas will result in significant changes to the ecosystems found there.").

[55] Nitrous oxide and methane are two potent greenhouse gases that tend to form when organic matter decomposes at depth. A 1999 study in the Southern Ocean found that "between 6 and 12 percent of the cooling effect from the iron addition was annulled by increased emissions of nitrous oxide." Hugh Powell, Part 3: *What Are the Possible Side Effects?, The Uncertainties and Unintended Consequences of Manipulating Ecosystems* (2008), *available at* http://www.whoi.edu/oceanus/viewArticle.do?id=35668.

[56] Bertram, *supra* note 52, at 3; Ocean acidification, which is caused by an increased concentration of carbon dioxide in the oceans, "has slowed growth rates in calcium carbonate based organisms such as coral reefs and crustaceans." Dean, *supra* note 32, at 331.

2.2.3 Monitoring Challenges

There are several factors that complicate monitoring the assessment of effects from OIF experiments. Adding the iron requires a twelve-hour zigzagging cruise across a theoretical square of water whose boundaries shift constantly in the ocean currents. In the weeks of monitoring that follow, a ship typically spends twelve hours every day mapping out the boundaries of the bloom.[57] Algae blooms are difficult to monitor because the added iron rapidly dilutes, sinks, and reacts with seawater, causing it to be virtually undetectable after a few days.[58] The eventual size of blooms from small iron additions can span 1,000 square kilometers or more, extend to depths of up to 100 meters, and drift hundreds of kilometers from their starting positions.[59]

As difficult as it can be to monitor the results of OIF experiments, verification of the carbon sequestered is even more vexing. If the objective of OIF is to claim "credit"[60] for enhanced sequestration of carbon, then verification must include measurement-based estimates of the amount of carbon sequestered in OIF experiments, which entails:

- monitoring of changes in the downward carbon export in both the fertilized areas and adjacent areas that were not fertilized, but were otherwise similar; and
- "long-term (months to years) and far-field monitoring to determine if there are subsequent rebound effects that might offset some of the initial change or might have negative impacts."[61]

Monitoring and verification of the short-term and long-term effects of OIF experiments are not only challenging to conduct accurately, they are also expensive because of the cutting-edge technology required. "Effective monitoring of the short-term, near-field intended effects of large scale fertilization will itself be costly. In the opinion of several scientists who have been involved in past iron fertilization experiments, adequate verification cannot yet be achieved with currently available observing capabilities."[62]

Proponents of OIF have advocated for OIF to be incorporated into the carbon credit systems established under the Kyoto Protocol. They maintain that satellites can accurately monitor and verify phytoplankton blooms from space; however, these

[57] Powell, *supra* note 25.
[58] *Id.*
[59] Powell, *supra* note 25.
[60] A carbon "credit" is a value that has been assigned to a reduction or offset of greenhouse gas emissions, which can be sold or used by a given entity. In some instances, environmental groups have purchased and retired these "credits" to achieve a net reduction in greenhouse gas emissions in a given area.
[61] UNESCO, *supra* note 13, at 3.
[62] *Id.*

satellites merely assess or interpret the amount of chlorophyll present in the very top layer of the ocean.[63] This technology is limited by its inability to (1) "identify which types of plankton are present in each bloom or if the composition changes with depth,"[64] and (2) "detect the amount of carbon that is re-released back into the atmosphere through phytoplankton respiration."[65]

3. THE INTERNATIONAL AND DOMESTIC LEGAL FRAMEWORK FOR OIF

OIF projects present daunting legal challenges because multiple existing sources of international and domestic law potentially apply to these activities. OIF needs to be addressed at the global level because its negative effects have the potential to be felt worldwide. Furthermore, the most promising locations for iron fertilization are found in the high seas and thus can only be regulated through an international forum.[66]

Three treaty regimes apply most directly to OIF: the United Nations Convention on the Law of the Sea (UNCLOS),[67] the London Convention and Protocol,[68] and the Convention on Biological Diversity.[69] The fragmented applicability of these treaty regimes creates two major problems. First, to the extent OIF is subject to regulation under these regimes, there are potentially conflicting mandates that must be reconciled. Second, much of OIF remains unregulated, and there needs to be either a new treaty regime or a harmonized version of existing treaty regimes to regulate it properly.

3.1 *UNCLOS*

UNCLOS regulates activities on, over, and under the world's oceans and defines countries' jurisdiction over, and rights of access to, the oceans and their resources.[70] Under Article 194, parties must take measures that are necessary to prevent, reduce,

[63] Dean, *supra* note 32, at 328.
[64] *Id.*
[65] *Id.*
[66] Abate & Greenlee, *supra* note 10, at 572.
[67] U.N. Convention on the Law of the Sea, Dec. 10, 1982, 1833 U.N.T.S. 397 [hereinafter UNCLOS].
[68] Convention on the Prevention of Marine Pollution by Dumping of Wastes and Other Matter, Dec. 29, 1972, 26 U.S.T. 2403, 1046 U.N.T.S. 138; 1996 Protocol to the Convention on the Prevention of Marine Pollution by Dumping of Wastes and Other Matter, 1972, Nov. 7, 1996, S. TREATY DOC. NO. 110–5, 36 I.L.M. 1 [hereinafter London Convention and Protocol].
[69] Convention on Biological Diversity, June 5, 1992, 1760 U.N.T.S. 79 [hereinafter CBD].
[70] Kelsi Bracmort, et al., Cong. Research Serv., R41371, *Geoengineering: Governance and Technology* 32 (2010), *available at* http://www.fas.org/sgp/crs/misc/R41371.pdf.

and control *pollution* of the marine environment from any source.[71] Article 1(1)(4) of UNCLOS defines pollution as "the introduction by man directly or indirectly, of substances or energy into the marine environment ... which results or is likely to result in such deleterious effects as harm to living resources and marine life, hazards to human health, hindrances to marine activities, including fishing and other legitimate uses of the sea, impairment of quality for use of sea water and reduction of amenities.[72]

OIF projects could trigger the UNCLOS framework for pollution regulation[73] if such projects have a negative effect on the marine ecosystem. However, OIF proponents could argue that the definition of "pollution" in UNCLOS would not cover their activities because iron enrichment of the oceans occurs naturally and the dire results predicted by some are largely based on modeling that operates under the assumption of worst case scenarios.[74] Moreover, OIF proponents also may argue that OIF could improve the marine ecosystem because the phytoplankton blooms stimulate the base of the food chain.[75]

Alternatively, OIF could be deemed to have a negative impact on the marine ecosystem by fitting within the definition of "dumping" under UNCLOS. "Dumping" is defined in Article 1(5)(a) as "any deliberate disposal of wastes or other matter from vessels, aircraft, platforms or other man-made structures at sea."[76] The addition of iron to the ocean in fertilization projects is indeed "deliberate"; however, it is unclear whether added iron is "waste." Article 1(5)(b) excludes from the definition of dumping the "placement of matter for a purpose other than the mere disposal thereof, provided that such placement is not contrary to the aims of this Convention."[77] OIF activities fit within the first clause of the exclusion. However, OIF opponents argue that because of negative impacts on the marine environment and the obligation established in Article 192, OIF activities fail to satisfy the second clause of the exclusion. However, OIF proponents argue that OIF activities would fail the second clause only if OIF activities are determined to be harmful, which is still unclear based on the current state of the science. Accordingly, UNCLOS could prohibit OIF activities, but it does not do so explicitly under its definitions of pollution and dumping.[78] In addition, more conclusive evidence from the scientific community that OIF activities are indeed harmful to the marine environment

[71] UNCLOS, *supra* note 67, art. 194 (emphasis added).
[72] *Id.* art. 1(1)(4).
[73] *E.g.*, arts. 56 and 238–241.
[74] Bracmort, *supra* note 70, at 32.
[75] Abate & Greenlee, *supra* note 10, at 573–74.
[76] UNCLOS, *supra* note 67, art. 1(5)(a).
[77] *Id.* art. 1(5)(b).
[78] Dean, *supra* note 32, at 334–35.

will be necessary before OIF activities may be deemed "contrary to the aims" of UNCLOS.

3.2 *London Convention and Protocol*

The relationship between OIF and its regulation under international marine dumping guidelines can be analyzed under the London Convention and Protocol.[79] The 1972 London Convention promotes the effective control of all sources of marine pollution and governs the deliberate disposal at sea of wastes or other matter.[80] A separate agreement, the 1996 London Protocol, updated the London Convention and eventually will replace it.[81]

The dumping of all wastes is prohibited except those listed in Annex 1 of the Convention.[82] This list contains two categories of waste that could be interpreted to incorporate the iron dust used in fertilization projects. Annex 1(1.5) allows for the at-sea disposal of "inert, inorganic geologic material," a category into which the particulate iron used in fertilization experiments fits.[83] In addition, Annex 1(1.7) expressly permits the disposal of iron, albeit in the form of a "bulky" structure.[84] Although iron dust is not "bulky," its chemical composition is comparable to that of "bulky" forms.[85]

The 1996 London Protocol includes a general obligation that parties shall "apply a precautionary approach to environmental protection from dumping of wastes or other matter whereby appropriate preventative measures are taken when there is reason to believe that wastes or other matter introduced into the marine environment are likely to cause harm even when there is no conclusive evidence to prove a causal relation between inputs and their effects."[86] Under each of these formulations, the regulatory response that is "appropriate" necessarily depends on the level of knowledge, the scale of the potential risks, and the benefits to be derived from the activity. In the case of OIF, there is no basis for applying these factors in a way that would lead to a prohibition. Instead, cautious regulation should be undertaken

79 *Id.* at 335.
80 Hugh Powell, Part 5: *Dumping Iron and Trading Carbon: Profits, Pollution, and Politics All Will Play Roles in Ocean Iron Fertilization*, OCEANUS (2008), *available at* http://www.whoi.edu/oceanus/viewArticle.do?id=35826§ionid=1000.
81 *Id.*
82 London Convention and Protocol, *supra* note 68, art. 3(1) ("dumping" is defined in Article 3(1) with the same language and exclusions as found in UNCLOS Article 1(5)).
83 Dean, *supra* note 32, at 336.
84 *Id.*
85 *Id.*
86 London Protocol, art. 1.

to ensure that OIF activities are conducted responsibly while additional data is collected to inform future regulatory decisions.[87]

The parties to the London Convention and Protocol have recently considered amendments and resolutions to these agreements to address some geoengineering technologies more explicitly. First, the 2006 amendments to the London Protocol provide guidance on the means by which sub-seabed geological sequestration of carbon dioxide can be conducted, stating that carbon dioxide streams may only be considered for dumping if "(1) disposal is into a sub-seabed geological formation; (2) the substances dumped consist overwhelmingly of carbon dioxide; and (3) no wastes or matter are added for the purpose of disposing of those wastes or other matter."[88]

Second, in a nonbinding resolution negotiated in October 2008 under the umbrella of the International Maritime Organization (IMO), the parties concluded that OIF projects should not be carried out except for careful scientific research, which in turn should be subject to member state permission.[89] Two years later, the IMO developed an assessment framework for conducting scientific research on OIF experiments.[90]

The status of OIF science remains unsettled. On March 5, 2010, the IMO held a meeting that addressed the progress made on OIF science.[91] The IMO mandated that a Correspondence Group be assembled to review the final text of the CBD report, "Scientific Synthesis on the Impacts of Ocean Fertilization on Marine Biodiversity,"[92] to assess its adequacy and provide recommendations to address the gaps in this report.[93] Members of the Correspondence Group failed to reach consensus on whether this report adequately summarizes the current state of scientific knowledge on OIF.[94]

87 Leinen, *supra* note 15, at 22–23.

88 *See* World Resources Institute, *London Protocol Adopts Amendment Allowing for Sub-Seabed Carbon Dioxide Storage*, *available at* http://pdf.wri.org/css_06_12_08_london_protocol.pdf.

89 *See* ANNEX SIX RESOLUTION LC-LP.1 ON THE REGULATION OF OCEAN FERTILIZATION (2008), http://www.imo.org/includes/blastDateOnly.asp/data_id%3D24337/LC-LP1(30).pdf.

90 *Assessment Framework for Scientific Research Involving Ocean Fertilization Agreed*, International Maritime Organization, Oct. 20, 2010, *available at* http://www.imo.org/mediacentre/pressbriefings/pages/assessment-framework-for-scientific-research-involving-ocean-fertilization-agreed.aspx.

91 International Maritime Organization, Scientific Group of the London Convention, Report of the Ocean Fertilization Correspondence Group 1 (Mar. 5, 2010), *available at* http://www.imo.org/includes/blastDataOnly.asp/data_id%3D27823/2–1.pdf [hereinafter IMO Report].

92 SECRETARIAT OF THE CONVENTION ON BIOLOGICAL DIVERSITY, SCIENTIFIC SYNTHESIS OF THE IMPACTS OF OCEAN IRON FERTILIZATION ON MARINE BIODIVERSITY (2009), *available at* http://www.cbd.int/doc/publications/cbd-ts-45-en.pdf.

93 IMO Report, *supra* note 91, at 1.

94 *Id.* at 2. As of this writing, the Correspondence Group is working on a document that will provide a comprehensive report on the current state of knowledge on OIF, which is scheduled for release in April 2011; *see* IMO, Scientific Group of the London Convention, *Ocean Fertilization: Development*

3.3 *Convention on Biological Diversity*

Apart from being regulated as marine pollution or dumping, OIF activities also may be regulated under agreements concerning biological conservation and integrity. The Convention on Biological Diversity (CBD) implements an ecosystem-based approach to biodiversity. This holistic regulatory scope is relevant in considering the potential cascade effects of OIF when determining whether an OIF project is appropriate.[95] Moreover, like UNCLOS and the London Convention and Protocol, the CBD endorses the precautionary principle.[96] Accordingly, the lack of long-term studies and observations on the effects of OIF does not necessarily prohibit regulation of OIF under the CBD.[97]

The CBD provides limited authorization to regulate OIF. Article 7(c) requires each party to "identify processes and categories of activities which have or are likely to have significant adverse impacts on the conservation and sustainable use of biological diversity," but requires only that the activities be identified and then "monitored."[98] As such, it does not effectively ban the use of detrimental processes and thus would still allow OIF activities.[99] Moreover, the CBD's regulatory scope is generally limited to waters in the territorial seas.[100] As the high seas are the regions in which OIF implementation is most likely, the CBD would not be the most effective framework to regulate OIF.[101]

In 2008, the Ninth Meeting of the Conference of the Parties to the CBD noted the work of parties to the London Convention and Protocol regarding OIF. The COP requested that its own parties "act to ensure that OIF activities do not take place until either there is adequate scientific basis on which to justify such

of Science Overviews on Ocean Fertilization, 2, IMO Doc. LC/SG/ES.2/1 (July 30, 2010), *available at* http://www.imo.org/includes/blastDataOnly.asp/data_id%3D29393/2–1.pdf.

95 OIF regulation also could be addressed from a biological conservation perspective using the Antarctic Treaty System. The Antarctic Treaty of 1959 and its 1991 Madrid Protocol seek to protect "the Antarctic environment and dependent and associated ecosystems and the intrinsic value of Antarctica" through methods similar to those used in the CBD. However, given that OIF experiments are not limited to the Southern Ocean, these instruments will not provide a comprehensive foundation for the international law framework governing OIF. Dean, *supra* note 32, at 337.

96 *Id.*

97 *Id.*

98 CBD, art. 7(c).

99 Dean, *supra* note 32, at 337; *see* CBD, art. 14 (article 14 of the CBD elaborates on these terms by requiring notification and measures to minimize damages, in addition to an environmental impact assessment requirement. These mandates would apply to the effects of OIF activities beyond a state's coastal waters).

100 Dean, *Id.*

101 *Id.*

activities or the activities are small-scale scientific research studies within coastal waters."[102]

In 2011, the Tenth Meeting of the Conference of the Parties to the CBD adopted Decision X/33, which addresses climate geoengineering and its relationship to the fulfillment of CBD objectives.[103] The decision requests "available scientific information, and views and experiences of indigenous and local communities and other stakeholders, on the possible impacts of geo-engineering techniques on biodiversity."[104] In preparation for the Sixteenth Meeting of the Subsidiary Body on Scientific, Technical, and Technological Advice in May 2012, the Liaison Group will prepare proposals on definitions of climate geoengineering and evaluate known, expected, and potential impacts of climate geoengineering on biodiversity.[105]

3.4 *U.S. Law*

In addition to the international law regulatory framework, certain U.S. laws potentially govern OIF activities. As of this writing, no federal law has been enacted expressly to govern OIF activities. Although this section addresses federal initiatives in the United States to address OIF and geoengineering regulation, these regulatory challenges and initiatives are not limited to the United States.

Two U.S. environmental laws, the Marine Protection, Research and Sanctuaries Act (MPRSA)[106] and the National Environmental Policy Act (NEPA),[107] may apply to domestic OIF practices and their impacts, depending on the type, location, and sponsor of the activity.[108] The American Power Act,[109] pending before the Senate

[102] Abate & Greenlee, *supra* note 10, at 576–77. In 2009, the CBD Secretariat released a report assessing the science underlying OIF. Secretariat of the Convention on Biological Diversity, *Scientific Synthesis of the Impacts of Ocean Fertilization on Marine Diversity* (2009), http://www.cbd.int/doc/publications/cbd-ts-45-en.pdf; The IMO has evaluated this report in making its determinations on the status of OIF science. *See supra* notes 86–88 and accompanying text.

[103] Secretariat of the Convention on Biological Diversity *Liaison Group Meeting on Climate-Related Geo-Engineering as It Relates to the Convention on Biological Diversity* (2011), http://www.cbd.int/doc/meetings/cc/lgcrg-eng-01/official/lgcrg-eng-01–01-en.pdf

[104] *Id.*

[105] *Id.*

[106] Bracmort, *supra* note 70, at 24.

[107] U.S. GOV'T ACCOUNTABILITY OFFICE, GAO-10–546T, CLIMATE CHANGE: PRELIMINARY OBSERVATIONS ON GEOENGINEERING SCIENCE, FEDERAL EFFORTS, AND GOVERNANCE ISSUES: TESTIMONY BEFORE THE COMMITTEE ON SCIENCE AND TECHNOLOGY 12 (2010) [hereinafter GAO Report] (statement of Frank Rusco, Director, Natural Resources and Environment), *available at* http://www.gao.gov/new.items/d10546t.pdf.

[108] Bracmort, *supra* note 70, at 24.

[109] *See* American Power Act (discussion draft), S. 1733, 111th Cong. (2010), *available at* http://kerry.senate.gov/imo/media/doc/APAbill3.pdf.

as of this writing, would address climate change regulation in the United States. If enacted, it would be the first U.S. law to expressly govern OIF activities.

The MPRSA is the implementing legislation in the United States for the London Convention. The Act prohibits persons from dumping material, including material for ocean fertilization, into the ocean without a permit from EPA.[110] The MPRSA exempts the placement of certain materials in the ocean "for a purpose other than disposal."[111] Title I of the MPRSA prohibits unpermitted ocean dumping by any U.S. vessel or other vessel sailing from a U.S. port in ocean waters under U.S. jurisdiction.[112] EPA may issue permits if it determines that the dumping will not unreasonably degrade or endanger human health, welfare, the marine environment, ecological systems, or economic potentialities.[113]

The MPRSA's jurisdictional reach is limited, however, to "disposition of materials for fertilization by vessels or aircraft registered in the United States, vessels or aircraft departing from the United States, federal agencies, or disposition of materials for fertilization conducted in U.S. territorial waters, which extend 12 miles from the shoreline or coastal baseline."[114] As a result of this loophole, "a domestic company could conduct ocean fertilization outside of EPA's regulatory jurisdiction and control if, for example, the company's fertilization activities took place outside U.S. territorial waters from a foreign-registered ship that embarked from a foreign port."[115]

The most direct response to OIF regulation in the United States is contained in proposed climate change legislation, the American Power Act (APA). Introduced by Senators John Kerry and Joseph Lieberman, the APA would severely restrict OIF activities. Language in the APA essentially bans iron and urea fertilization, the dumping of iron ferrites or urea to stimulate blooms of carbon-capturing plankton "as a means to mitigate climate change."[116]

The APA would impose limits on greenhouse gas emissions consistent with the administration's climate change goals. Based on the EPA's analysis of this draft legislation, the APA would significantly limit U.S. greenhouse gas emissions that contribute to increasing global temperature and other climate changes. As a part of

[110] GAO Report, *supra* note 107, at 13.

[111] Powell, *supra* note 80.

[112] Bracmort, *supra* note 70, at 24.

[113] *Id.* at 28–29.

[114] The National Environmental Policy Act (NEPA) also could apply to OIF activities. NEPA requires federal agencies to evaluate the likely environmental effects of certain major federal actions by using an environmental assessment or, if the projects likely would significantly affect the environment, a more detailed environmental impact statement. An OIF project could constitute a major federal action requiring a NEPA analysis. *See* GAO Report, *supra* note 107, at 11–12.

[115] *Id.* at 13.

[116] *Iron Fertilization Dead in the Water? Controversial Geoengineering Proposal Banned in US Climate Change Legislation*, UNDERWATER TIMES, May 26, 2010, http://www.underwatertimes.com/news.php?article_id=20716358491.

international efforts to limit global warming, the APA would substantially reduce the risk of dangerous global temperature changes by encouraging energy efficiency and the use of low-carbon technologies, such as nuclear power and renewable energy.[117]

Participants in this new market would include electric utilities, petrochemical refiners, and manufacturing and heavy industry.[118] When regulations would be scheduled to take effect in 2013, many of the pollution allowances would be given away for free, and the rest would be auctioned. The APA would regulate the price at which these pollution credits can be traded between industries, with a floor of $12 and a ceiling of $25 per ton of carbon. To help stabilize the price of pollution permits within that price range, the APA authorizes EPA to buy and sell allowances.[119] However, climate change regulation is just one dimension of the overlapping and potentially conflicting sources of international and domestic law that need to be reconciled to regulate OIF effectively.

4. RECOMMENDATIONS FOR OIF GOVERNANCE

The flurry of proposed regulatory responses to OIF at the international and domestic levels within the past few years underscores the need to establish an effective international law framework to regulate OIF. Moreover, existing international agreements such as UNCLOS, the London Convention and Protocol, the CBD, and the Kyoto Protocol need to be coordinated with respect to OIF regulation.

4.1 *Establish an Independent Treaty to Address Geoengineering*

A new treaty that addresses geoengineering specifically, or new technologies more generally, could be developed as a governance option for OIF. This type of treaty could take two forms. First, it could strictly regulate how research on geoengineering strategies, including OIF, may proceed. Second, such a treaty also could be patterned after the UN Environmental Modification Convention (ENMOD), which prohibits governments from using weather or climate as a weapon against other states.[120]

In evaluating independent treaty governance options for OIF, it is important as a threshold matter to make a distinction between the different types of OIF projects:

[117] U.S. Environmental Protection Agency, EPA's Analysis of the American Power Act (APA) (2010), http://www.epa.gov/climatechange/economics/pdfs/EPA_APA_Analysis_6-14-10.pdf.

[118] Sarah Terry-Cobo, *The American Power Act: Cap and Trade 2.0*, FRONTLINE/WORLD, June 14, 2010, http://www.pbs.org/frontlineworld/stories/carbonwatch/2010/06/the-american-power-act-cap-and-trade-20.html.

[119] *Id.*

[120] Bracmort, *supra* note 70, at 34.

(1) moderate-scale research projects to study the effects of OIF,[121] and (2) large-scale projects that would be necessary as a carbon mitigation strategy.[122] The concerns that have been raised about possible environmental harm from OIF typically are associated with potential harm from large-scale deployment of OIF, not from moderate-scale research experiments.[123] Moderate-scale research experiments can help scientists and policy makers understand the potential of large-scale OIF projects and design these projects in a manner that could safely remove large amounts of atmospheric CO_2. Improved research models will also be crucial to determine the level of large-scale OIF that "balances the carbon sequestration benefit with acceptable environmental risk."[124]

Regulation of these moderate-scale experiments could be conducted in a few ways. One option would be to establish a governance system for geoengineering research.[125] In a related vein, a new governance structure could be established in a separate treaty to evaluate new technologies under which OIF and other new and emerging technologies could be assessed, monitored, and regulated.[126] Such a new treaty could be focused exclusively on assessing, monitoring, and regulating climate geoengineering strategies.[127] Geoengineering research also could proceed under the direction of international research consortia.[128] The Human Genome Project and the European Organization for Nuclear Research are examples of this approach. This research consortia approach would apply precaution in considering emerging technologies while enabling expeditious research of such technologies through a new or revised treaty regime.[129]

4.2 *Harmonize Existing Treaties*

Harmonizing relevant treaties could promote one of two outcomes: (1) cautious regulation of OIF activities, or (2) a moratorium on OIF activities on a commercial scale. To promote cautious regulation of OIF, an inter-treaty body could be established to harmonize UNCLOS, CBD, and the London Convention and Protocol.[130]

[121] These projects range from 100 × 100 km to 200 × 200 km. Leinen, *supra* note 15, at 15.
[122] *Id.*
[123] *Id.*
[124] *Id.*
[125] Abate, *supra* note 8, at 14 (internal citations omitted).
[126] *Id.* at 14 (internal citations omitted) (citing DIANA BRONSON ET AL., SWEDISH SOCIETY FOR NATURE CONSERVATION, RETOOLING THE PLANET? CLIMATE CHAOS IN THE GEOENGINEERING AGE 40–41 (2009), *available at* http://www.etcgroup.org/upload/publication/pdf_file/Retooling%20the%20Planet.final_.pdf.
[127] Abate, *supra* note 8, at 14.
[128] *Id.*
[129] *Id.* at 14–15.
[130] Abate & Greenlee, *supra* note 10, at 589–91.

The IMO could serve as the implementing body for such inter-treaty coordination because it already oversees several other treaties, including the London Convention and Protocol, that regulate activities in the oceans.[131] Precedent for such inter-treaty coordination exists in international environmental law. For example, to address a potentially damaging regulatory gap in the climate change regulation context, "the Kyoto and Montreal Protocol regimes are working together to address the regulation of hydrofluorocarbons (HFCs), an ozone-depleting substance and a potent greenhouse gas."[132] Without this coordination, the use of HFCs to address the ozone depletion problem could have severely undermined efforts to regulate climate change.

Alternatively, or in addition to the inter-treaty body, the London Convention and the Kyoto Protocol could be amended to institute a limited moratorium on the use of OIF on a commercial scale. Under this approach, scientific exploration of the effects of OIF could still be explored on a smaller scale; however, such experiments would be limited to compiling information about the dynamics of OIF and its impact on the environment rather than allowing exploration of its effects on mitigating climate change.[133] To achieve this objective, Annex 1 of the London Convention could be amended to exclude particulate iron from the accepted list.[134] Moreover, the Kyoto Protocol and related guidance documents regulating the operation of the carbon markets and the CDM also could be modified to exclude OIF credits from the second commitment period.[135] This would eliminate any financial motivation for the continued use of OIF, and could be accomplished either by retaining the language of Article 3(3), which allows sink credits only from land-based sources, or by including a clause specifically excluding ocean sink sources.[136] Similar restrictions should be incorporated into carbon market guidance documents.[137] Petitions for OIF projects under the CDM also should be prohibited.[138]

Any effort to harmonize existing treaties governing OIF would have implications for domestic law coordination. The federal government could expand the regulatory scope of existing laws to specifically address OIF activities or develop new laws.[139]

[131] *Id.* at 590.

[132] Abate, *supra* note 8, at 14 (citing ENVIRONMENTAL INVESTIGATION AGENCY, A CLIMATE BRIEFING: THE MONTREAL PROTOCOL MUST WORK IN COLLABORATION WITH THE CLIMATE TALKS TO REGULATE HFCS TO PREVENT EXACERBATION OF GLOBAL CLIMATE CHANGE WHILE RESTORING THE OZONE LAYER 2 (2008), *available at* http://www.eia-global.org/PDF/report – Climate – Jan09.pdf).

[133] Dean, *supra* note 32, at 339.

[134] *Id.* at 340.

[135] *Id.*

[136] *Id.*

[137] *Id.*

[138] *Id.*

[139] Bracmort, *supra* note 70, at 24.

In addition, administrative agencies could interpret their statutory authority to authorize new rules explicitly addressing OIF activities.[140]

As of this writing, there is no coordinated federal strategy for geoengineering, including guidance on how to define federal geoengineering activities or efforts to identify and track federal funding related to geoengineering.[141] Officials from federal offices coordinating federal responses to climate change reported that they do not currently have a coordinated geoengineering strategy or position.[142] "In its recent study, Advancing the Science of Climate Change, NRC acknowledged the lack of consensus regarding what constitutes geoengineering in relation to widely accepted practices that remove CO_2 from the atmosphere."[143] Moreover, even though certain federal agencies share information about OIF as part of a working group for international regulation of the ocean, there is no working group to share information or coordinate geoengineering research more broadly, because such an action would require a decision from the administration to pursue geoengineering research on a larger scale.[144] Although this discussion has focused on the status of federal regulation of OIF and climate geoengineering in the United States, other countries face similar challenges in developing a regulatory framework for these activities.

Carbon sequestered in commercial OIF activities would generate credits for use in carbon markets. As of this writing, such credits would have to be sold on voluntary markets, a segment of the carbon market that is rapidly growing in pace with the public's rising concern over climate change.[145] In voluntary markets, concerned individuals and companies buy carbon credits despite having no obligation to reduce emissions. Because projects in voluntary markets are not subject to the strict regulations that govern projects under the Kyoto Protocol, concerns have been raised about improper or inaccurate accounting in these markets. However, voluntary markets are also much smaller than regulated markets, which limits the opportunity for OIF to realize significant commercial success.[146]

Voluntary markets are also strongly influenced by the perceptions of the public.[147] Consequently, OIF projects could suffer from a poor image in voluntary markets "because of the publicity concerning the potential adverse effects from such projects."[148] Nevertheless, as long as the legal status of OIF is not clearly defined and CO_2 emissions keep growing with increasing momentum, company interests

[140] *Id.*
[141] GAO Report, *supra* note 107, at 23.
[142] *Id.*
[143] *Id.* at 23–24.
[144] *Id.* at 26.
[145] Powell, *supra* note 80.
[146] *Id.*
[147] *Id.*
[148] *Id.*

in selling carbon offsets generated through OIF projects are likely to remain strong. Strict regulations would be necessary if these offsets were to be integrated into a post-Kyoto agreement or regulated carbon markets in the future.[149]

Two recent developments regarding carbon credits from reducing emissions from deforestation and degradation (REDD) projects offer some hope for a potential future for OIF to enter mandatory carbon markets. First, an international agreement on REDD was reached at COP 16 in Cancun, Mexico in 2010.[150] Second, also in December 2010, California approved a mandatory carbon credits scheme that authorizes use of carbon credits generated from avoided deforestation projects in Brazil and Mexico.[151]

5. CONCLUSION

OIF is a potentially effective, albeit controversial, climate change mitigation strategy. The regulation of OIF has reached a critical juncture in many respects. More experiments will need to be conducted to determine the effectiveness of OIF as a carbon sequestration strategy, and to diminish concerns about possible adverse environmental consequences. These experiments will need to be regulated through an international law framework, either in the form of an independent treaty to regulate climate geoengineering, or through a coordinated framework of existing international law regulation under UNCLOS, London, CBD, and Kyoto. Moreover, the lack of clarity regarding what international climate change regulation framework will be in place in the post-2012 era, if any, has promoted active consideration of the role OIF may play in a carbon credits system within or apart from that regime.

ACKNOWLEDGMENTS

The author gratefully acknowledges the assistance of Carla Nadal and Farheen Jahangir in preparing this chapter.

[149] Bertram, *supra* note 52, at 5.

[150] *See* Biello, *supra* note 3. Similarly, the use of carbon capture and storage credits was approved for projects under the Kyoto Protocol's Clean Development Mechanism as part of the "Cancun Agreements" adopted at COP 16. KPMG, *Post-Cancun – What it Means*, Jan. 13, 2011, http://www.kpmg.com/global/en/issuesandinsights/articlespublications/kpmg-cop16-cancun/pages/post-cancun-what-it-means.aspx

[151] *See California Approves Cap-and-Trade under Global Warming Law*, MONGABAY.COM, Dec. 17, 2010, http://news.mongabay.com/2010/1217-ca_ab32_cap-and-trade.html.

11

Ocean Iron Fertilization

Time to Lift the Research Taboo

Kerstin Güssow, Andreas Oschlies, Alexander Proelss, Katrin Rehdanz, and Wilfried Rickels

1. INTRODUCTION

Today, most countries have accepted a 2 degrees Celsius temperature increase above preindustrial levels as the maximum tolerable limit for global warming. An exceedance probability of below 20 percent for this limit implies an emission budget of less than 250 GtC from 2000 until 2049, however, extrapolating from current global CO_2 emissions, this budget will only last until 2024.[1] This sobering math should wake us up to the reality that all options, including climate engineering, need to be considered to address climate change.[2] Climate engineering options can be classified broadly into two categories: solar radiation management and carbon dioxide removal measures. Solar radiation management schemes seek to decrease the incoming solar radiation or to increase the reflection of incoming solar radiation. These approaches can generate fast climate responses, but do not immediately address the cause of the problem. Carbon dioxide removal measures seek to decrease atmospheric carbon concentrations by enhancing or substituting natural carbon sinks. The terrestrial carbon sink can be enhanced by means of forestation; the oceanic sink may, in some regions, be enhanced by means of fertilization, for example by artificially enhanced upwelling of macronutrients or by purposeful addition of the micronutrient iron; the mineral carbon sink can be enhanced by means of chemically accelerated weathering. Some analysts have expressed doubts about the potential of mitigating climate change by sink enhancement, because of concerns about whether carbon can be stored permanently.[3] Nevertheless,

[1] Malte Meinshausen et al., *Greenhouse-Gas Emission Targets for Limiting Global Warming to 2°C*, 458 NATURE 1158, 1158 (2009).

[2] Ken O. Buesseler et al., *Ocean Iron Fertilization Moving Forward in a Sea of Uncertainty*, 319 SCI. 162, 162 (2008).

[3] Miko Kirschbaum, *Temporary Carbon Sequestration Cannot Prevent Climate Change*, 11 MITIGATION & ADAPTION STRATEGIES FOR GLOBAL CHANGE 1151, 1158 (2006); Malte Meinshausen & Bill Haare,

terrestrial vegetation sinks have entered the Kyoto Protocol[4] (KP) as offsets for anthropogenic greenhouse gas emissions, but ocean sinks have not.

However, the uncertainty about undesired adverse effects of purposeful iron fertilization on marine ecosystems and biogeochemistry has led to attempts to ban commercial and, to some extent, scientific experiments aimed at a better understanding of the processes involved, potentially precluding further consideration of this option. Some authors argue that research, and in particular large-scale experiments on Ocean Iron Fertilization (OIF), should not be further pursued.[5] We challenge this view and argue that further research about the geoengineering potential of OIF is, indeed, necessary. The reality is that even if emissions would be cut significantly, uncertainty remains as to whether current levels of atmospheric carbon concentration are already sufficiently high to lead to irreversible climate change.[6] Carbon dioxide removal techniques directly decrease atmospheric carbon concentration and thus, in principle, could facilitate the removal of past emissions. Given the large uncertainties in current climate sensitivities, we cannot rule out situations in which large-scale sink enhancement projects may be justifiable and may be applied to prevent dangerous atmospheric peak concentrations. OIF could be a viable approach in this context, and thus needs to be explored in a timely manner.[7] Therefore, it is important to conduct a comprehensive analysis of the potential of OIF that looks at scientific, economic, and legal issues.

In this chapter, we will: (1) initially review the potential of OIF from an oceanographic perspective, (2) summarize findings of our analyses that investigate the economic potential of OIF in the context of an international climate agreement,[8] and (3) examine what public international law says today on the issue of OIF and what it should say in the future.

Temporary Sinks Do Not Cause Permanent Climatic Benefits. Achieving Short-Term-Emission Reduction Targets at the Future's Expense, GREENPEACE BACKGROUND PAPER, Sept. 2000, at 6.

4 Kyoto Protocol to the United Nations Framework Convention on Climate Change, Dec. 11, 1997, 2303 U.N.T.S. 148.

5 Aaron Strong et al., *Ocean Fertilization: Time to Move On*, 461 NATURE 347, 348 (2009).

6 Susan Solomon et al., *Irreversible Climate Change due to Carbon Dioxide Emissions*, 106 PNAS 1704, 1704 (2009).

7 Caroline Kousky et al., *Responding to Threats of Climate Change Mega-Catastrophes*, DISCUSSION PAPER, Nov. 2009, at 24.

8 Wilfried Rickels, Katrin Rehdanz & Andreas Oschlies, *Economic Prospects of Ocean Iron Fertilization in an International Carbon Market*, WORKING PAPER OF THE KIEL INSTITUTE FOR THE WORLD ECONOMY (2009); Andreas Oschlies et al., *Side Effects and Accounting Aspects of Hypothetical Large-Scale Southern Ocean Iron Fertilization*, BIOGEOSCIENCES DISCUSSIONS 2949 (2010); Wilfried Rickels, Katrin Rehdanz & Andreas Oschlies, *Methods for Greenhouse Gas Offsets Accounting: A Case Study of Ocean Iron Fertilization*, 69 ECO. ECON. 2495 (2010).

2. OCEAN IRON FERTILIZATION: THE OCEANOGRAPHIC PERSPECTIVE

Beginning with the experimental work of Martin and Fitzwater,[9] iron has been recognized for more than two decades as an important micronutrient regulating marine productivity and associated biogeochemistry over large ocean areas. This insight led Gribbin[10] to suggest that adding iron compounds to the ocean might be a practical way of removing carbon dioxide from the atmosphere. Meanwhile, a number of in-situ OIF experiments have confirmed that phytoplankton growth is limited by iron in the three major High Nutrient Low Chlorophyll (HNLC) regions, that is, the Southern Ocean,[11] the eastern equatorial Pacific,[12] and the subarctic North Pacific.[13] All experiments have revealed a significant increase in phytoplankton biomass and an associated decrease in the partial pressure of CO_2 (pCO_2) in the surface water, with enhanced particle export having been observed at the end of one experiment.[14] However, the experiments conducted so far did not focus on the potential prospects for carbon sequestration, but instead were aimed at a more genuine scientific understanding of the role of iron in marine ecology and biogeochemistry. Such an understanding is required for a number of reasons, including to better assess impacts of past and likely future changes in iron supply by dust or icebergs. Time and space scales of the experiments carried out so far have precluded a clear assessment of the export and fate of the extra carbon taken up as a result of fertilization.

Clear observational evidence for an iron-induced enhancement of carbon export has been obtained from programs targeting natural OIF at the Kerguelen Plateau and Crozet Islands in the Southern Ocean. At both sites, seasonal export fluxes were found to be more than three times higher than in adjacent non-fertilized regions.[15] Both estimates differ by an order of magnitude, however, in the inferred ratio of carbon export to iron supply. The reason for this difference is not yet understood and requires further study.[16]

9 John H. Martin & Steve E. Fitzwater, *Iron Deficiency Limits Phytoplankton Growth in the North-East Pacific Subarctic*, 331 NATURE 341, 341 (1988).

10 John Gribbin, *Any Old Iron?*, 331 NATURE 570, 570 (1988).

11 Philip W. Boyd et al., *A Mesoscale Phytoplankton Bloom in the Polar Southern Ocean Stimulated by Iron Fertilization*, 407 NATURE 695, 695 (2000).

12 John H. Martin *et al.*, *Testing the Iron Hypothesis in Ecosystems of the Equatorial Pacific Ocean*, 371 NATURE 123, 123 (1994).

13 Atsushi Tsuda et al., *A Mesoscale Iron Enrichment in the Western Subarctic Pacific Includes a Large Centric Diatom Bloom*, 300 SCI. 958, 958 (2003).

14 James K.B. Bishop et al., *Robotic Observations of Enhanced Carbon Biomass and Export at 55°S during SOFeX*, 304 SCI. 417, 419 (2004).

15 Stéphane Blain et al., *Effect of Natural Iron Fertilization on Carbon Sequestration in the Southern Ocean*, 446 NATURE 1070, 1072 (2007); Raymond T. Pollard et al., *Southern Ocean Deep-Water Carbon Export Enhanced by Natural Iron Fertilization*, 457 NATURE 577, 577 (2009).

16 Pollard et al., *supra* note 15, at 579.

To what extent the enhanced export of particulate carbon leads to a net drawdown of atmospheric CO_2 depends on the fertilization region. Model studies suggest that the carbon sequestration potential of OIF is essentially limited to the Southern Ocean, with very limited impact in the HNLC regions of the equatorial or subpolar North Pacific.[17] Information on the magnitude of the CO_2 sequestration potential of large-scale OIF comes from a combination of numerical ocean models and paleo records. Among these continental Antarctic ice core data of dust and of atmospheric CO_2 across glacial-interglacial cycles[18] and compilations of Southern Ocean sea-floor sediment records[19] suggest that enhanced glacial atmospheric iron supply led in the past to carbon sequestration of about 100 GtC. This atmospheric CO_2 drawdown, however, took several thousand years, and therefore occurred at a rate several orders of magnitude smaller than current anthropogenic CO_2 release. Conversely, it is not known to what extent the glacial dust supply was sufficient to fully relieve Southern Ocean iron limitation and whether sequestration rates could have been higher for a more intense iron supply.

Estimates of the sequestration potential of large-scale iron fertilization on centennial time scales, so far, essentially rely on numerical modeling studies. These have suggested that large-scale Southern Ocean iron fertilization may sequester some 70 to 180 GtC within one hundred years.[20] Even the lower end of the large range is far from negligible and amounts to about one "stabilization wedge" as proposed by Pacala and Socolow.[21]

However, there are also significant perturbations of marine biogeochemistry and ecology. In fact, some alteration of the function of pelagic ecosystems is the very objective of carbon sequestration by OIF. Any assessment of OIF therefore has to account for both intended and unintended consequences.[22] Unintended consequences identified so far include a downstream reduction of nutrients and

[17] Jorge L. Sarmiento & James C. Orr, *Three Dimensional Simulations of the Impact of Southern Ocean Nutrient Depletion on Atmospheric CO_2 and Ocean Chemistry*, 36 LIMNOLOGY & OCEANOGRAPHY 1928, 1937 (1991); Anand Gnanadesikan, Jorge L. Sarmiento & Richard D. Slater, *Effects of Patchy Ocean Fertilization on Atmospheric Carbon Dioxide and Biological Production*, 17 GLOBAL BIOGEOCHEMICAL CYCLES 1050, 1064 (2003); Olivier Aumont & Laurent Bopp, *Globalizing Results from Ocean in Situ Iron Fertilization Experiments*, 20 GLOBAL BIOGEOCHEMICAL CYCLES 1, **13** (2006).

[18] Andrew J. Watson et al., *Effect of Iron Supply on Southern Ocean CO_2 Uptake and Implications for Glacial Atmospheric CO2*, 407 NATURE 730, 730 (2000).

[19] Karen E. Kohfeld et al., *Role of Marine Biology in Glacial-Interglacial CO2 Cycles*, 308 SCI. 74, 74 (2005).

[20] Sarmiento & Orr, *supra* note 17, at 1948; Aumont & Bopp, *supra* note 17, at 12.

[21] Stephen Pacala & Robert Socolow, *Stabilization Wedges: Solving the Climate Problem for the Next 50 Years with Current Technologies*, 305 SCI. 968, 968. (2004).

[22] John J. Cullen & Philip W. Boyd, *Predicting and Verifying the Intended and Unintended Consequences of Large-Scale Ocean Iron Fertilization*, 364 MAR. ECO. PROGRESS SERIES **295**, 295. (2008).

productivity,[23] expansion of anoxic areas,[24] increased production of the greenhouse gas nitrous oxide,[25] and changes in species composition.[26] Interestingly, a model study of Southern Ocean OIF shows that volumes of low oxygen waters and associated production of N_2O may eventually decrease in response to downstream reduction in nutrients, fueling production above the tropical oxygen minimum zones.[27] Further studies are needed to obtain a robust assessment of the currently known potential consequences and to evaluate these against the potential consequences of leaving CO_2 in the atmosphere. Although we acknowledge that Garrett Hardin's first law of ecology, "we can never do merely one thing,"[28] does apply to iron fertilization, we have to bear in mind that it applies equally well to emitting CO_2 into the atmosphere.

3. OCEAN IRON FERTILIZATION: THE ECONOMIC PERSPECTIVE

Exploring the potential of OIF requires not just considering its effectiveness in increasing phytoplankton growth and carbon sequestration, but also measuring how much it costs to accomplish this environmental objective (economic efficiency) and comparing its carbon sequestration potential to other abatement options. We assume that carbon sequestration by OIF is realized within an international project as part of a post-Kyoto climate agreement. Without international coordination its usage would be inefficiently low, and it would be more difficult to establish a mechanism that addresses side effects.[29] Exploration of the economic potential of OIF in the context of an international treaty on climate change requires answers to the following questions: How many carbon credits are generated? How are they assigned? And, can they be used for compliance? The Kyoto Protocol (KP) established corresponding criteria for Clean Development Mechanism (CDM) and Joint Implementation (JI) projects. Carbon credit issue has to be based on an approved methodology, carbon credits can only be assigned to emission reductions or carbon storage that occur in addition to existing regulations or technical improvements and that are permanent, the credits have to be verified by a third party, and the

[23] Gnanadesikan et al., *supra* note 17, at 156.

[24] Sarmiento & Orr, *supra* note 17, at 1928.

[25] Xin Jin & Nicolas Gruber, *Offsetting the Radiative Benefit of Ocean Iron Fertilization by Enhancing N_2O Emissions*, 30 GEO. RESEARCH LETTERS, 2249, 2249 (2003).

[26] Sallie W. Chisholm & François M.M. Morel, *What Controls Phytoplankton Production in Nutrient Rich Areas of the Open Sea?*, 36 LIMNOLOGY & OCEANOGRAPHY, U1507, U1508 (1991).

[27] Oschlies et al., *supra* note 8, at 2959.

[28] GARRETT HARDIN, FILTERS AGAINST FOLLY 58 (1985).

[29] Kousky et al., *supra* note 7, at 15.

number of carbon credits needs to be corrected for leakage.[30] Leinen[31] discusses the fulfillment of these criteria for carbon sink enhancement through OIF, identifying the issues of permanence and leakage as the most critical ones. The degree of fulfillment of both criteria determines the number of carbon credits assigned to the sink enhancement project.

If we address the issue of permanence first, for terrestrial sinks various carbon accounting methodologies have been proposed to assess the value of different temporary storage projects that could be applied to OIF.[32] These methodologies commonly define permanence for assessment purposes as a time period of one hundred years, following the IPCC's definition of permanence for sequestration projects.[33] Four carbon accounting methods exist that assign permanent carbon credits (permanent methods): the net method, the average storage method, the discounting method, and the equivalence method. The net method, for example, measures the overall effect of OIF for a given period of time, generally one hundred years regardless of when the carbon fluxes take place within that period. Two carbon accounting methods exist that assign temporary carbon credits (temporary methods): the short-term method and the long-term method. Another method exists that assigns both permanent and temporary carbon credits (mixed method). Temporary carbon credits used for compliance have to be replaced at some point in time; permanent carbon credits need not be. Under the KP, two of the assignment options described earlier in the paragraph are applied, the permanent and the temporary methods, but terrestrial sink enhancement projects can generate temporary carbon credits only. Papers discussing the effectiveness of OIF implicitly apply the net method, because

[30] MICHAEL GRUBB, CHRISTIAAN VROLIJK & DUNCAN BRACK, THE KYOTO PROTOCOL: A GUIDE AND ASSESSMENT 53 (1999).

[31] Margaret Leinen, *Building Relationships between Scientists and Business in Ocean Iron Fertilization*, 364 MARINE ECOLOGY PROGRESS SERIES 251, 252. (2008). Following her line of reasoning, the criteria regarding methodology and additionality are easily fulfilled by OIF. The criterion of verification by a third party does apply in particular to projects between single firms or single countries in the context of CDM and JI.

[32] Michael Dutschke, *Fractions of Permanence – Squaring the Cycle of Sink Carbon Accounting*, 7 MITIGATION & ADAPTATION STRATEGIES FOR GLOBAL CHANGE 381, 386 (2002); Philip M. Fearnside, Daniel A. Lashof & Pedro Moura-Costa, *Accounting for Time in Mitigating Global Warming through Land-Use Change and Forestry*, 5 MITIGATION & ADAPTATION STRATEGIES FOR GLOBAL CHANGE 239, 242 (2000); Philip M. Fearnside, *Why a 100 Year Time Horizon Should Be Used for Global Warming Mitigation Calculations*, 7 MITIGATION & ADAPTATION STRATEGIES FOR GLOBAL CHANGE 19, 20. (2002); Gregg Marland, Kristy Fruit & Roger Sedjo, *Accounting for Sequestered Carbon: The Question of Permanence*, 4 ENVTL. SCI. & POL'Y 259, 262 (2001); Pedro Moura-Costa & Charlie Wilson, *An Equivalence Factor between CO_2 Avoided Emissions and Sequestration – Description and Applications in Forestry*, 5 MITIGATION & ADAPTATION STRATEGIES FOR GLOBAL CHANGE 51, 53. (2000).

[33] UNFCCC, *Report of the Conference of the Parties on Its Third Session*, TECHNICAL REPORT 31 (1997). The choice of one hundred years is not based on scientific rationale but was rather policy driven. *See* Leinen, *supra* note 31, at 252.

they do not distinguish when carbon uptake takes place within the permanence period.

Rickels et al.[34] are the first to discuss all of these accounting methods and to apply them to OIF. The results indicate that overall, and from an economic perspective, the short-term method is most appropriate for temporary OIF. Based on this method the largest amount of carbon credits is provided at an early state. Also, the fraction that is permanently provided until the end of the crediting period is larger compared to the fraction provided by the other methods. From an environmental perspective, the short-term method also seems most appropriate, as no additional carbon emissions will be released, because all credits have to be replaced at some point in time. Instead, even permanently stored carbon has to be replaced, so the application of the short-term method would provide extra climate benefits by reducing the atmospheric carbon concentration.

In addressing the issue of leakage, all potential offsets have to be taken into account to obtain the net amount of carbon credits. Potential offsets arise due to (1) carbon emissions outside the enhancement region of the OIF projects (spatial leakage), and (2) changes in emissions of GHGs other than carbon (GHG leakage).[35] To account for spatial leakage, Rickels et al.[36] apply the accounting methods to global data for oceanic carbon uptake rather than to local data. To account for GHG leakage, they introduce a discount factor. The discount factor reduces the gross amount of carbon credits to a net amount that can then be used for compliance. To offset GHG leakage, the upper and lower bounds for discount factors vary between the various accounting methods and the various experiments, ranging overall from 0.23 to 13.3 percent.

These ranges indicate that the potential of OIF cannot be determined with great accuracy. However, within an international treaty, such as the KP, a discount rate could be chosen that is large enough to compensate for this lack of knowledge and to take into account uncertainties. Considering offsets by other greenhouse gases as well as carbon emissions from ship operations, Rickels et al.[37] suggest an upper bound of 15 percent for the discount factor. Applying this discount factor to the net method, they find a range of 0.4 to 2.2 GtC for annual oceanic carbon uptake for OIF in the Southern Ocean, if OIF is implemented for seven years. Increasing the duration of implementation to one hundred years, the range narrows to between 0.5 to 1.4 GtC/yr. In comparison to OIF, enhancing terrestrial carbon sinks by forestry activities has entered the KP as offsets for anthropogenic carbon emissions, but the potential is uncertain as well. In a recent study, the annual potential of global forestry activities,

[34] Rickels, Rehdanz & Oschlies, *Methods for Greenhouse Gas Offsets, supra* note 8.

[35] In the context of OIF, GHG leakage occurs basically through a change in N_2O emissions. *See* Oschlies et al., *supra* note 8, at 2962.

[36] Rickels, Rehdanz & Oschlies, *Methods for Greenhouse Gas Offsets, supra* note 8.

[37] *Id.*

including reforestation, forest management, expanded use of forest products, and reduced deforestation, for carbon uptake is estimated between 0.4 and 0.8 GtC until 2030 assuming carbon prices of between 20 and 100 USD per ton CO_2.[38] The share of reforestation is approximately one-third.[39] Extending the time horizon to 2100, the range for reforestation enlarges and amounts to an annual carbon uptake of 0.2 to 1.1 GtC.[40] These figures indicate that the potential of forestation also cannot be determined with much accuracy. Van Kooten and Sohngen[41] show that there is a great inconsistency across forestry activity studies in how carbon uptake and costs are measured, so that costs of creating carbon credits through forestry activities vary widely. They conclude that the commonly held notion that these activities are a low-cost means for reducing atmospheric CO_2[42] needs to be reassessed.

As discussed above, another relevant issue for determining the effectiveness of a project is leakage, which is often ignored in bottom-up forestry activities analyses[43]. Forest management regimes including drainage systems[44] might lead to higher emissions of other greenhouse gases, in particular CH_4 and N_2O.[45] Estimates vary widely between 5 to 93 percent.[46] Leakage also arises if the stored carbon is intentionally or unintentionally released. In particular, the unintended release because of naturally occurring events such as fires, pest, droughts, or hurricanes imposes a risk on long-term storage prospects.[47] The likelihood of such naturally occurring risks may increase in the future because of global warming and would make terrestrial carbon sinks even less attractive.[48]

[38] Gert Jan Nabuurs et al., *Forestry*, *in* CLIMATE CHANGE 2007: MITIGATION. CONTRIBUTING OF WORKING GROUP III TO THE FOURTH ASSESSMENT REPORT OF THE INTERGOVERNMENTAL PANEL ON CLIMATE CHANGE, 541, 562 (Bert Metz et al. eds., 2007); Josep G. Canadell & Michael R. Raupach, *Managing Forests for Climate Change Mitigation*, 320 SCI. 1456, 1456 (2008).

[39] Nabuurs *et al.*, *supra* note 35, at 543.

[40] Jayant Sathaye et al., *GHG Mitigation Potential, Costs and Benefits in Global Forests: A Dynamic Partial Equilibrium Approach*, Multi-Greenhouse Gas Mitigation and Climate Policy Special Issue, ENERGY J. 95, 98. (2006); Brent Sohngen & Robert Sedjo, *Carbon Sequestration in Global Forests under Different Carbon Price Regimes*, ENERGY J. 109, 119 (2006); Bart J. Strengers, Jelle G. van Minnen & Bas Eickhout, *The Role of Carbon Plantations in Mitigating Climate Change: Potentials and Costs*, 88 CLIMATE CHANGE 343, 347 (2006).

[41] G. Cornelis van Kooten & Brent Sohngen, *Economics of Forest Ecosystem Carbon Sinks: A Review*, 1 INT'L REV. ENVTL. & RESOURCE ECON. 237, 258 (2007).

[42] *Id.* at 263

[43] *Id.*

[44] Drainage systems control the removal of surface and subsurface water in soils to allow optimal management of biomass production.

[45] Jane Ellis, *Forestry Projects: Permanence, Credit Accounting and Lifetime*, OECD AND IEA INFORMATION PAPER 10 (2001).

[46] Brian C. Murray, *Economics of Forest Carbon Sequestration*, *in* FOREST IN A MARKET ECONOMY 221, 239, (Erin O. Sills & Karen Lee Abt eds., 2003).

[47] ROYAL SOCIETY, THE ROLE OF LAND CARBON SINKS IN MITIGATING GLOBAL CLIMATE CHANGE 5 (2001).

[48] Ellis, *supra* note 45, at 30.

Using recent sequestration efficiency ratios from patch OIF experiments, Boyd estimates that the costs are between 8 and 80 USD per t CO_2 sequestered. For large-scale OIF, no cost estimates exist. However, OIF will not be used as a mitigation option if its costs exceed its benefits. Regarding the still existing uncertainty as to the volume of and costs of OIF, Rickels et al.[49] seek to determine the critical cost levels and the critical amounts of carbon credits from OIF for OIF to be competitive with forestry or CDM activities. Applying short-term OIF model experiments for the durations of one, five, and seven years and for different assumed fertilization efficiencies, they obtain critical unit costs for the upper level between 95 and 119 USD per t CO_2 and between 22 and 23 USD per t CO_2 for the lower level. The upper level of the estimates indicates whether OIF could be considered an abatement option compared to the current status of the KP, including all existing abatement options. For the lower level, it is assumed that the current restrictions regarding the use of carbon credits generated in low-cost countries is completely relaxed. The lower level of the estimates therefore indicates whether OIF would be comparable to options that achieve a given emission reduction target at lowest costs. OIF should at least generate the same efficiency gains as extending existing options, such as losing limitations with respect to usage of CER (Certified Emission Reduction) credits from CDM and Emission Reduction Units (ERU) credits from JI and Removal Unit (RMU) credits from forestation.

A comparison of this range of cost estimates to those of Boyd[50] for patch OIF experiments indicates that the upper and lower levels of those estimates are below the corresponding range of the upper and lower levels of the estimates of Rickels et al.[51] However, it must be noted that these cost estimates might not be representative of large-scale OIF.[52] In a comparison of this range of cost estimates to the range of estimates for forestation projects, they are in the same order of magnitude. However, OIF may well provide more carbon credits. Rickels et al.[53] demonstrate that seven years of large-scale OIF in the area of 30 degrees South can provide the same amount of credits as a global forestation project for the duration of twenty years. Therefore, we conclude, that current knowledge of OIF's potential as well as its costs militate against excluding it as a potential abatement option in the future.

[49] Rickels, Rehdanz & Oschlies, *Economic Prospects*, *supra* note 8, at 9.

[50] Philip W. Boyd, *Implications of Large-Scale Iron Fertilization of the Oceans – Introduction and Synthesis*, 364 MARINE ECOLOGY PROGRESS SERIES 214, 217. (2008).

[51] Rickels, Rehdanz & Oschlies, *Economic Prospects*, *supra* note 8, at 11.

[52] Christine Bertram, *Ocean Iron Fertilization in the Context of the Kyoto Protocol and the Post-Kyoto Process*, 38 ENERGY POL'Y 1130, 1133. (2010).

[53] Rickels, Rehdanz & Oschlies, *Economic Prospects*, *supra* note 8, at 11.

4. OCEAN IRON FERTILIZATION: THE PUBLIC INTERNATIONAL LAW PERSPECTIVE

The preceding economic analysis has shown that the comparison to efficiency criteria established by existing abatement options, in particular by existing sink enhancement options, militates against excluding OIF as a possible abatement option. Consequently, the inclusion of OIF activities in future global or regional emissions trading schemes could result in considerable economic benefits. This conclusion demands that calls for prohibiting or restricting any such activity on the grounds of inconsistency with public international law be meticulously justified. Although the 1992 Rio Declaration on Environment and Development[54] states that "[i]n order to achieve sustainable development, environmental protection shall constitute an integral part of the development process and cannot be considered in isolation from it," the very same Principle 4 of the Declaration implicitly acknowledges that economic development, indeed, constitutes one of the three central pillars (the other two being environmental protection and intergenerational justice) on which the concept of sustainable development is founded.[55] Thus, economic considerations at least should be taken into account (even though not necessarily given priority) whenever a certain activity is assessed by the competent fora in respect to whether it should be accepted.

In consideration of the fact that the concept of sustainable development is not a binding principle of international law, but rather a political concept,[56] its purpose is to provide a framework for national and international decision making.[57] The need to consider the economic impacts of OIF during such a decision-making process arises from the scientific uncertainty connected with its potentially negative effects on the marine environment and the novel character of the underlying legal questions, as well as the epochal challenge posed by global warming. However, as will be discussed later in the chapter, global political developments seem to point in the opposite direction (i.e., the imposition of a complete moratorium on OIF).

[54] 1992 Rio Declaration on Environment and Development, Datum, 31 ILM 874 (1992).

[55] Principles 11, 12, and 16 refer to economic aspects that ought to be taken into account. Principles 4, 24, and 25 deal with environmental protection, and Principle 3 mentions the needs of future generations equally next to the needs of present generations.

[56] The United Nations Conference on Environment and Development (UNCED) adopted three nonbinding instruments: the Rio Declaration on Environment and Development, the Agenda 21, and the UNCED Forest Principles. Besides those instruments, UNCED developed two treaties to be signed by states: the Convention on Biological Diversity (CBD) and the United Framework Convention on Climate Change (UNFCCC); *cf.* Philippe Sands, *Principles of International Environmental Law* 52 *et seq.* (2d ed. 2003).

[57] *See, e.g.*, Guy Beaucamp, *Das Konzept der zukunftsfähigen Entwicklung im Recht* 109 (2002).

The Conference of the Parties to the London Convention and Protocol[58] – one of the competent international bodies in this matter – adopted a catalog of numerous and strict criteria that should be fulfilled prior to the commencement of scientific OIF experiments. If lack of a scientific basis on which to justify a certain potentially harmful activity is used to strengthen the case against scientific research on the very same subject matter, though, it is difficult to argue that such a course of conduct is sustainable. Sustainability requires a balanced decision-making process. Referring strictly to potential negative side effects neglects the other pillars of the concept of sustainable development. Both OIF as a measure of climate engineering and global warming itself have the potential of risky side effects. We submit that from a legal perspective the development of a moratorium on OIF is not, in fact, based on an accurate reading of the precautionary principle, because it does not address the issue of potential negative impacts on the marine environment in an isolated manner, but instead assumes that these impacts must, again, be weighed in light of the global challenges deriving from climate change.[59]

4.1 *Relevant International Agreements*

In examining the rules of public international law applicable to OIF, the 1982 United Nations Convention on the Law of the Sea (UNCLOS)[60] should be referred to first. This framework treaty, a "constitution of the seas," was concluded according to its preamble with the objective "to promote [...] the study, protection and preservation of the marine environment," as specified by Part XII of the Convention.[61] Given that OIF conducted in certain marine areas could constitute "dumping,"[62] Art. 210 UNCLOS is the initially relevant norm. Paragraph 1 requires the contracting parties "[to] adopt laws and regulations to prevent, reduce and control pollution of the marine environment by dumping."[63] The reference to "global rules and standards" contained in this norm is generally understood as a reference to the London

[58] 1972 Convention on the Prevention of Marine Pollution by Dumping of Wastes and Other Matter, Nov. 29, 1972 ("London Convention" or "LC"), 1046 U.N.T.S. 120, *available at* http://www5.imo.org/SharePoint/blastDataHelper.asp/data_id%3D16925/LC1972.pdf; 1996 Protocol to the 1972 Convention on Prevention of Marine Pollution by Dumping of Wastes and Other Matter, Nov. 7, 1996 ("London Protocol" or "LP"), ILM 36 [1996], 1, *available at* http://www.imo.org/OurWork/Environment/SpecialProgrammesAndInitiatives/Pages/London-Convention-and-Protocol.aspx.

[59] Similarly Gareth Davies, *Law and Policy Issues of Unilateral Geoengineering: Moving to a Managed World* (Jan. 29, 2009), at 4 *et seq.*, *available at* http://ssrn.com/abstract=1334625.

[60] United Nations Convention on the Law of the Sea (UNCLOS), Dec. 10, 1982, 1833 U.N.T.S. 397, *available at*http://www.un.org/Depts/los/convention_agreements/texts/unclos/unclos_e.pdf.

[61] *See id.*, Recital 4 of the preamble, Art. 192 et seq.

[62] *Id.* Art. 210.

[63] *Id.* Art. 210(1).

Convention and the London Protocol, which supersedes the LC for acceding parties. Both treaties specifically apply to pollution by dumping.[64]

The concept of "dumping" is defined in Art. III (1)(a) LC and Art. 1 No. 4.1.1 LP (as well as in Art. 1 (5)(a) UNCLOS) as follows:

(i) any deliberate disposal at sea of wastes or other matter from vessels, aircraft, platforms or other man-made structures at sea;
(ii) any deliberate disposal at sea of vessels, aircraft, platforms or other man-made structures at sea.[65]

Even if iron filings introduced into the marine environment were not classified as "wastes," they would still be classified as "other matter." As they will remain in the ocean, "disposal" appears to be occurring.[66] However, this alone does not lead to the conclusion that OIF constitutes "dumping." Art. III (1)(b)(ii) LC and Art. 1 (4) No. 2.2 LP (as well as Art. 1 (5)(b)(ii)) UNCLOS) contain the following qualification to the definition of "dumping":

> "Dumping" does not include: [. . .] (ii) placement of matter for a purpose other than the mere disposal thereof, provided that such placement is not contrary to the aims of this Convention.[67]

As the goal of OIF is the stimulation of the primary production of phytoplankton in order to scientifically examine this process and its consequences with a view to potential increases in the uptake of CO_2, an objective other than the mere disposal of iron filings is being pursued. Under either applicable convention, OIF does not result in dumping so long as the placement of the iron filings are not "contrary to the aims" of the conventions.

The question remains whether OIF activities are contrary to the aims of the LC and the LP. As the purpose of these treaties is to prevent the pollution of the oceans through the dumping of wastes and other substances, the objectives of the conventions would seem to be contravened only when the substances introduced have a potentially damaging effect on human health, living resources, and/or marine life.[68]

[64] K. Russell LaMotte, *Legal Posture of Ocean Iron Fertilization under International Law*, II(1) Int'l Envtl. Law Committee Newsletter 8 (2009).

[65] London Convention, *supra* note 58, Art. III (1)(a); London Protocol, *supra* note 58, Art. 1 No. 4.1.1; UNCLOS, *supra* note 60, Art. 1 (5)(a).

[66] Rosemary Rayfuse et al., *Ocean Fertilisation and Climate change: The Need to Regulate Emerging High Sea Uses*, 23 INT'L J. MARINE & COASTAL L. 297, 312 (2008); David Freestone & Rosemary Rayfuse, *Ocean Iron Fertilization and International Law*, 364 MARINE ECO. PROGRESS SERIES 227, 229 (2008).

[67] London Convention, *supra* note 58, Art. III (1)(b)(ii); London Protocol, *supra* note 58, Art. 1 (4) No. 2.2; UNCLOS, *supra* note 60, Art. 1 (5)(b)(ii)).

[68] *See* London Convention, *supra* note 58, Art. I, in connection with London Protocol, *supra* note 58, Art. 2 Art. 1.6.10.

As shown above, it is currently not possible to rule out negative consequences of OIF for marine life or for human beings.[69] Our having said that, it should not be ignored that the main purpose of OIF experiments is not the mere stimulation of primary production in the ocean, but instead to investigate the potential for stimulating phytoplankton blooms under specific conditions and the consequences of such stimulation, as well as to achieve a more general understanding of the role of iron in marine ecology and biogeochemistry. To determine whether the effects of pursuing this goal contradict the objectives of the LC and the LP, Resolution LC-LP.1 on the Regulation of Ocean Fertilization can be consulted. According to this, the development of evaluation standards by the scientific groups of the LC and the LP is foreseen. As long as these standards have not been adopted, paragraph 6 of the Resolution requires the exercise of the utmost caution and the use of existing information for the best possible evaluation of research proposals in regard to the protection of the marine environment according to the LC and LP. This conclusion strongly militates in favor of accepting that not *all* scientific OIF experiments are contrary to the aims of the LC and the LP.

In this respect, we must note that the relevant issue here is also addressed by other international treaties, which potentially overlap with the aforementioned law of the sea instruments. In particular, reference to the primary agreement relevant to climate change, the 1992 United Nations Framework Convention on Climate Change (UNFCCC)[70] and its 1997 KP, is pertinent. The ultimate aim of the UNFCCC is to achieve a stabilization of greenhouse gas concentrations in the atmosphere at a level that would prevent dangerous anthropogenic interference with the climate system,[71] but it contains only comparatively weak obligations of a mainly procedural nature such as the duty to gather and share information on greenhouse gas emissions, national policies, and best practices. In contrast, the KP obliges industrialized states (Annex I States) to ensure that their greenhouse gas emissions do not exceed their individually assigned limitation and reduction commitments inscribed in Annex B. It is generally recognized that the ocean is a natural CO_2 sink in terms of the KP. Against this background, we might well ask whether an isolated interpretation of the aims of the LC and LP might, ultimately, not obviate the objectives of the climate change regime.

[69] Sallie W. Chisholm et al., *Dis-Crediting Ocean Fertilization*, SCI. 294, 309, 310 (2001); Ken L. Denman, *Climate Change, Ocean Processes and Ocean Iron Fertilization*, 364 MARINE ECO. PROGRESS SERIES 219, 223 *et seq*. (2008); Richard S. Lampitt et al., *Ocean Fertilization: A Potential Means of Geoengineering?*, 366 PHIL. TRANSACTIONS ROYAL SOC'Y A 366, 3919, 3930 *et seq*. (2008).

[70] United Nations Framework Convention on Climate Change, May 9, 1992, 1771 U.N.T.S. 107, *available at* http://unfccc.int/resource/docs/convkp/conveng.pdf.

[71] *See* UNFCCC, *supra* note 56, Art. 2.

4.2 *Current Developments*

The uncertainty of the legal status of OIF has recently led several competent international regimes to address this issue. As regards the legality of OIF activities under the Convention on Biological Diversity, the 9th Conference of the Parties (COP) to the Convention adopted Decision IX/16 on "Biodiversity and Climate Change" in May 2008, whose relevant part reads:

> *The Conference of the Parties,* [...]
>
> 4. *Bearing in mind* the ongoing scientific and legal analysis occurring under the auspices of the London Convention (1972) and the 1996 London Protocol, *requests* Parties and *urges* other Governments, in accordance with the precautionary approach, to ensure that ocean fertilization activities do not take place until there is an adequate scientific basis on which to justify such activities [...]; with the exception of small scale scientific research studies within coastal waters.[72]

As small-scale scientific research studies within coastal waters are not suitable for such experiments,[73] Decision IX/16 amounts, in substance, to a moratorium on OIF activities, including scientific experiments. In particular, a restriction to experiments within coastal waters seems to be inappropriate, because the marine productivity there is mostly not restricted by micronutrients such as iron. Furthermore, by increasing a research area, effects on the fertilized patch of mixing with water that has not been fertilized becomes less important.[74] However, in October 2010 the 10th COP integrated several cross-references to Decision IX/16 in their Decisions on Marine and Coastal Biodiversity and on Biodiversity and Climate Change,[75] which confirmed the conditions established by Decision IX/16. At the same time, COP 10 adopted a modification of the exception for OIF experiments:

> *The Conference of the Parties* [...]
>
> 8. *Invites* Parties and other Governments, according to national circumstances and priorities [...] to consider the guidance below [...]:

[72] UNEP/CBD/COP/DEC/IX/16, Decision Adopted by the Conference of the Parties to the Convention on Biological Diversity at Its Ninth Meeting – IX/16. Biodiversity and Climate Change, Oct. 9, 2008, Part C, para. 4, *available at* http://www.cbd.int/doc/decisions/cop-09/cop-09-dec-16-en.pdf.

[73] Denman, *supra* note 69, at 221 *et seq*.; Lampitt et al., *supra* note 69, at 3938; Jorge L. Sarmiento & Nicolas Gruber, *Sinks for Anthropogenic Carbon*, 55(8) PHYSICS TODAY 30, 31 (2002).

[74] Intergovernmental Oceanographic Commission, Statement of the IOC Ad Hoc Consultative Group on Ocean Fertilization, June 14, 2008, II.2., III.

[75] Advanced unedited text reflecting the decision as adopted on the basis of document UNEP/CBD/COP/10/L.42, paras. 13 lit. (e), 57 ff.; advanced unedited text reflecting the decision as adopted on the basis of document UNEP/CBD/COP/10/L.36, paras. 8 lit. (w), 9 lit. (o), (p).

> (w) Ensure, in line and consistent with decision IX/16 C, on ocean fertilization and biodiversity and climate change, in the absence of science based, global, transparent and effective control and regulatory mechanisms for geo-engineering, and in accordance with the precautionary approach and Article 14 of the Convention, that no climate-related geo-engineering activities that may affect biodiversity take place, until there is an adequate scientific basis on which to justify such activities and appropriate consideration of the associated risks for the environment and biodiversity and associated social, economic and cultural impacts, with the exception of small scale scientific research studies that would be conducted in a controlled setting in accordance with Article 3 of the Convention, and only if they are justified by the need to gather specific scientific data and are subject to a thorough prior assessment of the potential impacts on the environment.[76]

The restriction to coastal waters is not mentioned in the new passage. The reason for the omission can easily be explained by pointing to the reference made in the decision to all possible geoengineering techniques. Instead, four new conditions were established: the conduct in a controlled setting, a reference to Article 3 of the CBD, which requires states to address potential impacts of their activities on other states or the global commons; the need to gather specific data; and the need to conduct a environmental impact assessment.

In October 2007, the Meeting of the Parties (MOP) to the LC and the LP released a Statement of Concern regarding OIF. In this document, it was stated that,

> recognizing that it was within the purview of each State to consider proposals on a case-by-case basis in accordance with the London Convention and Protocol, urged States to use the utmost caution when considering proposals for large-scale ocean fertilization operations. The governing bodies took the view that, given the present state of knowledge regarding ocean fertilization, such large-scale operations were currently not justified.[77]

One year later, in November 2008 (i.e., after the adoption of CBD Decision IX/16), the same body adopted Resolution LC-LP.1 (2008) on the regulation of OIF. According to paragraph 8 of this document, OIF activities are contrary to the objectives of the London regime if and to the extent to which they cannot be qualified as legitimate scientific research:

> The Thirtieth Meeting of the Contracting Parties to the London Convention and the Third Meeting of the Contracting Parties to the London Protocol [...]

[76] Advanced unedited text reflecting the decision as adopted on the basis of document UNEP/CBD/COP/10/L.36, para. 8 lit. (w) (footnote omitted).

[77] LC 29/17, Report of the Twenty-Ninth Consultative Meeting and Second Meeting of Contracting Parties, Dec. 14, 2007, para. 4.23.

8. AGREE that, given the present state of knowledge, ocean fertilization activities other than legitimate scientific research should not be allowed. To this end, such other activities should be considered as contrary to the aims of the Convention and Protocol and not currently qualify for any exemption from the definition of dumping in Article III.1(b) of the Convention and Article 1.4.2 of the Protocol.[78]

It was pointed out by Proelss[79] that neither the CBD Decisions nor the Statement of Concern and the resolution LC-LP.1 are by themselves legally binding. However, as Resolution LC-LP.1 (2008) directly examines the question of whether OIF should be categorized as dumping under the LC and LP, it can be referred to as an aid in the interpretation of the scope of the respective Conventions. The conclusion is that *legitimate* OIF experiments cannot be considered as prohibited dumping.

As provided for in Resolution LC-LP.1 (2008), the Scientific Groups of the LC and LP developed an Assessment Framework for scientific research involving OIF. This Assessment Framework was adopted by the Contracting Parties during their last meeting in October 2010.[80] The framework contains a detailed catalog (approximately twenty pages) of strict criteria for evaluating whether an OIF experiment constitutes legitimate scientific research in terms of the Resolution. An initial assessment, as well as a detailed and often interdisciplinary risk analysis, will be required, effectively recognizing that an OIF experiment is likely to pose a serious, if not unrealizable, challenge for scientists.

Careful scientists will in any case take into account the risks of potential adverse effects of a certain activity on the marine environment. However, some components of the requisite data will simply not be available before a fertilization experiment is undertaken. This data can be determined only by conducting the experiment itself. Arguably, requiring a scientist to carry out individual parts of the experiment in advance and to bear the risk of not receiving an approval for it afterward, restricts marine scientific research in a way that does not seem to be compatible with the spirit of resolution LC-LP.1. Against this background, the course of action taken by the Scientific Group threatens to undermine the decision that legitimate scientific research shall be considered as being lawful. Additionally, in light of the economic benefits described above, it is at least doubtful whether any such implementation of Resolution LC-LP.1 (2008) can be held to be sustainable.

Our having said that, it should be noted that the pertinent paragraph of the assessment framework that governs the decision-making process to be followed by

[78] Resolution LC-LP.1 (2008) on the Regulation of Ocean Fertilization, Oct. 31, 2008, para. 8.

[79] Alexander Proelss, *Legal Opinion on the Legality of the LOHAFEX Marine Research Experiment under International Law* (2009), at 9, 15, *available at* http://www.internat-recht.uni-kiel.de/veranstaltungen/opinions/Legal%20Opinion%20LOHAFEX%20(Proelss).pdf.

[80] Resolution lc-lp.2 (2010) on the Assessment Framework for Scientific Research Involving Ocean Fertilization, Oct. 14, 2010.

the national competent authority was amended during the negotiation process.[81] Although the first drafts of the paragraph did not at all clarify under what conditions an experiment is "contrary to the aims of the Convention and Protocol,"[82] the text that was subsequently adopted substantiates the decision-making process by referring to the precautionary principle:

> [i]f the risks and/or uncertainties are so high as to be deemed unacceptable, with respect to the protection of the marine environment, taking into account the precautionary approach, then a decision should be made to seek revision of or reject the proposal.[83]

This wording suggests – in connection with the requirement to minimize environmental disturbance and to maximize scientific benefits[84] – an assessment based on balancing risks and uncertainties.

4.3 *Impact of the Precautionary Principle*

It is submitted that further clarification can be achieved by reference to the precautionary principle. Notwithstanding a considerable degree of uncertainty as to its normative content and validity,[85] it is well established that Principle 15 of the Rio Declaration contains the most widely known formulation of the precautionary principle:

> In order to protect the environment, the precautionary approach shall be widely applied by all States according to their capabilities. Where there are threats of serious or irreversible damage, lack of full scientific certainty shall not be used as a reason for postponing cost-effective measures to prevent environmental degradations.[86]

By explicitly referring to cost-effective measures, the precautionary principle requires a careful analysis of the economic impact of a decision.[87] It does not provide an

[81] *See* London Convention, *supra* note 58, Art. III (1)(b)(ii); London Protocol, *supra* note 58, Art. 1.4.2.2.

[82] *Cf.* LC/SG 32/15, Draft, Assessment Framework for Scientific Research Involving Ocean Fertilization, June 29, 2009, Annex 2, para. 9.

[83] Assessment Framework for Scientific Research Involving Ocean Fertilization, *supra* note 82, para. 4.3.

[84] *Id.*, para. 4.1.

[85] Malgosia Fitzmaurice, CONTEMPORARY ISSUES IN INTERNATIONAL ENVIRONMENTAL LAW 1 *et seq.* (2009); David Freestone & Ellen Hey, *Origins and Developments of the Precautionary Principle*, *in* THE PRECAUTIONARY PRINCIPLE AND INTERNATIONAL LAW – THE CHALLENGE OF IMPLEMENTATION 3 (David Freestone & Ellen Hey eds., 1996); Simon Marr, *The Precautionary Principle in the Law of the Sea* 7 (2003); Rüdiger Wolfrum, *Precautionary Principle*, *in* NEW TECHNOLOGIES AND LAW OF THE MARINE ENVIRONMENT 203 (Jean-Pierre Beurier, Alexandre Kiss & Said Mahmoudi eds., (1999).

[86] Rio Declaration, *supra* note 54, Principle 15.

[87] Sergio A. Hasbun, *The Precautionary Principle in the SPS Agreement*, 12 ZEITSCHIFT FÜR EUROPARECHTLICHE STUDIE 455, 464 (2009).

authorization to act, but shall be considered whenever states exercise their rights and obligations under public international law.

The precautionary principle constitutes the common denominator of virtually all of the pertinent legal instruments, including the LC and LP, and may, arguably, be used as a balancing tool to measure the environmental benefits arising out of a certain activity against its potentially negative impacts on another part of the environment.[88] Additionally, it is commonly held to be one of the cornerstones of the concept of sustainable development.[89] Although it is true that assessment frameworks constitute one of the means of implementation of the precautionary principle, one might ask whether the catalog of criteria discussed within the context of the LC and LP is consistent with its requirements.

If one attempts to explore the relevance of the precautionary principle in the context at hand, recourse to the differentiation between rules and principles appears to be helpful. Dealing with Hart's concept of positivistic legal theory,[90] Dworkin developed his famous principle paradigm.[91] According to his concept of law, principles are characterized as "optimizing commands."[92] They express certain values, but do not require a specific behavior of the respective subject of law. By contrast, rules are structured in the pattern of fact and legal consequence and are applicable in an "all-or-nothing-fashion."[93] They are specific in their requirements and consequences. If a rule is valid, it prescribes a definitive legal consequence by permitting, forbidding, or commanding something. If it is not valid, it has no influence on the decision. Principles can be realized to varying degrees subject to the legal possibilities, that is, the extent to which a certain principle can be implemented depends on the existence and scope of competing principles. Thus, the application of legal principles generally results in a fair balance of values.

As regards the precautionary principle, its elements are characterized by a degree of indetermination. This becomes particularly manifest in the element "lack of full scientific certainty" contained in Principle 15 of the Rio Declaration.[94] It is exactly this vagueness that shows that the precautionary principle must be qualified as a legal principle.[95]

[88] Alexander Proelss & Monika Krivickaite, *Marine Biodiversity and Climate Change*, 3(4) CARBON & CLIMATE L. REV. 437, 445 (2009).

[89] CHRISTINA VOIGT, SUSTAINABLE DEVELOPMENT AS A PRINCIPLE OF INTERNATIONAL LAW 48 (2009). *Cf.* Davies, *supra* note 59, at 4, who argues to consider the risky side effects of geoengineering and of global warming by weighing them against each other.

[90] HERBERT L. A. HART, THE CONCEPT OF LAW 89 (1961).

[91] RONALD DWORKIN, TAKING RIGHTS SERIOUSLY 14 (1982).

[92] ROBERT ALEXY, THE ARGUMENT FROM INJUSTICE – A REPLY TO LEGAL POSITIVISM 70 (2002).

[93] DWORKIN, *supra* note 91, at 24.

[94] Rio Declaration, *supra* note 54, Principle 15.

[95] Marr, *supra* note 85, at 13.

If one applies this classification to the case of OIF, one must note that on the basis of an isolated reading of the relevant provisions of the law of the sea,[96] the precautionary principle seems to militate in favor of the protection of the marine environment. On the other hand, Art. 3(3) UNFCCC demands that the lack of full scientific certainty of mitigation measures should not be used as a reason for postponing such measures where there are threats of serious or irreversible damage. Consequently, within the context of global warming the precautionary principle supports permitting OIF activities. Against this background, and keeping in mind the nature of the precautionary principle as a principle of law, the precautionary principle ought to be used to balance the risks arising out of scientific OIF activities (which are likely to contradict the aims contained in the CBD) with the potential advantages relevant to the objectives of the UNFCCC and the KP.

One has to measure the potential negative impacts of OIF on the marine environment against the global dangers resulting from rising CO_2 concentrations in the atmosphere. It is submitted that a proper application of the precautionary principle can only lead to the conclusion that further scientific research must be permitted to explore the sequestration potential of OIF. In conclusion, this concept might either be rejected or integrated into the flexible mechanisms contained in the KP or a successor agreement. This is even more so with a view to the potential economic benefits of OIF examined in this chapter. That being said, whether commercial activities should be permitted by inclusion of OIF in the flexible Kyoto mechanisms depends on the outcome of experiments addressing the potential negative impacts of OIF on the marine environment.

5. CONCLUSION

Our scientific and economic analysis leads us to the conclusion that OIF should be considered as a carbon dioxide removal option. Likewise, we have demonstrated that public international law does not require imposing a complete moratorium on OIF. To the contrary, as far as scientific research experiments are concerned, a proper analysis of the pertinent agreements, as well as an adequate reading of the precautionary principle, results in a clear presumption in favor of permitting such activities.

Against the background of an ever-declining carbon emission budget on the one hand, and widespread reluctance to accept meaningful global reduction targets on the other, including OIF in a post-Kyoto climate agreement might provide new incentives for the negotiation process. Rickels et al.[97] show that countries with high

[96] *See* UNCLOS, *supra* note 60, Art. 1(1) No. 4, Art. 194(1).

[97] Rickels, Rehdanz & Oschlies, *Economic Prospects*, *supra* note 8, at 13

abatement costs are expected to be more or less indifferent as to the choice between the option of extending the share of carbon credits traded with CDM countries and the option of including OIF, presuming that only countries with positive reduction targets are included in the allocation of OIF carbon credits. CDM countries such as China are expected to favor the first option. Consequently, a third option could be considered. It would incorporate both other options, extending the share of CDM carbon credits and including OIF, with allocation of OIF carbon credits to CDM countries that would accept emission reduction targets in a future commitment period.

However, only discussing OIF as a potential geoengineering option tends to provoke public resistance, which in the case of the German-Indian LOHAFEX experiment[98] resulted in antiscientific propaganda by individual nongovernmental organizations, political struggle between different German government authorities, and calls for the implementation of a complete ban on commercial and, to some extent, scientific experiments.[99] These views and attempts, based on statements about uncertain side effects and consequences, reveal an attitude that emphasizes continuity above alteration.

If we disregard our cultural reservation against non–fishery-related ocean change, a valuation might be more complicated. For example, how do we value the likelihood of enhanced marine production in the Southern Ocean that may turn out to be beneficial for many species, including severely depleted whale populations?[100] How do we account for the possibility that large-scale Southern Ocean OIF might, via downstream reduction of macronutrients, lead to reduced oxygen minimum zones and associated nitrous oxide emissions in the tropical oceans?[101]

We have to acknowledge that we will never have full knowledge or forethought with respect to all the risks associated with OIF – or of the risks associated with

[98] LOHAFEX was conducted from January to March 2009 in the Southern Ocean and was the largest OIF experiment so far. The LOHAFEX team fertilized a 300-square-kilometer patch of ocean using about six tons of dissolved iron. *Cf.* Press Release, Alfred Wegener Institute Ending of the Lohafex expedition (Mar. 23, 2009), *available at* http://www.awi.de/en/news/selected_news/2009/lohafex/press_releases/.

[99] *Cf.* Statement of the German Minster of Education and Research, *available at* http://www.bmbf.de/press/2453.php; Statement of the German Minster of the Environment, *available at* http://www.bmu.de/pressearchiv/16_legislaturperiode/pm/42974.php; ETC Group News Release, Jan. 13, 2009 LOHAFEX Update: Geo-Engineering Ship Plows on as Environment Ministry Calls for a Halt, *available at* http://www.etcgroup.org/upload/publication/712/01/nretc_lohafexupdate13jan09_final.pdf; ETC Group News Release Jan. 28, 2009 LOHAFEX Update: Throwing Precaution (and Iron) to the Wind (and Waves), *available at* http://www.etcgroup.org/upload/publication/719/01/etcnr_lohafexupdate28jan09.pdf.

[100] Victor Smetacek & Syed W. Naqvi, *The Next Generation of Iron Fertilization Experiments in the Southern Ocean*, 366 PHIL. TRANS. R. SOC. A 3947, 3947 (2008).

[101] Oschlies et al., *supra* note 8, at 2959.

foregoing OIF research in the face of burgeoning CO_2 emissions. Given the multi-sectoral and overwhelmingly serious challenges posed by climate change (a truly global phenomenon), as well as the difficulties in achieving global agreement on a sufficient degree of emissions reductions, there is, indeed, no alternative to the further exploration of engineering options such as large-scale OIF.

ACKNOWLEDGMENTS

The chapter is a revision of the article "Ocean Iron Fertilization: Why Further Research Is Needed" by the authors, published in *Marine Policy* 34 (2010), pp. 911–918. The DFG provided financial support through the Excellence Initiative Future Ocean.

12

Remaking the World to Save It

Applying U.S. Environmental Laws to Climate Engineering Projects

Tracy Hester

The long-running struggle over climate change policy may ultimately fall under the shadow of a much larger concern: What if our best strategies and legal measures to control greenhouse gas (GHG) emissions and adapt to climate change, in the end, are simply not enough?

The question is becoming increasingly important. Although U.S. regulatory and policy efforts have picked up new momentum, federal legislative efforts in the United States have ebbed after Congress' failure to pass a comprehensive climate change bill.[1] International efforts to limit GHG emissions have not yet achieved significant reductions or even appreciably slowed the rate of increase in emissions.[2]

[1] The history of prior climate change legislative and regulatory initiatives is complex and fast-moving, and it lies beyond the scope of this chapter. Significant milestones include President Obama's decision to focus his first Oval Office speech on the need to move away from fossil fuels and to reduce GHG emissions through fostering renewable energy technologies. Barack Obama, President, United States, Remarks by the President to the Nation on the BP Oil Spill (June 15, 2010), *available at* http://www.whitehouse.gov/the-press-office/remarks-president-nation-bp-oil-spill. The U.S. Environmental Protection Agency (EPA) issued its long-pending endangerment finding under the federal Clean Air Act that GHG emissions threaten human health and the environment. National Emission Standards for Hazardous Air Pollutants for Source Categories: Gasoline Distribution Bulk Terminals, Bulk Plants, and Pipeline Facilities; and Gasoline Dispensing Facilities, 74 Fed. Reg. 66,495 (Dec. 15, 2009) (codified at 40 C.F.R. pts. 9 & 63). The EPA's finding, even though it faces numerous petitions for judicial review, has already triggered a cascade of regulations to control industrial GHG emissions. *See, e.g.*, N. Richardson, Art Fraas & Dallas Burtraw, *Greenhouse Gas Regulation under the Clean Air Act: Structure, Effects, and Implications of a Knowable Pathway*, 41 ENVTL. L. REP. (Envtl. Law. Inst.) 10,098, 10,100 (Feb. 2011), *available at* http://www.elr.info/articles/vol41/41.10098.pdf. Several states have also acted to limit GHG emissions in their jurisdictions, and their efforts have helped to form regional compacts to lay the groundwork for future GHG trading and controls. *See infra* note 23 and accompanying text (discussing the Regional Greenhouse Gas Initiative and the Western Climate Initiative).

[2] When the Kyoto Protocol was set to expire at the end of 2012, a last minute agreement was reached in December to extend the Protocol until 2020. This was after many years of failed attempts to achieve a post-Kyoto agreement putting in place significantly greater reductions in greenhouse gases. Specifically, the Sixteenth Conference of the Parties to the United Nations Framework Convention

Anthropogenic GHG emissions[3] remain at historically high levels,[4] and the growing use of fossil fuels by developing economies virtually guarantees large increases in future emissions.[5] Given the political and logistical challenges to widespread implementations of alternative fuels, petroleum will likely remain the primary source of

for Climate Change in Cancun, Mexico, announced on Dec. 11, 2010, a set of agreements that outlined voluntary commitments to provide financing for green energy development and to reduce GHG emissions. *Cancun Climate Outcome "Consistent with U.S. Objectives,"* ENVTL. NEWS SERV. (Dec. 14, 2010), http://www.ens-newswire.com/ens/dec2010/2010–12–14–02.html. *Id.* The parties at the Fifteenth Conference of the Parties in 2009 in Copenhagen failed to reach any binding agreement that would significantly limit future GHG emissions. *See id.* (noting that attempts during the 2009 Copenhagen Conference were "not fruitful"). A small subgroup (including the United States, China, and India) instead agreed to examine steps to limit the rate of growth of GHG emissions, and the remaining body of delegates desultorily "took notice" of the new Copenhagen Accords. United Nations Framework Convention on Climate Change, Report of the Conference of the Parties on its Fifteenth Session, Copenhagen, Denmark, Dec. 7–9, 2009, *Copenhagen Accord*, Decision 2/CP.15, ¶ 6, U.N. Doc. FCCC/CP/2009/11/Add.1 (Mar. 30, 2010), *available at* http://unfccc.int/resource/docs/2009/cop15/eng/11a01; *see also* John M. Broder, *Climate Goal Is Supported by China and India*, N.Y. TIMES, Mar. 9, 2010, at A1, *available at* http://www.nytimes.com/2010/03/10/science/earth/10climate.html. More important, some initial assessments of the Cancun Agreement have concluded that it did not include sufficient emission reduction pledges to keep global temperature increases below a target of 2 degrees Celsius or less. CLAUDINE CHEN ET AL., CLIMATE ACTION TRACKER, CANCUN CLIMATE TALKS – KEEPING OPTIONS OPEN TO CLOSE THE GAP 2, Jan. 10, 2011, *available at* http://www.climateactiontracker.org/briefing_paper_cancun.pdf.

[3] Anthropogenic GHG emissions are GHG releases caused by human activities. These activities can include industrial operations, farming activities, transportation emissions, and alterations to natural ecosystem emissions caused by human activities.

[4] "The radiative forcing of the climate system is dominated by the long-lived GHGs … Global GHG emissions due to human activities have grown since pre-industrial times, with an increase of 70% between 1970 and 2004." INTERGOVERNMENTAL PANEL ON CLIMATE CHANGE, CLIMATE CHANGE 2007: SYNTHESIS REPORT 36 (2007), *available at* http://www.ipcc.ch/pdf/assessment-report/ar4/syr/ar4_syr.pdf [hereinafter SYNTHESIS REPORT]. The Intergovernmental Panel on Climate Change (IPCC) report further notes that "[g]lobal atmospheric concentrations of CO_2 [carbon dioxide], CH_4 [methane],and N_2O [nitrous oxide] have increased markedly as a result of human activities since 1750 and now far exceed pre-industrial values determined from ice cores spanning many thousands of years. The atmospheric concentrations of $CO_2$and CH_4in 2005 exceed by far the natural range over the last 650,000 years." *Id.* at 37 (citation to figures omitted). *But cf.* U.S. ENERGY INFO. ADMIN., DOE/EIA-0573(2008), EMISSION OF GREENHOUSE GASES IN THE UNITED STATES 2008, at 1 (2009), *available at* ftp://ftp.eia.doe.gov/pub/oiaf/1605/cdrom/pdf/ggrpt/057308.pdf (finding total U.S. GHG emissions decreased by 2.2 percent from 2007 to 2008).

[5] In 2009, the International Energy Agency predicted China and India would account for 53 percent of the increase in global demand for energy between 2009 and 2030 and that these two nations will predominantly rely on GHG-emitting technologies to reach that position. INT'L ENERGY AGENCY, WORLD ENERGY OUTLOOK 2009 FACT SHEET 1 (2009), *available at* http://www.iea.org/weo/docs/weo2009/fact_sheets_WEO_2009.pdf. On November 2, 2010, Indian Prime Minister Manmohan Singh said demand for hydrocarbons in his country will increase by 40 percent over the next decade. Walid Mazi, *Indian Energy Firms Advised to Expand Amid Soaring Fuel Demand*, ARABNEWS.COM (Nov. 2, 2010), http://arabnews.com/economy/article177746.ece.

energy for transportation for decades and will further swell GHG emissions.[6] And even if these accelerating GHG sources could be slowed, the atmosphere has already received sufficient anthropogenic GHGs to assure that climate change effects will grow during the next century or even accelerate as self-reinforcing warming processes take root.[7] The risk of self-reinforcing feedback processes has also heightened concerns over abrupt and disruptive climate change.[8]

Against this pessimistic backdrop, some scientists have begun to seriously study direct actions to modify the earth's climate in ways that would offset anthropogenic global warming. These strategies, discussed further in Section 2, include: releasing sulfur dioxide aerosols into the upper stratosphere to reflect solar radiation back into space, enhancing the reflectivity of clouds in the polar oceans, constructing and distributing millions of mechanical units to filter ambient air and remove carbon dioxide (CO_2), using reflective satellites to control solar radiation reaching the earth's surface, and seeding oceans with iron to enhance phytoplankton growth and draw large quantities of CO_2 out of the atmosphere.[9] These ideas, collectively labeled "climate engineering" or "geoengineering,"[10] are polarizing and controversial, but their

[6] The transportation sector is the largest growth segment of total oil demand; by 2030, oil demand in developing countries will exceed oil demand in countries in the Organization of Economically Developed Countries. Jacqueline L. Weaver, *The Traditional Petroleum-Based Economy: An "Eventful" Future*, 36 CUMB. L. REV. 505, 528 (2006) (discussing energy use projections by major energy corporations and U.S. agencies).

[7] For example, some scientists have argued that arboreal soils and permafrost may release large amounts of CO_2 as they thaw in a warming climate. E. Schuur et al., *The Effect of Permafrost Thaw on Old Carbon Release and Net Carbon Exchange from Tundra*, 459 NATURE 556 (2009). Such soils contain significantly more carbon than the amount of CO_2 already present in the atmosphere. As a result, those increased CO_2 emissions may in turn magnify climate change effects and enhance ambient temperature increases, which would then accelerate continuing CO_2 emissions from the soils. *See, e.g.*, Eric A. Davidson & Ivan A. Janssens, *Temperature Sensitivity of Soil Carbon Decomposition and Feedbacks to Climate Change*, 440 NATURE 165 (2006).

[8] Some climatologists have concluded that geologic records show that earth's climate can change significantly and abruptly over a time span as short as ten years. Under this model, earth's climatic system can shift quickly and unpredictably from one stable state into another without gradual or cumulative changes. For example, if increased levels of fresh water in the North Atlantic lead to a disruption or cessation of the Gulf Stream component of the ocean currents that convey warmer waters toward northern Europe and Africa, those regions could see dramatic drops in temperatures and changes in precipitation over a short time span. R. Gagosian, President, Woods Hole Oceanic Institute, Presentation to Davos Summit: Abrupt Climate Change: Should We Be Worried? (Feb. 10, 2003), *available at* http://www.whoi.edu/page.do?cid=9986&pid=12455&tid=282; Wallace S. Broecker, *Thermohaline Circulation, the Achilles' Heel of Our Climate System: Will Man-Made CO_2 Upset the Current Balance?*, 278 SCI. 1582, 1584 (1997). The U.S. National Academy of Sciences noted in 2002 that "available evidence suggests that abrupt climate changes are not only possible but likely in the future, potentially with large impacts on ecosystems and societies." U.S. NAT'L ACAD. OF SCIS., ABRUPT CLIMATE CHANGE: INEVITABLE SURPRISES, at v (2002).

[9] *See* discussion *infra* Section 2.

[10] In keeping with the developing trend, this chapter uses the term "climate engineering" instead of "geoengineering." The term "geoengineering" can also apply to large-scale earth moving operations,

rapid emergence as "Plan B" for climate change strategies will ultimately put federal and state environmental laws squarely in the middle of contentious fundamental disputes over the future direction of U.S. and global climate change policy.

If climate engineering someday becomes a component of U.S. and global climate change policy, U.S. environmental laws will almost certainly be used to challenge demonstrations of climate engineering technologies conducted by U.S. corporations and citizens, or those demonstrations in territories or airspace under U.S. jurisdiction. Environmental advocates have frequently turned to U.S. environmental laws to slow or stop the implementation of arguably risky or unexamined technologies. For example, critics of novel technologies used U.S. environmental laws to challenge the deployment of genetically modified organisms into the environment, the distribution of nanomaterials into the workplace and commerce, and the siting of certain renewable energy technologies.[11] Ironically, if climate engineering proves an essential component of federal climate change policy to control or minimize climatic disruptions, environmental law may play an instrumental role in limiting

and some groups have begun to use "climate engineering" as a clearer term. COMM. ON SCI. & TECH., 111TH CONG., ENGINEERING THE CLIMATE: RESEARCH AND STRATEGIES FOR INTERNATIONAL COORDINATION 13 (Comm. Print 2010); J. SHEPHERD, THE ROYAL SOCIETY, GEOENGINEERING THE CLIMATE: SCIENCE, GOVERNANCE AND UNCERTAINTY 30 (2009), *available at* http://royalsociety.org/uploadedFiles/Royal_Society_Content/policy/publications/2009/8693.pdf [hereinafter Royal Society Study].

[11] For example, the Foundation on Economic Trends turned to the National Environmental Policy Act to obtain an injunction halting the deployment of genetically modified tomatoes outside the laboratory setting. Found. for Econ. Trends v. Heckler, 587 F. Supp. 753 (D.D.C. 1984), *aff'd in part*, 756 F.2d 143 (D.C. Cir. 1985). Current environmental groups continue to use environmental statutes to challenge expanded use of genetically modified organisms. *See, e.g.*, Complaint, Ctr. for Food Safety v. Vilsack, No. CV11–1310 (filed Mar. 18, 2011), *available at* http://www.centerforfoodsafety.org/wp-content/uploads/2011/03/1-Complaint.pdf (challenging deregulation of Roundup Ready Alfafa genetically engineered to tolerate glycophosphate-based pesticides). Environmental groups have also repeatedly urged EPA to regulate nanoscale materials more aggressively under the Toxic Substances Control Act. *See, e.g.*, Natural Res. Def. Council, Comments on EPA Proposed Voluntary Pilot Program for Nanomaterials, Docket ID: OPPT-2004–0122 (July 20, 2005), *available at* http://www.nanoaction.org/doc/OPPT-2004–0122–0037.pdf (urging the EPA to regulate nanoscale materials as new chemicals under TSCA). Environmental groups and wastewater system operators recently petitioned the EPA to promulgate rules to control the potential release of nanoscale silver to wastewater treatment systems and the environment. *See, e.g.*, Letter from Michele Pla, Exec. Dir., Bay Area Clean Water Agencies to Nathanael R. Martin, Office of Pesticide Programs, U.S. EPA, (Mar. 19, 2009), http://bacwa.org/Portals/0/Committees/BAPPG/Archive/BACWA%20Comments%20on%20Petition%20for%20rulemaking%20to%20regulate%20nonsilver%20as%20pesticide%203–09.pdf. Last, environmental groups have also seized upon environmental statutes to oppose renewable energy projects. For example, opponents of the Cape Wind project to place turbines offshore of Massachusetts filed complaints alleging that the wind turbines would kill endangered and threatened species protected under the Endangered Species Act. *See, e.g.*, Complaint, Pub. Emps. For Envtl. Responsibility v. Bromwich, No. 1:10-cv-01067-RMU (D.D.C. filed June 25, 2010), *available at* http://www.marinelog.com/PDF/capewindcomplaint.pdf.

options available to address one of the most daunting environmental challenges of our time.

If existing U.S. environmental laws become the initial battleground for disputes over climate engineering research and test projects, those fights may yield surprises for litigants on both sides. U.S. environmental laws could extend an unexpectedly long and broad reach over novel climate engineering technologies. The federal courts have allowed administrative agencies, including the EPA, a considerable degree of flexibility and freedom to interpret current statutes to cover emerging environmental threats and concerns.[12] Beyond this statutory malleability, the federal judiciary may provide a more hospitable forum for climate engineering litigation than it has offered to climate change tort claims under federal common law.[13] Climate engineering litigation can sidestep some of the jurisprudential traps that have waylaid other climate change courtroom initiatives by presenting a reversed image of earlier climate change public nuisance lawsuits: rather than attempting to hold innumerable defendants liable for greenhouse gases emitted throughout the globe over extended periods of time, climate engineering lawsuits would target a small number of defendants for projects expressly designed to yield measurable contemporaneous changes to climate.

As a result, climate engineering litigation may provide an unexpected opportunity for U.S. courts to clarify threshold issues on the judicial branch's ability to hear lawsuits over global climate change. Although federal climate change nuisance lawsuits have garnered the most immediate attention,[14] legal battles over climate engineering projects may ultimately offer a faster, clearer, and more compelling avenue for the U.S. courts to define their role in the developing law of climate change control and liability.

This chapter examines how U.S. environmental laws might apply to climate engineering research and how the U.S. courts would review disputes over those projects. Section 1 surveys the development and background of climate change policy and explains how climate engineering fits into that structure. Section 2 outlines specific technologies and techniques used in climate engineering. The attributes of climate engineering itself will define the likely parties involved in future legal actions as well as the likely initial strategies and approaches to these legal issues. Section 3 examines how challenges to climate engineering might avoid, or fall prey to, roadblocks that have impeded efforts to bring environmental lawsuits under federal

[12] Chevron U.S.A., Inc. v. Natural Res. Def. Council, 467 U.S. 837 (1984) (using a deferential standard to review an agency determination within the area of expertise provided to that agency by the statute, where underlying federal statute did not convey congressional intent in unambiguous language). *See also* Am. Elec. Power Co. v. Connecticut, 131 S. Ct. 2527 (2011) (discussed *infra* at note 178).

[13] *See* discussion *infra* Section 3.2 (barriers to climate change public nuisance litigation).

[14] *Id.*

environmental statutes and tort law targeting governmental or private entities for their contributions to global climate change effects. Some of these litigation pitfalls include doctrines on standing, justiciability, proof of causation, and limitations on remedies that a court can impose. This chapter concludes by pointing out how this new type of environmental litigation may provide an opportunity for U.S. courts to address climate change issues in a context better suited to their institutional role and limits, and offers suggestions on how the federal government might best respond to these challenges.

1. CURRENT CLIMATE CHANGE LEGAL STRATEGIES: CONTROLLING EMISSIONS AND MITIGATING DAMAGES

Existing international and U.S. regulatory strategies to ameliorate climate change – with some important exceptions – focus largely on either mitigation[15] or adaptation.[16] These approaches generally seek to limit future climate disruption by either reducing current or future emissions of GHGs through regulatory controls, incentives, and sequestration activities, or by helping societies or ecosystems to adapt to an environment with higher temperatures.[17] From the U.N. Framework Convention on Climate Change[18] to the Kyoto Protocol[19] to the Cancun Agreement,[20] almost every international agreement has incorporated these two approaches. Although the UNFCC and its implementing instruments also offer other compliance options that

[15] Mitigation strategies focus on reducing or modifying activities that emit anthropogenic GHGs. These efforts frequently focus on reducing current and future GHG emissions per unit of output. SYNTHESIS REPORT, *supra* note 4, at 84.

[16] Adaptation strategies focus on modifying human societies and natural ecosystems so that they can continue to function in warmer climates. Adaptation strategies do not attempt to minimize climate change itself. *Id.* at 76. For a survey of potential strategies that large urban centers may use to deal with higher temperatures, see MATTHEW E. KAHN, CLIMATOPOLIS: HOW OUR CITIES WILL THRIVE IN THE HOTTER FUTURE (2010).

[17] SYNTHESIS REPORT, *supra* note 4, at 76, 84.

[18] United Nations Conference on Environment and Development, Rio de Janiero, Braz., June 3–14, 1992, *United Nations Framework Convention on Climate Change*, U.N. Doc. FCCC/INFORMAL/84 (1992) [hereinafter UNFCCC], *available at* http://unfccc.int/resource/docs/convkp/conveng.pdf. After 166 countries ratified the UNFCCC, it entered into force on March 21, 1994. *Status of Ratification of the Convention*, UNITED NATIONS, http://unfccc.int/essential_background/convention/status_of_ratification/items/2631.php (last visited Nov. 29, 2010). Currently, 194 countries have ratified the UNFCCC. *Id.*

[19] Kyoto Protocol to the United Nations Framework Convention on Climate Change, Dec. 11, 1997, 2303 U.N.T.S. 148.

[20] United Nations Framework Convention on Climate Change, Report of the Conference of the Parties on Its Sixteenth Session, Cancun, Mexico, Nov. 29–Dec. 10, 2010, *The Cancun Agreements: Outcome of the Work of the Ad Hoc Working Group on Long-Term Cooperative Action under the Convention*, Decision 1/CP.16, U.N. Doc. FCCC/CP/2010/7/Add.1 (Mar. 15, 2011), *available at* http://unfccc.int/resource/docs/2010/cop16/eng/07a01.pdf#page=2 [hereinafter The Cancun Agreements].

would arguably reduce ambient GHG levels through afforestation or agricultural activities, these alternatives generally concentrate on generating credits or allowances that can offset GHG emissions from other activities.[21] Individual efforts by other nations primarily adopt mitigation and adaptation techniques as well.[22]

In particular, U.S. legislative initiatives and state programs have focused primarily on such mitigation and adaptation. For example, both the Regional Greenhouse Gas Initiative (RGGI) and California's Assembly Bill 32 statutory program establish cap-and-trade programs that seek to limit future emissions of carbon dioxide and other greenhouse gases and thereby reduce the growing amount of greenhouse gases in the atmosphere.[23] Although this generalization admittedly excludes some projects that actively remove greenhouse gases from the atmosphere (for example, carbon sequestration through afforestation), the majority of climate change strategies focus on either reducing the flow of gases into the atmosphere, promoting or protecting natural processes that absorb GHGs, or planning to adapt to an altered global climate.

A growing group of researchers now believe, however, that efforts to curb current and future GHG emissions may not be sufficient to keep the concentrations of GHGs in the atmosphere below the critical threshold.[24] These researchers base their

[21] The mechanisms under the Kyoto Protocol are emissions trading, the Clean Development Mechanism (CDM), and Joint Implementation (JI). Article 17 of the Kyoto Protocol governs emissions trading. Article 12 defines CDM, which allows an Annex B party under the Protocol to implement an emission-reducing program in a developing country and thereby earn certified emission reduction credits equal to one ton of carbon dioxide. The JI falls under Article 6 and allows an Annex B party to earn emission reduction units from emission-reducing or emission removal projects in other Annex B countries. For more information, see *The Mechanisms under the Kyoto Protocol: Emissions Trading, the Clean Development Mechanism, and Joint Implementation*, UNITED NATIONS, at http://unfccc.int/kyoto_protocol/mechanisms/items/1673.php (last visited 11/8/11).

[22] Brazil, for example, has used a mixture of energy efficiency, renewable electricity, cogeneration, and biofuels to reduce the country's annual emissions by 10 percent. WILLIAM CHANDLER ET AL., PEW CENTER ON GLOBAL CLIMATE CHANGE, CLIMATE CHANGE MITIGATION IN DEVELOPING COUNTRIES: BRAZIL, CHINA, INDIA, MEXICO, SOUTH AFRICA, AND TURKEY, at iii (2002). The study also notes that deforestation in Brazil is a major contributor to climate change, and the government has done very little to abate that problem. *Id.* at 5.

[23] For a comprehensive description of regional initiatives against climate change and a fifty-state survey of state climate change laws, regulations, and policies, see GLOBAL CLIMATE CHANGE AND U.S. LAW 315–419 (M. Gerrard ed., 2007).

[24] For example, Dr. James Hansen, a well-known and influential scientist advocating aggressive action to constrain GHG emissions, has argued that "[t]he dangerous level of carbon dioxide, at which we will set in motion unstoppable changes, is at most 450 parts per million, but it may be less ... We must make significant changes within a decade to avoid setting in motion unstoppable climatic change." James Hansen, *Tipping Point: Perspective of a Climatologist*, *in* STATE OF THE WILD 2008–2009: A GLOBAL PORTRAIT OF WILDLIFE, WILDLANDS, AND OCEANS 6–15 (E. Fearn ed., 2008); *see also* James Hansen & Makiko Sato, *Paleoclimate Implications for Human-Made Climate Change* (2011), *available at* www.columbia.edu/~jeh1/mailings/2011/20110118_MilankovicPaper.pdf (arguing similar tipping point effects in sea level rise from GHG levels under business-as-usual scenarios).

concerns on the physical properties of some GHGs and the sheer volume of GHGs already in the atmosphere. One estimate of the longevity of atmospheric CO_2 perturbations concluded that the atmosphere would still retain 40 percent of its peak CO_2 concentration enhancement over preindustrial values as a quasi-equilibrium state even after a thousand years.[25] The decay rate of the remaining CO_2 would fall to even slower rates for years after the thousand-year mark.[26] Although preindustrial concentrations of carbon dioxide in the atmosphere were approximately 280 parts per million (ppm), the existing atmospheric loads of CO_2 have already reached 388.92 ppm.[27] This CO_2 burden will not cycle out of the atmosphere for several hundred years even if all industrial activities halted immediately.[28] In effect, significant climate changes because of elevated ambient GHG levels may have already happened. We are simply waiting for the full ramifications of changes that will result from prior activities. The risk of self-reinforcing processes that release GHGs and the prospect of abrupt climate change have only heightened these concerns.

Given these daunting challenges, some engineers and scientists began to call for strategies to directly alter climate change processes. The idea of this type of climate engineering is not new. The advent of advanced weather radar systems after World War II raised hopes that the practice of planet-wide climate modification was within reach.[29] In a well-publicized and controversial early climate engineering effort known as "Project Cirrus," General Electric (GE) attempted to modify the strength and path of an Atlantic hurricane.[30] Although the storm originally was drifting away from land into the eastern Atlantic, the storm reversed course after

[25] S. Solomon et al., *Irreversible Climate Change due to Carbon Dioxide Emissions*, 106 PROC. OF THE NAT'L ACAD. OF SCI. 1704, 1705 (2008), *available at* http://www.pnas.org/content/early/2009/01/28/0812721106.full.pdf+html; H. Damon Matthews & Ken Caldeira, *Stabilizing Climate Requires Near-Zero Emissions*, GEOPHYSICAL RES. LETTERS Vol. 35 L04705, Feb. 27, 2008. By some estimates, 25 percent of CO_2 emitted currently will remain in the atmosphere after five thousand years. A. Montenegro et al., *Long Term Fate of Anthropogenic Carbon*, GEOPHYSICAL RES. LETTERS 1 (Oct. 2007).

[26] Solomo et al., *supra* note 25, at 1705.

[27] The monthly mean CO_2 levels for October 2011 reached 388.92 ppm at the National Oceanic and Atmospheric Administration's monitoring post at Mauna Loa, Hawai'i. Nat'l Oceanic & Atmospheric Admin., *Mauna Loa CO2 Monthly Mean Data* (Nov. 16, 2011, 2:45 PM), ftp://ftp.cmdl.noaa.gov/ccg/co2/trends/co2_mm_mlo.txt.

[28] Solomon et al. *supra* note 25, at 1705. Although emissions of other GHGs such as methane NH_4 or N_2O oxides can affect climate change over a time period of decades or centuries, they do not persist in the atmosphere on the same timescales as CO_2. *Id.*; *see also* Piers Forster et al., *Changes in Atmospheric Constituents and in Radiative Forcing, in* CLIMATE CHANGE 2007: THE PHYSICAL SCIENCE BASIS 747–845 (S. Solomon et al. eds., 2007).

[29] For an illuminating review of the colorful prior attempts to modify the weather, see JAMES R. FLEMING, FIXING THE SKY: THE CHECKERED HISTORY OF WEATHER AND CLIMATE CONTROL (2010).

[30] GEN. ELEC., REPORT NO. RL-758, JULY 1952 GENERAL ELECTRIC RESEARCH LABORATORY, HISTORY OF PROJECT CIRRUS 61–64 (Barrington S. Havens ed., 1952), *available at* http://ia700402.us.archive.org/14/items/historyofprojectoohave/historyofprojectoohave.pdf.

GE's seeding effort and eventually struck the Georgia coast where it inflicted serious damage.[31] GE subsequently abandoned its hurricane program,[32] but discussions of weather engineering have continued to circulate through the climate community. These efforts included renewed but unsuccessful attempts to modify hurricanes in "Project Stormfury" from 1962 through 1983.[33]

Climate engineering and other adaptation strategies have historically drawn opposition out of concerns that they would simply distract popular attention and political will from necessary GHG emission control strategies.[34] That resistance shifted significantly in 2006. After long reluctance to seriously scrutinize climate engineering strategies, several climate scientists stepped forward to urge new efforts to study these alternatives as a fallback strategy to control climate change if current greenhouse gas emission control strategies failed. In particular, Paul Crutzen, a Nobel laureate in atmospheric science studies, published a keynote paper that assessed the feasibility of releasing aerosol particles into the upper atmosphere to reduce the amount of

[31] FLEMING, *supra* note 29, at 152–53. Of course, the storm's change in direction after seeding did not directly or conclusively establish that GE's efforts actually steered the hurricane.

[32] *Id.* at 153. GE's decision to withdraw from hurricane modification and weather manipulation research after Project Cirrus may reflect in part its earlier concerns over potential lawsuits over damage from weather events that the technologies allegedly affected. *Id.* at 148–49; C. PARKINSON, COMING CLIMATE CRISIS? CONSIDER THE PAST, BEWARE THE BIG FIX 206 (2010) ("[n]o one knows whether this shift [of the hurricane's direction] was or was not influenced by the cloud seeding, but in any event the changed direction resulted in the hurricane's slamming into the coast of the state of Georgia, with the quite undesired further effect of a flurry of lawsuits against General Electric..."). Weather modification (predominantly rain making) has a long legal history where courts or legislatures have attempted to allocate liabilities for damages allegedly caused by cloud seeding or other technologies. These laws are surveyed in the work of the late Professor Ray Jay Davis, who was a recognized expert in weather modification law. R. Davis, *Real Property Issues in Weather Control, in* 8–71 THOMPSON ON REAL PROPERTY § 71.06 (David A. Thomas, ed. 2004).

[33] FLEMING, *supra* note 29, at 177–79; *see also* Nat'l Atmospheric & Oceanic Administration, *Hurricane Research Division*, www.aoml.noaa.gov/hrd/hrd_sub/sfury.html (last visited Nov. 30, 2011). Despite doubts that the Cirrus Project showed that cloud seeding or other weather modification techniques could affect the course of a hurricane or other large storm system, the U.S. government undertook even more ambitious attempts to modify hurricane formation and direction in 1962 through 1983 under Project Stormfury. This effort led to seedings of several hurricanes with silver iodide or dry ice from 1963 through 1971. Ultimately, the project failed to yield unequivocal data to demonstrate that the hurricanes' behavior reflected human intervention rather than normal climatic processes. *See* FLEMING, *supra*, at 177–79 ("Frustration mounted as Stormfury scientists began to realize that their hurricane-seeding hypotheses were flawed. First of all, hurricanes contain very little of the super-cooled water that is necessary for effective silver iodide seeding. Also, the effects of seeding were so small that they were impossible to measure. Morale plummeted when Stormfury scientists learned that the navy intended to weaponize their research").

[34] *See, e.g.*, JEFF GOODELL, HOW TO COOL THE PLANET: CLIMATE ENGINEERING AND THE AUDACIOUS QUEST TO FIX THE EARTH'S CLIMATE 13 (2010) ("Although the dream of manipulating the weather is almost as old as civilization itself, the idea of studying ways of deploying technology to manage the earth's climate was seen by some scientists as politically incorrect, dangerous, or just downright silly.").

sunlight reaching the earth's surface.[35] Crutzen concluded that this strategy could yield substantial temperature reductions on a global scale, but he also pointed out that there are large areas of uncertainty and undesirable effects that this strategy might cause. For example, he noted that these techniques would not reduce damages attributable to increased rain acidification or answer the unchecked acidification of ocean waters.[36] Crutzen also spearheaded a symposium at Harvard University in 2008 to discuss potential climate engineering strategies.[37]

After the Harvard symposium, the discussion of climate engineering proposals steadily grew in scientific journals and spilled over into more mainstream sources and policy reports.[38] The British Royal Society released a comprehensive study of climate engineering options that highlighted the most viable technologies for countering the effects of climate change and concluded that "further research and development of climate engineering options should be undertaken to investigate whether low risk methods can be made available if it becomes necessary to reduce the rate of warming this century."[39] The same sentiment is also seen in the IPCC's meeting in June 2011 to consider the scientific basis for climate engineering as well as its costs and impacts.[40] Additionally, the U.S. Congress has held hearings to assess the implications of what committee members called "large-scale climate intervention."[41]

The burst of interest in climate engineering has already sparked efforts to limit research and demonstration projects. Most of the early attention has focused on ocean fertilization because at least thirteen experiments have already occurred on

[35] Paul J. Crutzen, *Albedo Enhancement by Stratospheric Sulfur Injections: A Contribution to Resolve a Policy Dilemma?*, 77 CLIMATIC CHANGE 211 (2006).

[36] *Id.* at 217; *see generally A Rational Discussion of Climate Change: The Science, the Evidence, the Response: Hearing Before the Subcomm. on Energy and Env't of the H. Comm. on Sci. and Tech.*, 111th Cong. (2010) (testimony of Richard A. Feely, PhD, Office of Oceanic and Atmospheric Research) (overview of general effects of ocean acidification because of elevated atmospheric levels of carbon dioxide).

[37] *See generally* Eli Kintisch, *Tinkering with the Climate to Get Hearing at Harvard Meeting*, 318 SCI. 551 (Oct. 26, 2007) (discussing Crutzen's upcoming workshop at Harvard).

[38] ELI KINTISCH, HACK THE PLANET: SCIENCE'S BEST HOPE – OR WORST NIGHTMARE – FOR AVERTING CLIMATE CATASTROPHE 13–14 (2010) ("Since the Harvard meeting, almost every forum relevant to the climate crisis has reached out to embrace, if tentatively, the former pariah called geoengineering.").

[39] ROYAL SOCIETY STUDY, *supra* note 10, at 57.

[40] *See also* Alyson Kenward, *Scientists Consider Whether to Cause Global Cooling*, CLIMATE CENTRAL (Oct. 19, 2010), http://www.climatecentral.org/news/newsscientists-consider-whether-to-cause-global-cooling/; Jeff Tollefson, *Geoengineering Faces Ban*, NATURENEWS (NOV. 2, 2010), http://www.nature.com/news/2010/101102/full/468013a.html.

[41] The House Science & Technology Committee held hearings in 2009 to explore the technological background, risks, benefits, and governance issues surrounding potential climate engineering schemes. *Geoengineering: Assessing the Implications of Large-Scale Climate Intervention: Hearing Before the H. Comm. on Sci. and Tech.*, 111th Cong. (2009); *see also* Press Release, H. Comm. on Sci. and Tech., Climate Engineering Research Needed, Members Hear (Nov. 5, 2009), http://archives.democrats.science.house.gov/press/PRArticle.aspx?NewsID=2676.

the high seas.[42] One particular proposal by Planktos, Inc., a commercial venture group seeking to generate tradable carbon credits, created controversy because it planned to release one hundred tons of iron ore dust into the Pacific Ocean near the Galapagos Islands in August 2007.[43] The experiment aimed to investigate marine phytoplankton blooms as a potential tool to sequester CO_2 in deep waters.[44] After strong environmentalist opposition – including a permanent patrol vessel by Greenpeace to intercept and halt any attempt by Planktos to release the iron – Planktos abandoned the project in February 2008.[45]

In response to the controversy, the EPA notified Planktos that the iron seeding might require a permit under the Marine Protection, Research and Sanctuaries Act (MPRSA).[46] EPA also submitted a statement of concern on behalf of the United States to the parties to the London Convention.[47] An International Marine

[42] United Nations Educational, Scientific and Cultural Organization, OCEAN FERTILIZATION: A SCIENTIFIC SUMMARY FOR POLICY MAKERS 3 (2011) [hereinafter OCEAN FERTILIZATION].

[43] R. Abate & A. Greenlee, *Sowing Seeds Uncertain: Ocean Iron Fertilization, Climate Change, and the International Environmental Law Framework*, 27 PACE ENVTL. L. REV. 555, 558 (2010), *available at* http://digitalcommons.pace.edu/pelr/vol27/iss2/5; Int'l Maritime Org., Scientific Group of the London Convention & Scientific Group of the London Protocol, June 18–22, 2007, LC/SG 30/INF.28 (June 1, 2007) ("[i]t is the understanding of the United States Government that the United States-based for-profit company Planktos, Inc., plans to dissolve up to 100 tons of iron dust in a 100 km by 100 km area approximately 350 miles west of the Galapagos Islands in June 2007 in order to stimulate phytoplankton blooms. Because this iron release project will not be done by vessels flagged in the United States or by vessels leaving from the United States, the United States Government does not have jurisdiction to regulate this project under its law implementing the London Convention").

[44] *Id.* at 558; Raphael Sagarin et al., *Iron Fertilization in the Ocean for Climate Mitigation: Legal, Economic and Environmental Challenges* 7–8 (Duke University Institute for Environmental Policy Solutions, Working Paper 07–07, 2007), *available at* http://nicholasinstitute.duke.edu/oceans/marinees/iron-fertilization-in-the-ocean-for-climate-mitigation-legal-economic-and-environmental-challenges/at_download/paper.

[45] *Planktos Kills Iron Fertilization Project due to Environmental Opposition*, MONGABAY.COM (Feb. 19, 2008), http://news.mongabay.com/2008/0219-planktos.html; *see also Planktos Is a No-Show in the Galapagos*, SEA SHEPHERD (Aug. 10, 2007), http://www.seashepherd.org/news-and-media/news-070810–1.html.

[46] The EPA notified Planktos that MPRSA might apply to the experiment if it took place in waters under U.S. jurisdiction or if Planktos undertook the project from a United-States-flagged vessel. Planktos responded that it would not use a United-States-flagged vessel for the experiment. *See* discussion, *infra* notes 134–139 of potential MPRSA requirements for climate engineering projects.

[47] The London Convention, an international organization consisting of eighty-six member states, is charged with implementation of the London Convention of 1972. This Convention controls the discharge of pollutants into the high seas. The London Protocol was agreed to in 1996 to modernize the Convention and, eventually, supersede its substantive provisions. The Protocol prohibits all dumping except for potentially acceptable wastes on the so-called reverse list. It entered into force on March 24, 2006, and thirty-eight states have joined the Protocol. The United States has joined the London Convention, but it has not acceded to the London Protocol. INTERNATIONAL MARITIME ORGANIZATION [IMO], THE LONDON CONVENTION AND PROTOCOL: THEIR ROLE AND CONTRIBUTION TO PROTECTION OF THE MARINE ENVIRONMENT (2008) *available at* http://www.imo.org/KnowledgeCentre/

Organization committee then adopted a resolution that included a "scientific statement of concern" and called for a halt to ocean fertilization projects unless they constituted legitimate scientific research.[48] The Convention subsequently adopted another resolution containing an assessment framework for scientific research into ocean fertilization.[49] These resolutions effectively declared that the Convention parties prohibited ocean fertilization projects conducted for commercial or non-scientific purposes; even scientific research could proceed only on a case-by-case basis. By 2012 the Convention intends to promulgate regulations governing ocean fertilization research.[50]

Other governmental entities have also taken action. In 2009, the German federal government ordered a team of researchers from the Alfred Wegener Institute for Polar and Marine Research to halt a test of iron seeding in the Southern Ocean in response to complaints that the iron releases constituted prohibited marine pollution.[51] Although the German government quickly withdrew its order,[52] legal opposition to climate engineering projects escalated. Most notably, the latest Conference of Parties to the Convention on Biological Diversity adopted a resolution that called for a limited moratorium on climate engineering activities "until there is an adequate scientific basis on which to justify such activities."[53] Despite division of opinions on

ShipsAndShippingFactsAndFigures/TheRoleandImportanceofInternationalShipping/IMO_Brochures/Documents/6%20page%20flyer%20London%20Convention.pdf.

[48] IMO, *Resolution on the Regulation of Ocean Fertilization*, Res.LC-LP.1 (Oct. 31, 2008).

[49] IMO, *Assessment Framework for Scientific Research Involving Ocean Fertilization*, Resolution LC-LP.2 (2010).

[50] IMO, Information on Work on Carbon Capture and Storage in Sub-Seabed Geological Formation and Ocean Fertilization under the London Convention and London Protocol 2, 16th Conf. of the UNFCCC (Nov. 2010), *available at* http://www.imo.org/OurWork/Environment/PollutionPrevention/AirPollution/Documents/COP%2016%20Submissions/IMO%20note%20on%20LC-LP%20matters.pdf.

[51] *Who Ate All the Algae? Using Phytoplankton to Capture Carbon Dioxide Hits a Snag*, ECONOMIST (Mar. 26, 2009), *available at* http://www.economist.com/node/13361464.

[52] *Id.*

[53] Conference of the Parties to the Convention on Biological Diversity, *Biodiversity and Climate Change: Draft Decision Submitted by the Chair of Working Group I* (Oct. 29, 2010). The original draft text included language that might have supported a blanket ban on climate engineering research projects, but the final text limited the prohibition to climate engineering projects that might affect biodiversity and that lacked transparent and effective governance mechanisms. The final language also included important exceptions for small-scale scientific research as well as a working definition of "geoengineering." *Compare* Convention on Biological Diversity, *Conference of Parties 10 Decision X/33 on Biodiversity and Climate Change* (Oct. 29, 2010) (final text), *available at* http://www.cbd.int/decision/cop/?id=12299, *with* Biodiversity and Climate Change, Draft Decision Submitted by Chair of Working Group I, UNEP/CBD/COP/10/L.36 at 8(w), (Oct. 29, 2010), *available at* http://www.cbd.int/doc/meetings/cop/cop-10/in-session/cop-10-L-36-en.doc. The Convention on Biological Diversity has also established a liaison group to assess climate engineering technologies and how international legal obligations may apply to them. *See, e.g.*, Report of the Liaison Group Meeting on Climate-Related

geoengineering research among environmental groups,[54] some organizations have both at international conferences[55] and in independent policy statements actively advocated a moratorium on further climate engineering research.[56]

Tests have also begun on climate engineering technologies beyond ocean fertilization. Researchers in the United Kingdom, for example, announced that they intended to lift a hose measuring up to twenty-five kilometers long into the upper atmosphere with a weather balloon so that they could test technologies for potential large-scale dispersion of sulfate aerosols.[57] Russian scientists sprayed a small amount of aerosols into the atmosphere to measure their effect on incoming solar radiation.[58]

Geo-Engineering as It Relates to the Convention on Biological Diversity, UNEP/CBD/LGCRG/ENG/1/5 (July 1, 2011).

[54] Interestingly, some environmental groups have signaled their willingness to consider carefully controlled research into potential geoengineering strategies. These groups usually emphasize the need for a strong governance structure before significant additional climate engineering research can take place. *See, e.g.*, FRIENDS OF THE EARTH, BRIEFING NOTE: GEOENGINEERING 4–5 (Nov. 2009), *available at* http://www.foe.co.uk/resource/briefing_notes/geoengineering.pdf (although condemning the failure of rich nations to reduce GHG emissions and opposing geoengineering proposals to reduce solar radiation reaching the earth's surface, Friends of the Earth concludes that "[i]t is now clear that mitigation alone cannot keep global temperatures below a safer threshold of 1–1.5 degrees above preindustrial levels" and that "[l]arge amounts of chemical air capture of carbon and storage – funded and carried out by rich countries – will probably be necessary, as long as safe storage sites can be identified and governance issues addressed"); THE ROYAL SOCIETY ET AL., THE SOLAR RADIATION MANAGEMENT GOVERNANCE INITIATIVE (SRMGI): ADVANCING THE INTERNATIONAL GOVERNANCE OF GEOENGINEERING 1–3 (Oct. 2010), *available at* http://www.srmgi.org/files/2010/10/SRMGI-project-description.pdf (arguing, with Environmental Defense Fund and the Academy of Sciences for the Developing World, for focus on governance of solar radiation management approaches to geoengineering); STEPHEN BRICK, NAT'L RES. DEF. COUNCIL, BIOCHAR: ASSESSING THE PROMISE AND RISKS TO GUIDE U.S. POLICY at iv–v, 1, 11–12 (Nov. 2010), *available at* www.nrdc.org/energy/files/biochar_paper.pdf (recommending additional research on biochar and noting its possible use as a "climate mitigation tool" to sequester large amounts of carbon dioxide out of the atmosphere).

[55] News Release, Action Grp. on Erosion, Tech. and Concentration, Hands Off Mother Earth! Civil Society Groups Announce New Global Campaign against Geoengineering Tests (Apr. 21, 2010). Over sixty civil society groups announced a joint campaign to oppose climate engineering tests. *Id.*

[56] *Id.*; *see also* ACTION GRP. ON EROSION, TECH. AND CONCENTRATION, GEOPIRACY: THE CASE AGAINST GEOENGINEERING 39–40 (Oct. 2010), *available at* http://www.etcgroup.org/en/node/5217 (calling for ban on climate engineering research until governance framework in place).

[57] University of Cambridge, *Stratospheric Particle Injection for Climate Engineering (SPICE)*, http://www2.eng.cam.ac.uk/~hemh/SPICE/SPICE.htm (last visited Nov. 27, 2011). Although the research consortium originally intended to conduct its experiment in November 2011, it postponed the experiment until April 2012 because of criticism and objections from numerous parties. Bob Yirka, *SPICE Geoengineering Project Delayed due to Critics Issues* (Oct. 5, 2011), http://www.physorg.com/news/2011-10-spice-geoengineering-due-critics-issues.html (last visited Nov. 27, 2011). The experiment was cancelled again in April 2012 because some of the participants in the project had concerns about a patent application. Daniel Cressey, *Cancelled Project Spurs Debate Over Geoengineering Patents*, 485 NATURE 429, 429 (2012).

[58] Yu A. Israel et al., *Field Experiment on Studying Solar Radiation Passing through Aerosol Layers*, 34 RUSSIAN METEOROLOGY & HYDROLOGY 265, 266 (2009).

Another proposed project would assess technologies to reverse the effects of ocean acidification caused by elevated atmospheric CO_2 levels.[59]

Despite calls for a moratorium on climate engineering research, the comparatively low research costs have enticed private investors to take initial steps into the field. For example, Bill Gates has funded more than $4.5 million worth of research on reducing the amount of GHGs in the atmosphere through adaptation measures and climate engineering.[60] In 2010, Gates was part of a group providing funds to a Silicon Valley inventor's plan to make clouds whiter so that they more effectively reflect solar radiation.[61] Additionally, private companies such as Climos have formed to attract capital and to conduct research outside the realm of public subsidies or public policy statements.[62] If climate engineering projects ultimately yield tradable credits for reductions in GHG emissions, private investors will have even stronger incentives to become more actively involved in climate engineering research and projects.[63]

2. THE NEXT STEP: POSSIBLE CLIMATE ENGINEERING STRATEGIES

Several possible engineering strategies have surfaced to address global climate change effects. Surprisingly, initial evaluations of some of these strategies show that they might significantly reduce climate change effects caused by current GHG

[59] Michael Marshall, *Geoengineering Trials Underway*, NOVIM (Sept. 14, 2011), http://www.novim.org/resources/novim-news/121-geoengineering-trials-get-under-way ("[e]lsewhere, Ken Caldeira of the Carnegie Institution for Science in Stanford, California, has permission to add sodium hydroxide – an alkali – to a small patch of ocean to see if it can reverse the effects of ocean acidification").

[60] Eli Kintisch, *Bill Gates Funding Geoengineering Research*, SCIENCE INSIDER (Jan. 26, 2010, 2:10 PM), http://news.sciencemag.org/scienceinsider/2010/01/bill-gates-fund.html. Gates has already applied as a coinventor on a patent in 2008 to "sap hurricanes of their strength by mixing surface and deep ocean water." *Id.*

[61] Oren Dorell, *Can Whiter Clouds Reduce Global Warming?*, USA TODAY (June 11, 2010, 12:37 AM), http://www.usatoday.com/weather/research/2010–06–10-cloud-whitening_N.htm.

[62] Scant information is available regarding these companies, but for more information on Climos's funding and business model, see *Frequently Asked Questions about Ocean Fertilization*, CLIMOS, http://www.climos.com/faq.php#9 (last visited Nov. 8, 2011).

[63] As noted in the introduction to this chapter, some entrepreneurs have already undertaken ocean iron seeding projects in hopes of generating tradable carbon emission credits for profit. Other entrepreneurs will undoubtedly view climate engineering as a set of valuable marketable technical skills that they can provide to governments or individuals who wish to respond to or forestall climate events. Notably, the final version of the American Clean Energy and Security Act of 2009 specifically excluded ocean fertilization projects from the definition of CO_2 "sequestration" that could receive funding and tax credits. *See* H.R. 2454 (111st Congr., 1st Sess.) (placed on Senate Calendar, July 7, 2009) (section 312 adds new Section 700(44) that defines "Sequestered and Sequestration" as "the separation, isolation, or removal of greenhouse gases from the atmosphere, as determined by the Administrator. The terms include biological, geologic, and mineral sequestration, but do not include ocean fertilization techniques").

levels in the atmosphere. More research and information will be needed, however, where each of these techniques poses unique risks and areas of concern.

Controversy has already emerged over the definitions of "climate intervention" or "geoengineering." These disagreements arise largely from the fact that the definition of these terms could exclude some technologies from any future regulatory framework or treaty governing climate engineering. For example, some definitions would exclude techniques such as biochar management, carbon capture and sequestration, and albedo enhancement through white roofs and more reflective vegetation.[64] Most definitions, however, include three common elements: (1) the intentional intervention or manipulation (2) of environmental systems, including systems related to climate, (3) to reduce or offset the effects of anthropogenic global warming.[65] The technologies described below contain each of these concepts. This chapter will focus on technologies that, by consensus, squarely fall within the definition of climate engineering, but many of the legal issues raised below will also apply to techniques that might lie outside some definitions of the term.[66]

Most proposed climate engineering strategies seek either to remediate existing high stores of CO_2 in the ambient atmosphere or to intervene directly in climatic processes that generate global warming. For example, one category of climate engineering would modify the amount of solar radiation that reaches the earth's surface

[64] Biochar is a charcoal-like substance made from biomass such as crop wastes and dross, and it can serve as a soil conditioner or secondary energy source. BRICK, *supra* note 54, at iv (Nov. 2010), *available at* www.nrdc.org/energy/files/biochar_paper.pdf. One study has suggested that the production of biochar may also allow the long-term sequestration in soils of up to 12 percent of global emissions of carbon dioxide. *Id.* at 11–12; Dominic Woolf et al., *Sustainable Biochar to Mitigate Global Climate Change*, NATURE COMMS. (Aug. 10, 2010), http://www.nature.com/ncomms/journal/v1/n5/full/ncomms1053.html. Carbon capture and sequestration technologies could remove GHGs from industrial emissions and then store them in secure geological or engineered structures for long time periods. Federal Requirements under the Underground Injection Control (UIC) Program for Carbon Dioxide (CO_2) Geologic Sequestration (GS) Wells, 75 Fed. Reg. 77,230, 77,233–35 (Dec. 10, 2010) (to be codified at 40 C.F.R. pts. 124, & 144–147). Albedo enhancement reduces climate change caused by solar radiation influx by simply reflecting as much sunlight as possible away from the earth's surface and back into space. Some albedo enhancement techniques include the use of light-colored material in roofs or placing reflective materials over large areas of unoccupied land. David W. Keith, *Geoengineering the Climate: History and Prospect*, 25 ANN. REV. OF ENERGY & THE ENV'T 245, 264 (2000) (discussing geoengineering through surface albedo enhancement); PARKINSON, *supra* note 32, at 173–75, 190 (finding that a square mile of desert covered with light-reflective polyethylene sheets would offset the emissions of seven thousand sports utility vehicles over a fifteen-year period).

[65] *See, e.g.*, ROYAL SOCIETY STUDY, *supra* note 10, at 1 ("[g]eoengineering proposals aim to intervene in the climate system by deliberately modifying the Earth's energy balance to reduce increases of temperature and eventually stabilise temperature at a lower level than would otherwise be attained"); U.S. GOVERNMENT ACCOUNTABILITY OFFICE, CENTER FOR SCIENCE, TECHNOLOGY AND ENGINEERING, TECHNOLOGY ASSESSMENT CLIMATE ENGINEERING: TECHNICAL STATUS, FUTURE DIRECTIONS, AND POTENTIAL RESPONSES, GAO-11–71, at 3 (July 2011).

[66] Royal Society Study,*Id.*; *see also* R. Lal, *Sequestering Atmospheric Carbon Dioxide*, 28:3 CRIT. REV. PLANT SCI. 90, 90 (2009); Keith, *supra* note 64, at 265, 281 (discussing carbon uptake by genetically modified organisms).

(solar radiation management (SRM)). By contrast, other technologies spur uptake of GHG by marine, geological, or arboreal biological sources, or by mechanical devices to remove or directly reduce existing stocks of GHGs in the atmosphere (carbon dioxide removal (CDR)). Although SRM technologies tend to attract the most concern and legal attention (for reasons discussed below), even CDR technologies can pose nettlesome policy and legal issues. For example, the use of CDR may significantly affect delicate ecosystems where the technology is deployed.[67]

With this division in mind, some of the most imminently feasible climate engineering approaches include the following methods:

Reduce Solar Influx. Much of the initial scientific scrutiny and concern has centered on techniques that directly reduce the amount of sunlight reaching the earth's surface. Several different techniques can achieve this goal. In particular, Crutzen's proposal would use the dispersal of sulfate aerosol particles in the stratosphere to scatter and reflect sunlight back into space. According to his calculations, this approach can yield significant reductions in surface global temperatures on a wide scale for a comparatively small cost of $25 to $50 billion annually.[68] Other proposals would involve the use of space-based reflective particles or mirrors placed in low or geostationary orbit to directly scatter sunlight before it reaches the earth's atmosphere.[69]

Enhance Production of High-Albedo Cloud and Surface Cover. Because certain types of clouds reflect a significant percentage of sunlight back into space, several proposals have focused on using seeding techniques to generate wide swaths of cloud cover over ocean areas. These techniques rely on recent scientific data showing that

[67] For example, large-scale iron seeding to enhance algal blooms may deplete levels of oxygen in the water column or promote the production of algal toxins. Large- scale CO_2 capture devices may also generate large volumes of calcium carbonate waste streams and possibly create waste disposal issues. Howard Herzog, Assessing the Feasibility of Capturing CO_2 from the Air (Oct. 2003) (unpublished thesis, Mass. Inst. of Tech.), *available at* http://step.berkeley.edu/Journal_Club/paper1_02092010.pdf; Diego Alvarez et al., *Behavior of Different Calcium-Based Sorbents in a Calcination/Carbonation Cycle for CO_2 Capture,* 21 ENERGY FUELS 1534, 1540 (2007); Charles G. Trick et al., *Iron Enrichment Stimulates Toxic Diatom Production in High-Nitrate, Low-Chlorophyll Areas,* 107 PROC. OF THE NAT'L ACAD. OF SCI. USA 5887, 5889 (2010); Arthur J. Miller et al., *Global Change and Oceanic Primary Productivity: Effects of Ocean-Atmosphere-Biological Feedbacks, in* 73 GLOBAL CLIMATE CHANGE AND RESPONSE OF CARBON CYCLE IN THE EQUATORIAL PACIFIC AND INDIAN OCEANS AND ADJACENT LANDMASSES 473 (2007).

[68] Crutzen, *supra* note 35, at 213 (to counteract global warming effects, the project would need to inject one to two teragrams of sulfur particulates into the stratosphere each year; such an effort would cost $25 to $50 billion annually). Estimates of the cost of unabated climate change damages are notoriously difficult and controversial. *See, e.g.*, NICHOLAS STERN, STERN REVIEW ON THE ECONOMICS OF CLIMATE CHANGE (2006). By comparison, however, one study estimates that the State of Alaska alone will face costs of up to $10 billion over the next few decades to address damage to its infrastructure caused by rising global temperatures. P. LARSEN ET AL., UNIVERSITY OF ALASKA, ESTIMATING FUTURE COSTS FOR ALASKA PUBLIC INFRASTRUCTURE AT RISK FROM CLIMATE CHANGE (2007), *available at* www.iser.uaa.alaska.edu/publications/JuneICICLE.pdf.

[69] ROYAL SOCIETY STUDY, *supra* note 10, at 46–48.

boat and jet contrails can be surprisingly effective at generating persistent high-level cloud formation. Under these proposals, autonomous sailing craft equipped with solar-powered engines would pump seawater to create a fine mist that they would disperse above sea level. In theory, these mists would have the ability to seed subsequent cloud formations.[70]

Increase Formation of Sea Ice. To halt or reverse the rapid shrinkage of polar ice caps and sea-based ice shelves, some scientists have proposed the use of sea-based snow projection for ice manufacturing that would seed additional production of ice at polar latitudes.[71] This approach, which could also rely on wind and nuclear power to help generate the ice, would theoretically need sufficient sea ice to create an enhanced albedo that would reflect sunlight back into space and reduce surface temperatures.[72]

Direct CO2 Sequestration through Ocean Seeding. One frequently discussed method of climate engineering is the addition of trace elements such as iron to certain portions of the ocean to enhance blooms of algae.[73] Because certain portions of the ocean ecosystem are limited by the scarce amounts of iron, even a comparatively small addition of distributed iron particles can lead to a burst of phytoplankton growth that can absorb CO_2 from the atmosphere directly above the ocean's surface.[74] In theory, the phytoplankton would then die and precipitate downward with CO_2 locked in their body mass.[75] At the ocean floor, the phytoplankton and the CO_2 would be sequestered on a long-term basis.[76] According to some studies, this process

[70] *Id.* at 27–28.

[71] S. Zhou & P.C. Flynn, *Geoengineering Downwelling Ocean Currents: A Cost Assessment*, 71 CLIMATIC CHANGE 203, 220 (2005) (concluding that formation of thicker sea ice by pumping ocean water onto the surface of ice sheets is the least-expensive proposed method to enhance downwelling ocean currents that would remove GHGs from the atmosphere).

[72] *Id.* at 207, 211.

[73] As noted above, the Alfred Wegener Institute in Germany planned to conduct an iron seeding experiment in 2009. *See, e.g., supra* notes 51–56 and accompanying text. The research ship was loaded with twenty tons of iron and ready to sail when the German government ordered it to stop and for further research to be conducted before the experiment was conducted. Quirin Schiermeier, *Ocean Fertilization Experiment Suspended,* NATURENEWS (Jan. 14, 2009), http://www.nature.com/news/2009/090114/full/news.2009.26.html.

[74] OCEAN FERTILIZATION, *supra* note 42, at 1, 5, 7.

[75] R.S. Lampitt et al., *Ocean Fertilization: A Potential Means of Geoengineering?*, 366 PHIL. TRANSACTIONS OF THE ROYAL SOC'Y A 3919, 3922 (2008) (concluding that CO_2 absorbed by algae can be sequestered from atmosphere for over one hundred years). Some researchers have raised concerns that the sinking of large amounts of algae into the deep ocean could convey nutrients such as nitrogen and phosphorous that would alter ocean ecological systems in unpredictable ways. Aaron Strong et al., *Ocean Fertilization: Time to Move On*, 461 NATURE 347 (2009). In particular, these concerns include the risk that ocean fertilization on a global scale could cause oxygen starvation in large regions of the ocean. *Id.*

[76] One common criticism of ocean fertilization experiments (and, indeed, of climate engineering approaches in general) is that they do not address other serious consequences of elevated ambient

has already begun on a natural basis because of releases of particulate iron from receding glaciers that have enhanced polar phytoplankton blooms.[77] Recent proposals have noted that iron fertilization of the ocean can also have substantial regional effects on wind patterns and the albedo of clouds affected by the release of sulfates from the enhanced phytoplankton growth.[78]

Marine Heat Transfer. Many of the most problematic climate change effects arise from higher ocean surface temperatures. For example, some climate models show that a broader difference between ocean surface temperatures and ambient air temperatures may lead to the formation of stronger and potentially more destructive hurricanes.[79] Ocean temperatures at deeper levels, however, remain much less affected by higher ambient air temperatures or surface solar radiation. As a result, some researchers have suggested that ocean heat pumps could moderate these climate effects by exchanging cooler deep marine waters with warmer surface waters.[80] These ocean heat pumps would consist of a large number of floating columns that would rely on the energy of wave motions to transport cooler water to the surface.[81] Some models show that a significant number of these floating heat exchangers could arguably reduce ocean surface temperatures over a broad area and potentially mitigate processes that might exacerbate the risk of more severe hurricanes.[82]

Direct Air Capture. Another proposed strategy would tackle ambient CO_2 levels in a direct fashion by using a large number of mechanical devices to "scrub" the

CO_2 levels. For example, heightened CO_2 levels have contributed to growing acidification of ocean waters. Some researchers have suggested that some technologies could directly reduce or at least not increase ocean acidification on at least a regional level. Marshall, *supra* note 59, at 8; *see* K. House et al., *Electrochemical Acceleration of Chemical Weathering as an Energetically Feasible Approach to Mitigating Anthropogenic Climate Change*, 41 ENVTL. SCI. & TECH. 8464, 8464 (2007). Although these additional large-scale projects also likely qualify as climate engineering, this chapter will focus instead on projects directly aimed at either SRM or CDM.

77 Rob Raiswell et al., *Contributions from Glacially Derived Sediment to the Global Iron (Oxyhydr) oxide Cycle: Implications for Iron Delivery to the Oceans*, 70 GEOCHIMICA ET COSMOCHIMICA ACTA 2765–80 (2006) (concluding that delivery of iron nanoparticles through glacial shedding of icebergs may fertilize oceanic productivity and draw down atmospheric levels of carbon dioxide).

78 Nicholas Meskhidze et al., *Phytoplankton and Cloudiness in the Southern Ocean*, 314 SCI. 1419, 1420–21 (2006).

79 Kevin E. Trenberth, *Warmer Oceans, Stronger Hurricanes*, SCI. AM., June 14, 2007, at 44. Professor Kerry Emanuel from MIT was one of the first to publish research connecting these areas, and in 2008 Professor Emanuel released new findings further supporting his 2005 research. *See, e.g.*, *Interview: Exploring the Links between Hurricanes and Ocean Warming*, YALE ENV'T 360 (Sept. 15, 2010), http://e360.yale.edu/feature/exploring_the_links_between_hurricanes_and_ocean_warming/2318/.

80 Kelly Klima et al., *Does It Make Sense to Modify Tropical Cyclones? A Decision-Analytic Assessment*, 45 ENVTL. SCI. & TECH. 4242, 4242 (2011) (discussing a computer model that indicated that use of wind-wave pumps in path of tropical cyclone approaching South Florida "could reduce net losses from an intense storm more than hardening structures").

81 David Biello, *Halting Hurricanes*, SCI. AM., Nov. 2011, at 24.

82 *Id.*

CO_2 out of the air. This approach, if adopted on a large scale, would use liquid or dry sorbents[83] to capture CO_2 (typically in a carbonate), chemically release the CO_2 in a subsequent step, and then reuse the restored sorbent to collect more CO_2.[84] The captured CO_2 could either be sold for commercial use or geologically sequestered.[85] Under these scenarios, the global deployment of ten million CO_2 capture units could theoretically reduce ambient CO_2 levels by five parts per million per year, and the projected costs could drop to $30 per ton of CO_2 captured.[86] If it proves cost-effective, this technology could reduce ambient CO_2 levels with fewer side effects than other potential climate engineering techniques.[87]

As climate engineering studies continue to refine potential methods and techniques, some of the strategies described previously in this section may undergo significant revisions. For example, one suggested modification would use precisely engineered nanoparticles in place of sulfate aerosols to scatter sunlight from the upper atmosphere back into space.[88] The proposal notes that these particles could remain in the upper stratosphere for a much longer time than sulfate aerosols, and the nanoparticles can be engineered to cause them to aggregate in polar regions.[89] This type of regional climate engineering may offer an important step in protecting the environments facing the highest risks, such as the polar ice caps and the Great Barrier Reef,[90] but regional SRM climate engineering efforts may pose especially high risks of weather disruption and governance challenges.[91]

[83] A sorbent is "[a] material having the property of collecting molecules of a substance by sorption." Sorption in turn is "[t]he combined or undifferentiated action of adsorption [the adherence of specific gases, liquids or substances to the exposed surfaces of materials, usually solids, they are in contact with] and absorption [the swallowing up of items through their inclusion in or assimilation to something else]." OXFORD ENGLISH DICTIONARY (compact ed. 1987) (definitions of sorbent, absorption and adsorption).

[84] Klaus S. Lackner, *Washing Carbon out of the Air*, SCI. AM., June 2010, at 66, 66–69.

[85] *Id.* at 70.

[86] *Id.* at 65; David W. Keith et al., *Climate Strategy with CO2 Capture from the Air*, 74 CLIMATIC CHANGE 17–45 (Jan. 2006).

[87] Lackner, *supra* note 84, at 70–71 (refusing to classify direct removal of carbon dioxide from the atmosphere as geoengineering because it "does not change the natural dynamics of the earth or create a potential environmental risk," and "[a]ir capture simply withdraws the excess CO_2 from the atmosphere that humans are putting there").

[88] David W. Keith, *Photophoretic Levitation of Engineered Aerosols for Geoengineering*, 107 PROC. OF THE NAT'L ACAD. OF SCI. USA 16,428 (Sept. 21, 2010), *available at* www.pnas.org/cgi/doi/10.1073/pnas.1009519107.

[89] *Id.*

[90] COMM. ON SCI. & TECH., 111TH CONG., ENGINEERING THE CLIMATE: RESEARCH AND STRATEGIES FOR INTERNATIONAL COORDINATION 41 (Comm. Print 2010).

[91] *Staff of H. Comm. on Sci. and Tech., 111th Cong., Rep. on Geoengineering: Assessing the Implications of a Large Scale Climate Intervention: Hearing* (2009) (testimony by Dr. Shepherd of the Royal Society) (that "[i]t would … be generally undesirable to attempt to localize SRM methods, because any localized radiative forcing would need to be proportionally larger to achieve the same global effect, and this

Although the field is in its infancy, several striking characteristics of these various climate engineering techniques may affect future assessments of their legal status. First, all of these techniques offer the prospect of immediate and short-term moderation of climate change effects. This benefit, however, comes with a high degree of uncertainty about other potential costs and damages. For example, proposals to reduce solar influx through stratospheric distribution of aerosols have raised concerns that aerosol distribution might alter regional precipitation patterns, could delay recovery of the ozone layer and thereby increase skin cancer rates, will not address – and in fact may enhance – ocean acidification, could increase risk of damage to aircraft engines, and might cause particulates to precipitate onto surface environments in ways that affect human or ecological systems.[92]

The second notable common characteristic of these climate engineering approaches is that they can be performed unilaterally. As opposed to global emission control conventions that require participation from all of the significant players to yield any material effects, a single nation or even large corporation may have the resources to undertake one or many of these climate engineering projects.[93] For example, the cost of an aerosol distribution project could easily fall within the scope of one nation's resources.[94]

Third, every one of these climate engineering techniques will likely spark strong and impassioned opposition from potentially affected individuals and interest groups. Because of the large unknowns associated with each of these techniques and the risk

is likely to induce modifications to normal spatial patterns of weather systems including winds, clouds, precipitation and ocean currents and upwelling patterns").

[92] *See, e.g.*, Victor Brovkin et al., *Geoengineering Climate by Stratospheric Sulfur Injections: Earth System Vulnerability to Technological Failure*, 92 CLIMATIC CHANGE 243, 255 (2008) (DOI 10.1007/s10584-008-9490-1 (concluding that "stratospheric sulfur injections might be a feasible emergency solution for cooling the planet," but the injections would have to continue "for millennia unless future generations find a secure way to remove CO_2 from the atmosphere." The authors also point out that "[a] critical consequence of climate engineering is a possibility of extremely rapid warming in case the emissions are abruptly interrupted" leading to warming in polar regions that could exceed 10 degrees Celsius within a few decades).

[93] David Victor et al., *The Geoengineering Option: A Last Resort against Global Warming?*, FOREIGN AFFAIRS (Mar. /Apr. 2009), *available at* http://iis-db.stanford.edu/pubs/22456/The_Geoengineering_Option.pdf ("[b]y contrast, geoengineering is an option at the disposal of any reasonably advanced nation. A single country could deploy geoengineering systems from its own territory without consulting the rest of the planet ... Although governments are the most likely actors, some geoengineering options are cheap enough to be deployed by wealthy and capable individuals or companies").

[94] This prospect of unilateral climate engineering efforts by a major national power has already surfaced. In November 2005, the head of the Russian Global Climate and Energy Institute (and previously a vice-chair of the Intergovernmental Panel on Climate Change) urged Russian President Vladimir Putin that Russia should immediately discharge into the atmosphere 600,000 tons of sulfur aerosol particles. C. Brahic, *Hacking the Planet: The Only Climate Solution Left?*, 2697 NEW SCIENTIST 8, 10 (2009), *available at* http://www.newscientist.com/article/mg20126973.600-hacking-the-planet-the-only-climate-solution-left.html.

of unintentional damages that they pose, several environmental advocacy groups have already soundly denounced any approach that would use climate engineering.[95] Other groups and governments have opposed the use of climate engineering projects, or even investigations into their soundness, because they might detract from efforts to reduce ongoing emissions.[96] This opposition will likely grow over concerns that even relatively small additions of greenhouse gases to existing elevated atmospheric levels may cause large, unpredictable, or chaotic effects on climate. In other words, although the prospect of abrupt climate change might give climate engineering projects a sense of urgency, it also highlights the risk of unexpected catastrophic effects caused by those same technologies.

3. LEGAL PRINCIPLES FOR CLIMATE ENGINEERING DISPUTES

Most of the nascent legal challenges to climate engineering projects have focused on using existing international legal regimes to oppose or control test programs or demonstration efforts. This initial orientation appropriately reflects the global consequences of climate change issues and the planned location for climate engineering experiments, such as polar environments or on the high seas, which fall within the jurisdiction of international laws and treaties. Some observers and nongovernmental organizations urged delegates at the UNFCCC Conference of the Parties in Cancun, Mexico, to explore the regulation of climate engineering approaches, but the modest climate agreement from the conference did not expressly address this issue.[97]

3.1 *Potential Challenges under U.S. Environmental Laws to Climate Engineering Projects*

Approaches that would use domestic national laws to control unilateral climate engineering projects, by contrast, have received less attention.[98] In particular, U.S.

[95] Lauren Morello et al., *At U.N. Convention, Groups Push for Geoengineering Moratorium*, SCI. AM. (Oct. 20, 2010), http://www.scientificamerican.com/article.cfm?id=at-un-convention-groups-push. *But see* Keith, *supra* note 88 (discussing the division of opinion among environmental groups, and that some of the largest environmental advocacy organizations have signaled willingness to accept climate engineering research if performed with adequate controls and governance).

[96] For example, a large collection of environmental groups have banded together into a campaign named "Hands Off Mother Earth" (HOME). *See* http://www.handsoffmotherearth.org/about/ (last visited Nov. 29, 2010). The HOME coalition will advocate for an international prohibition or regulation of efforts to test or implement climate engineering technologies. *Id.*

[97] *See* The Cancun Agreements, *supra* note 20.

[98] U.S. GOV'T ACCOUNTABILITY OFFICE, CLIMATE CHANGE: A COORDINATED STRATEGY COULD FOCUS FEDERAL GEOENGINEERING RESEARCH AND INFORM GOVERNANCE EFFORTS, GAO-10-903, at 27 (Sept. 2010) ("EPA officials stated that the extent to which existing federal environmental laws apply to

courts will likely be the fora for some of the initial legal actions to fight climate engineering efforts that might cause environmental damage or large-scale unanticipated effects. The emergence of the United States as a central forum for litigation would reflect its developing key role in the nascent enterprise of climate engineering. Research projects on climate engineering have already received a high level of attention in the United States: U.S. citizens and corporations have provided significant early funding for climate engineering theoretical research.[99] Some early climate engineering projects will likely be directed by U.S. citizens or within U.S. territory, and domestic U.S. environmental statutes would offer attractive opportunities to challenge those first efforts. Federal and state courts may offer personal jurisdiction over U.S. citizens who undertake or participate in other climate engineering projects. U.S. courts and environmental laws may also provide opportunities for injunctive relief or damages that other national court systems might not grant as readily.[100]

For example, if a corporation with significant operations in the United States (or that had incorporated itself within a U.S. state) decides to undertake a climate engineering project within the United States, environmental groups could draw on many potential options under multiple federal environmental statutes to contest the project. Only some of those challenges are within the scope of this chapter. Most important, the specific facts surrounding each climate engineering project – including its location, type of technology, scale, and projected effects – will play a critical role in invoking the jurisdiction and application of particular federal or state

geoengineering is unclear, largely because detailed information on most geoengineering approaches and effects is not available"). This general statement by the EPA notably does not reflect either the EPA's assertion of MPRSA jurisdiction over ocean fertilization experiments by ships flying the U.S. flag. By comparison, the German federal government relied on domestic German law to temporarily restrict ocean fertilization experiments in the Southern Ocean in 2009. *See supra* notes 51–52.

[99] For example, Bill Gates has provided at least $4.5 million in funding on geoengineering research for many years, although none of those funds have gone to any field experiments. *See supra* note 60. Entrepreneurs have obtained funding for demonstration projects on ocean iron seeding. *See* discussion *supra*, notes 43–45 (initial seeding projects by Planktos) and notes 79–81 (wind-wave pump demonstration projects to dampen the effects of climate change on hurricane intensification); *see also* K. Jerch, *Capitalizing on Carbon*, *in* Alan Robock, 20 *Reasons Why Geoengineering May Be a Bad Idea*, 64 BULL. ATOMIC SCI. 16 (May/June 2008), *available at* http://www.thebulletin.org/files/064002006_0.pdf (Climos obtained $3.5 million in funding from Braemar Energy Ventures for ocean iron fertilization projects; other ocean fertilization ventures by Ocean Nourishment Corporation and Atmocean obtained funding).

[100] In addition to federal environmental laws, state laws also impose significant regulatory requirements on weather modification activities. These laws typically addressed efforts to make or control amounts of rainfall in a local region. *See supra* note 32. Although weather modification laws might provide a useful historical backdrop, these state and local laws ultimately will not likely play a significant role in legal challenges to climate engineering projects on a global or regional scale.

environmental statutes. Current nascent climate engineering proposals simply lack enough detail as yet to allow a fully focused assessment of the environmental statutory and regulatory duties that they might trigger.

Nonetheless, we can still forecast general principles and strategies for the application of federal U.S. environmental statutes to climate engineering efforts. First, and surprisingly, the United States may have already established – albeit unintentionally – a statutory framework to mandate reporting of any climate engineering projects. Although the federal government has left the substantive regulation of weather modification (predominantly cloud seeding and rainmaking ventures) to local and state authorities, Congress passed the National Weather Modification Policy Act of 1972 to track burgeoning weather modification activities.[101] Pursuant to this Act, the National Oceanic and Atmospheric Administration promulgated regulatory reporting requirements for such projects.[102] In particular, these regulations require persons who engage in weather modification to keep and preserve records of their activities and to report the results of their actions to the federal government.[103] The activities subject to this reporting requirement expressly include "[m]odifying the solar radiation energy exchange of the earth or clouds, through the release of gases, dusts, liquids or aerosols into the atmosphere"[104] – a definition that seems to apply readily to climate engineering technologies that employ solar radiation management. To date, no one has notified the federal government that it has undertaken a climate engineering project under this regulatory program.

[101] Weather Modification Reporting Act of 1972, Pub. L. No. 92–205, §3(a), 85 Stat. 735 (1971) (uncodified provisions where Congress declared a policy to establish a national policy for weather modification and appropriated $1,000,000 to the Secretary of the U.S. Department of Commerce to prepare a comprehensive study on the effects and potential of weather modification).

[102] 15 C.F.R. § 908 (2011).

[103] *Id.* §§ 908.4–908.9.

[104] *Id.* § 908.3(a)(3). The reporting requirement applies to broad categories of activities that might encompass other emerging climate engineering technologies, including "[s]eeding or dispersing of any substance into clouds or fog, to alter drop size distribution, produce ice crystals or coagulation of droplets, alter the development of hail or lightning, or influence in any way the natural development cycle of clouds or their environment"; "[m]odifying the characteristics of land or water surfaces by dusting or treating with powders, liquid sprays, dyes, or other materials"; "[r]eleasing electrically charged or radioactive particles, or ions, into the atmosphere"; "[a]pplying shock waves, sonic energy sources, or other explosive or acoustic sources to the atmosphere"; "[u]sing lasers or other sources of electromagnetic radiation"; and "other similar activities falling within the definition of weather modification as set forth in § 908.1." *Id.* §§ 903(a)(1)–(b). Although this reporting requirement does not apply to "activities of a purely local nature that can reasonably be expected not to modify the weather outside of the area of operation," this exemption will facially not apply to climate engineering projects. *See id.* § 908.3(c). The regulations also limit this exemption solely to the use of lightning rods, deployment of small heat sources to prevent frost damage, and religious activities and ceremonies seeking to alter the weather. *Id.*

Beyond this federal reporting requirement, a few key questions will guide the application of federal environmental statutes generally to climate engineering projects:

Who does the statute regulate? Most environmental statutes expressly define the "person" who falls within the statute's requirements. The definition of "person" in the Clean Air Act, for example, expressly includes individuals, corporations, states, and federal governmental agencies.[105] This broad scope of "person" means that virtually anyone sponsoring a climate engineering project – including state agencies or federal entities – could fall within the ambit of "persons" who must comply with Clean Air Act requirements.

Where does the statute apply? Although this analysis focuses on climate engineering projects occurring within U.S. territory, many initial projects may occur outside U.S. territory or on the high seas. If so, climate engineering litigation could pose difficult questions of extraterritorial application of federal environmental laws. The federal courts have generally disfavored a broad application of those laws outside U.S. borders without express congressional authorization.[106] Actions outside the United States that have direct effects within U.S. borders, however, have provided a basis for application of U.S. environmental laws to foreign actors.[107]

Will the court have jurisdiction over the defendants? Even if the federal courts upheld the extraterritorial application of U.S. environmental statutes, claimants would still need to satisfy minimum contacts required for the constitutional exercise of personal jurisdiction over persons or corporations acting entirely outside the United States.[108] The simple fact that a defendant may be an individual U.S. citizen or be incorporated in a U.S. state, by itself, may suffice given further statutory authorization or additional contacts with the U.S. forum.[109]

[105] 42 U.S.C. § 7602(e) (2011) (defining "person").

[106] *See, e.g.*, Corrosion Proof Fitting v. EPA, 947 F.2d 1201 (5th Cir. 1991) (holding that Canadian asbestos producers lacked standing to challenge EPA regulations because the Toxic Substances Control Act did not require the EPA to consider extraterritorial effect of domestic regulations); Arc Ecology v. U.S. Dep't. of the Air Force, 294 F. Supp. 2d 1152 (N.D. Cal. 2003) (dismissing a claim by two Filipino citizens seeking declaration that CERCLA could be applied to two former U.S. military bases in the Philippines). *Cf.* Lujan v. Defenders of Wildlife, 504 U.S. 555, 563 (1992) (holding that plaintiffs lacked standing to challenge rule that limited federal consultation requirements under Section 7 of the Endangered Species Act to actions within the United States or on the high seas).

[107] *See, e.g.*, Pakootas v. Teck Cominco Metals, Ltd., 452 F.3d 1066 (9th Cir. 2006) (holding that CERCLA liability does not reach beyond the U.S. border into Canada, but finding that a Canadian factory created a "facility" within the definition of CERCLA in the United States because its discharges flowed directly into a U.S. water body).

[108] World-Wide Volkswagen Corp. v. Woodson, 444 U.S. 286, 292 (1980).

[109] According to the *Restatement (Second) of Conflict of Laws*, "... a citizen of the United States is subject to the judicial jurisdiction of the United States even though he is domiciled abroad. National citizenship constitutes a basis for the exercise of judicial jurisdiction by the United States." The Restatement adds that "[i]n actions between private persons, the Congress of the United States has

Who is opposing the project? The identity of the persons challenging the climate engineering project can play a large role in determining which causes of action and remedies might be available. In particular, the U.S. government, a state entity, or a local governmental unit would have access to a broader array of potential actions and remedies than private parties in citizen suits. For example, the federal government can bring actions or issue administrative orders to respond to emergencies or to imminent threats to human health or the environment. Governmental entities, as trustees for natural resources, might also have the ability to seek compensation for any damage to natural resources caused by climate engineering projects.[110]

With these questions in mind, the federal environmental statutes that might first apply to climate engineering projects would probably include the Clean Air Act, Clean Water Act, Endangered Species Act, National Environmental Policy Act, Marine Protection, Research and Sanctuaries Act, and several other federal statutes that could regulate aspects of specific types of projects (e.g., projects that might affect migratory birds).

3.1.1 The Clean Air Act

The federal Clean Air Act[111] provides the most likely statutory basis to challenge climate engineering projects. Most notably, EPA has already determined that the

never authorized the federal courts to exercise jurisdiction on the basis of citizenship over citizens of the United States who are domiciled abroad. The federal courts, however, have statutory authority in certain situations to order an American citizen who resides abroad to return to the United States in order to give testimony in either a civil or criminal proceeding and to punish him for contempt if he fails to comply with this order." RESTATEMENT (SECOND) OF CONFLICT OF LAWS § 31 cmt b.

[110] For example, see discussion infra notes 168–170 on the United States' broad emergency authorities under the Comprehensive Environmental Response, Compensation and Liability Act to respond to releases of "pollutants" that pose an imminent and substantial threat to human health and the environment.

[111] 42 U.S.C. §§ 7401–7661 (2006). The CAA sets out complex interlocking requirements for facilities that emit sufficient amounts of specified air pollutants. In particular, the CAA requires owners and operators to obtain permits if their facilities (1) emit sufficient amounts of criteria air pollutants to qualify as major sources that need either Prevention of Significant Deterioration permits for areas that meet ambient air quality standards, or New Source Review permits for facilities in nonattainment areas, *id.* §§ 7470–7479, 7501–7503; (2) install maximum available control technology on sources in a facility that emit hazardous air pollutants, *id.* § 7412; (3) control emissions or leaks of certain substances that deplete stratospheric ozone, *id.* §§ 7671a–7671e; or (4) obtain tradable emission credits or limit emissions of sulfur dioxide (SO_2) that can contribute to the formation of acid rain, *id.* §§ 7651–7651o. The operators must include all of these controls in a comprehensive federal facility operating permit under Title V of the CAA, and they must submit a certified statement that verifies that the facility has either complied with its permit requirements or has listed all of its deviations from the permit. *Id.* §§ 7661–7661c. This cursory overview of the CAA obviously and intentionally overlooks the vast and rich body of complex statutory and regulatory requirements set out by the Act. *See* CLEAN AIR ACT HANDBOOK (Robert J. Martineau & David P. Novello eds., 2004) (providing further background on the CAA) [hereinafter CLEAN AIR ACT HANDBOOK].

Clean Air Act applies to GHG emissions and provides an appropriate statutory vehicle to address climate change.[112] EPA has relied on existing Clean Air Act authorities to undertake an ambitious regulatory initiative to require GHG emission controls. This effort has included a finding under the Clean Air Act that GHG emissions endanger public health and welfare, as well as a determination that major stationary sources of GHG emissions must obtain Prevention of Significant Deterioration or New Source Review permits for producing the emissions.[113] Given its willingness to regulate activities to reduce the effects of GHG emissions, the EPA might take an expansive view of the Clean Air Act's applicability to other activities that could alter climate processes or directly release aerosols or other compounds into the atmosphere to mitigate climate change effects.[114]

The Clean Air Act therefore offers obvious avenues for claimants who oppose certain types of climate engineering projects. For example, an environmental advocate might assert that the dispersion of a sulfate aerosol in the upper stratosphere constituted a release of an air pollutant that violates Clean Air Act prohibitions or requires a permit or authorization.[115] If so, that advocate could bring a citizen suit to compel the EPA Administrator to use her nondiscretionary duty to stop or control those emissions from the project.[116]

Such a citizen suit action, however, would throw a sharp light on potentially difficult jurisdictional questions evoked by applying the Clean Air Act to climate engineering projects. First, the citizen suit would have to wrestle with a threshold issue: Do stratospheric aerosols, when released to achieve a specific purpose, constitute a "pollutant" that would trigger Clean Air Act jurisdiction.[117] Second, the Clean Air

[112] By contrast, the U.S. government has not supported the use of other federal environmental statutes in other contexts to regulate activities that might affect climate change. *See* discussion *infra* notes – (discussing Interior Department's refusal to use its authority under the Endangered Species Act to designate critical habitat for threatened species as a basis to regulate activities that might contribute generally to climate change).

[113] Endangerment and Cause or Contribute Findings for Greenhouse Gases under Section 202(a) of the Clean Air Act, 74 Fed. Reg. 66,496 (Dec. 15, 2009) (to be codified at 40 C.F.R. Ch. I); Prevention of Significant Deterioration and Title V Greenhouse Gas Tailoring Rule, 75 Fed. Reg. 31,514 (June 3, 2010).

[114] We have not identified any instance where EPA has officially taken a position on whether it can exercise its existing authority under the Clean Air Act to regulate climate engineering research, demonstration projects, or full-scale deployment in the United States or by U.S. citizens.

[115] A significant portion of sulfate particulates released into the upper stratosphere might also fall down to the troposphere where it could directly contribute to aggravated acid deposition in rain or snow. Robock, *supra* note 100, at 16.

[116] The federal government would have different tools to fight a proposed climate engineering project, including enforcement actions for failure to comply with federal environmental statutes as well as administrative orders to abate imminent endangerments to human health or the environment. *See* discussion *infra* notes 169–170.

[117] The Clean Air Act applies only to releases of "pollutants" that meet statutory and regulatory criteria. 42 U.S.C. § 7602(g). For example, the intentional discharge of chemicals into the air to fight forest

Act has historically not applied to activities that promote healthier ambient atmospheric conditions through any means other than emission controls.[118] For example, prior efforts to reduce ambient particulate matter concentrations or directly reduce ambient ozone levels have not triggered Clean Air Act regulatory requirements.[119] Third, the Clean Air Act lacks an express regulatory framework for emission limitations on climate engineering projects that might not conveniently fall into the existing rules for industrial source categories, priority pollutants under Title I for ambient air quality standards, air toxics regulated under Title III, or even stratospheric ozone protection under Title VI.

These and other difficult questions will ultimately turn on the specific design of the proposed climate engineering technology. For example, proposals to reduce solar radiation influx through releasing sulfate aerosols in the upper stratosphere may open several legal challenges under the Clean Air Act. This particular technology could pose regulatory obligations under:

(i) Title I for nonattainment of national ambient air quality standards.[120] Sulfur dioxide is a criteria pollutant with a NAAQS level as well as extensive permitting

fires has not triggered a need for Clean Air Act permits. U.S. FOREST SERVICES, DECISION NOTICE AND FINDING OF NO SIGNIFICANT IMPACT: AERIAL APPLICATION OF FIRE RETARDANT (2008), *available at* http://www.fs.fed.us/fire/retardant/Aerial_Application_of_Fire_Retardant.pdf. Historical attempts to modify weather through cloud seeding or other rainmaking technology have fallen under separate state regulatory regimes rather than the Clean Air Act. *See* discussion *infra* note.

[118] Although they have received comparably little attention, other proposed technologies would directly remove or absorb criteria pollutants from the ambient atmosphere. For example, state environmental agencies have explored the use of certain catalytic coatings for mobile sources, concrete structures, high-volume air conditioning systems, and road surfaces to directly absorb ozone and its precursors. Such materials have been successfully introduced in Japan, Italy, and Great Britain as a method of controlling emissions, and since 2005 they have been proposed for use as part of the air pollution control strategy for the Dallas–Fort Worth area. *See* TEXAS COMMISSION ON ENVIRONMENTAL. QUALITY, DRAFT: AREA – POTENTIAL CONTROL STRATEGIES FOR DFW ATTAINMENT DEMONSTRATION (Oct. 10, 2005), http://www.tceq.state.tx.us/assets/public/implementation/air/sip/miscdocs/area_8–31–05.pdf.

Theoretically, EPA and delegated states could also authorize techniques that directly remove air pollutants from the ambient atmosphere as an appropriate technology to satisfy BACT requirements for Title I permitting purposes. We have not located any BACT approvals, however, that have authorized this approach.

[119] Although EPA has not used the Clean Air Act to regulate technologies that directly reduce ambient levels of criteria pollutants, states have sought EPA's approval of these techniques so that they could claim credit for pollutant reductions for SIP modeling purposes. *See id.* at 1–2 (discussing the Texas Commission on Environmental Quality's (TCEQ) proposal of an emission reduction increment for requiring catalytic coating for pavement and building surfaces to directly reduce nitrogen oxides and volatile organic compounds).

[120] Section 109 of Title I requires the EPA to identify National Ambient Air Quality Standards (NAAQS) for several air pollutants. EPA must design each NAAQS to assure that it protects public health and welfare with an adequate margin of safety. To date, the EPA has promulgated NAAQS for six air pollutants: ozone, particulate matter, sulfur dioxide, nitrogen dioxide, carbon monoxide, and lead. If the air within a designated geographic region exceeds the NAAQS level for any pollutant, either the

requirements for areas not in attainment with that standard.[121] In addition, sulfate aerosols could constitute a precursor to the formation of particulate matter that falls within either particulate matter NAAQS standard.[122] Some proposals for sulfate dispersion in the upper atmosphere would rely on large stationary generators that would then convey their sulfate emissions into the stratosphere through immensely long flexible tubes supported by high-altitude balloons.[123] These sources may arguably trigger Clean Air Act permitting requirements if the generators emit enough sulfur dioxide or PM to constitute a "major source."[124]

Title I also imposes restrictions on emissions from "major sources" that might impair visibility in mandatory "Class I areas."[125] It is unclear whether sulfate aerosol or other scattering media would potentially affect visibility or regional haze formation. If so, visibility "New Source Review" requirements might apply to climate engineering projects that qualify as stationary major sources.[126]

EPA or the state must provide an implementation plan designed to help the affected region attain compliance with the NAAQS. 42 U.S.C. §§ 7401–7431, 7501–7515 (2006).

[121] It is unclear whether the generation and dispersal of sulfate aerosols would require the direct emission of sulfur dioxides, which would fall within the sulfur dioxide NAAQS. In addition, sulfate aerosols may also come within air quality planning and permitting requirements under Title I if their emission would contribute to the formation or decomposition of compounds into sulfur dioxide in ambient environmental conditions.

[122] EPA has promulgated two NAAQS for particulate matter. In 1987, EPA changed the indicator for particles from Total Suspended Particulates to PM_{10}, including particles with a mean aerodynamic diameter less than or equal to ten micrometers not to be exceeded once per year. Revisions to the National Ambient Air Quality Standard for Particulate Matter, 52 Fed.Reg. 24,634 (July 1, 1987) (codified at 40 C.F.R. pt. 50). EPA later issued a second NAAQS that set a lower ambient concentration threshold for $PM_{2.5}$ – particulate matter with a mean aerodynamic diameter less than or equal to 2.5 micrometers – because the agency had concluded that ultrafine PM contributed to increased incidents of pulmonary disease and other human health effects. National Ambient Air Quality Standards for Particulate Matter, 62 Fed.Reg. 38,652 (July 18, 1997) (codified at 40 C.F.R. pt. 50).

[123] Philip Rasch et al., *An Overview of Geoengineering of Climate Using Stratospheric Sulphate Aerosols*, 366 PHILOSOPHICAL TRANSACTIONS ROYAL SOC'Y 4007, 4013 (2008).

[124] A "major source" is a stationary source that has the potential to emit at least one regulated pollutant under the New Source Review Program in amounts that exceed major source thresholds. The thresholds can vary from 250 tons per year or more of any pollutant subject to regulation in geographic regions that meet NAAQS standards down to 10 tons per year of ozone precursors if the geographic region is an extreme nonattainment area for ozone NAAQS. CLEAN AIR ACT HANDBOOK, *supra* note 111, at 143–44, 178. If these stationary source facilities were located in nonattainment areas, it would raise the interesting question of whether they fall under nonattainment emission limitations even though their ultimate discharge actually occurs far above or outside the nonattainment area itself.

[125] A Class I area is an area "of special national or regional value from a natural,scenic, recreational, or historic perspective." CLEAN AIR ACT HANDBOOK, *supra* note 111, at 171, 192 n.261, 196–97.

[126] *See* 42 U.S.C. § 7491 (2006); *see also* CLEAN AIR ACT HANDBOOK, *supra* note 111, at 203–08.

(ii) Title II.[127] Releases of large amounts of sulfates from aircraft flying in the upper atmosphere may invoke complex regulatory provisions that govern emissions from mobile sources and aircraft. The Clean Air Act's mobile source program may have limited application, however, because it largely targets emissions from the operation of engines rather than intentional releases conveyed by the mobile sources themselves.

Title II may have a more direct application to other climate engineering proposals that rely on solar radiation scattering by stratospheric aerosol particulates. Recent models have shown that aircraft contrails in the upper atmosphere can have a significant effect on climate systems.[128] As a result, at least one proposal has suggested that aircraft fuels could be formulated to enhance their scattering effect by promoting the creation of high-altitude contrails or by encouraging formation of particulates.[129] An aircraft operator who sought to use these fuels may have to assure that the fuel meets mobile source fuel standards set out by the EPA under Title II or obtain a waiver from the EPA.

(iii) Title IV.[130] Sulfur dioxide is also a regulated precursor to the formation of acid rain.[131] Although Title IV regulates stationary sources in specific industrial categories

[127] Title II requires EPA to promulgate emission standards for automobiles and other mobile sources of air pollutants. Under this program, EPA has also comprehensively regulated the content and distribution of fuels for mobile sources, including fuels from renewable energy sources such as ethanol. 42 U.S.C. §§ 7521–7574 (2006).

[128] Most computer models show that aircraft contrails from current high-altitude operations contribute to climate change effects because their radiative forcing traps significant energy in the atmosphere. *Contrails Warm the World More than Aviation Emissions*, 2806 NEW SCI. 16 (Apr. 2, 2011), *available at* http://www.newscientist.com/article/dn20304-contrails-warm-the-world-more-than-aviation-emissions.html; *see also* Ulrike Burkhardt & Bernd Karcher, *Global Radiative Forcing from Contrail Cirrus*, 1 NATURE CLIMATE CHANGE 54 (Mar. 29, 2011).

[129] The U.S. Patent Office issued a patent for the addition of metallic Welsbach materials to jet fuels to provide "a method of reducing atmospheric warming ..." U.S. Patent No. 5,003,186, at [1] (filed Apr. 23, 1990) (issued Mar. 26, 1991). Notably, none of these proposals has appeared in peer-reviewed journals, and the field has become mired in accusations of covert government action and conspiracy theories. *See, e.g.*, Traci Watson, *Conspiracy Theories Find Menace in Contrails*, USA TODAY (Mar. 7, 2001, 10:38 AM), http://www.usatoday.com/weather/science/2001-03-07-contrails.htm.

[130] Title IV requires EPA to control the emissions of precursor pollutants that might cause the formation of acid rain. Under this program, EPA has comprehensively regulated emissions of sulfur dioxide from power generation units, including large electrical power generation plants. 42 U.S.C. §§ 7651–7651O (2006). Most notably, EPA has relied on a market-based auction system to allocate credits that allow operators to emit sulfur dioxide from their power generation units. This approach has led to significant reductions in sulfur dioxide emissions at less cost than a traditional command-and-control regulatory approach. Dallas Burtraw & David Evans, *Tradable Rights to Emit Air Pollution* 4–5 Research for the Future, Discussion Paper No. 08–08, 2008, *available at* http://www.rff.org/Documents/RFF-DP-08-08.pdf ("The administrative performance of the SO_2 program has been nearly perfect, with virtually 100 percent compliance and unexpectedly little litigation."). Burtraw and Evans cite studies concluding the SO_2 trading system saved 43 to 55 percent in costs compared with a uniform standard. *Id.* at 5.

[131] *See* 42 U.S.C. § 7651(a) (2011) (Congressional finding that sulfur and nitrogen dioxides are the "principal sources of acidic compounds ... in the atmosphere").

(e.g., power plants), large-scale releases of sulfur aerosols that may affect the acidity of regional precipitation might lead to regulatory scrutiny.[132]

(iv) Title VI. The Clean Air Act empowers the EPA to regulate emissions of stratospheric ozone-depleting substances (ODS) to assure that the United States meets its obligations under the Montreal Protocol.[133] Under Title VI, the EPA can add certain compounds to the list of ODS if it concludes that they contribute to ozone depletion. Some scientists have raised concerns that the release of sulfur aerosols into the upper stratosphere may cause significant ozone depletion.[134] If so, the EPA may have regulatory authority to add these types of activities and substances to the list of ODS and implement controls on their distribution and use. To date, the EPA has not included stratospheric sulfate aerosols in the list of ODS under Title VI.

Importantly, the potential application of these Clean Air Act requirements to climate engineering projects does not necessarily deny the EPA the flexibility to modify these regulatory standards in certain circumstances. For example, the Clean Air Act's provisions and exemptions for research projects may provide the EPA with some degree of flexibility to handle initial rounds of climate engineering projects or experiments.[135] The EPA may also have the ability to modify some regulatory obligations through consent agreements or compliance schedules that provide supplemental pathways for satisfying Clean Air Act requirements.

3.1.2 Clean Water Act

Climate engineering projects that require the addition of substances to waters of the United States may require authorization under the federal Clean Water

[132] This regulatory scrutiny would arise from the possible effects that sulfate dispersion might have on regional precipitation. *See* Marshall, *supra* note 59 ("Spraying aerosols locally [into the stratosphere with a hose] allows the particles to clump together, making them less effective at reflecting sunlight and more likely to be swept down by rain"); P. Heckendorn et al., *The Impact of Geoengineering Aerosols on Stratospheric Temperature and Ozone*, ENVTL. RES. LETTERS, Nov. 2009, at 6–7, *available at* http://www.see.ed.ac.uk/~shs/Climate%20change/Stratospherics/Heckendorn_Et_al_ERL2009.pdf. But Title IV and its implementing regulations currently do not address emissions from any of the types of sources that might be used in climate engineering projects. As noted previously, the application of these requirements for SO_2 to sulfate aerosol projects might also depend on whether dispersal of sulfates requires the direct emission of SO_2 into the atmosphere or will contribute to heightened SO_2 ambient levels due to the decomposition of other compounds or through other atmospheric chemical processes. *See* discussion *supra* notes 115–132 and accompanying text.

[133] Montreal Protocol on Substances That Deplete the Ozone Layer, opened for signature, Sept. 16, 1987, reprinted in 26 I.L.M. 1541 (1987).

[134] *See generally* Heckendorn et al., *supra* note, 132 at 1 ("[t]herefore, geoengineering by means of sulfate aerosols is predicted to accelerate the hydroxyl catalyzed ozone destruction cycles and cause a significant depletion of the ozone layer even though future halogen concentrations will be significantly reduced."); Patricia Kenzelmann et al., *Geoengineering Side Effects: Heating the Tropical Tropopause by Sedimenting Sulphur Aerosol?*, IOP CONFERENCE SERIES: EARTH & ENVTL. SCI. 6 (2009), *available at* http://www.iac.ethz.ch/people/kenzelpa/EGU2008_quer_handout.pdf.

[135] 42 U.S.C. § 7403(2006) (discussing alternative permitting options for research projects).

Act.[136] This Act prohibits the discharge of any pollutant from a point source into navigable waters unless that discharger has a permit or other form of authorization.[137] For example, a project that disperses iron or other nutrients into U.S. marine waters for a fertilization demonstration project may constitute a discharge that requires a permit under either the National Pollutant Discharge Elimination System (NPDES) or a delegated state program.[138] Notably, the EPA has construed the definition of "pollutant" to include the addition of heat to water bodies.[139] If a climate engineering project involves the addition or alteration of heat levels within U.S. waters, those transfers of heat may trigger NPDES permitting requirements.[140]

The Clean Water Act may also directly affect climate engineering projects that require alterations to land use or geographic features. For example, some climate engineering proposals would encourage the placement of highly reflective materials onto large swaths of land to increase surface albedo. By reflecting more sunlight back into space, these projects would reduce solar influx and ultimately reduce projected climate change effects.[141] Other projects would encourage large-scale CO_2 sequestration through the construction of artificial wetlands or restrictions on land uses that release trapped carbon.[142] If these efforts would involve alterations or placement of materials into wetlands within the jurisdictional reach of the Clean Water Act, the project operators may have to obtain authorization or permits from the U.S. Army Corps of Engineers.

Water quality concerns could also indirectly affect climate engineering projects. For example, direct CO_2 capture will generate a large volume of CO_2 in either a

[136] 33 U.S.C. § 1251 (2006).

[137] 33 U.S.C. § 1311(a) (2006) (prohibiting the discharge of any pollutant by any person into waters of the United States except in compliance with requirements of the Clean Water Act).

[138] *Id.* The prospects for significant field tests of iron fertilization in U.S. waters is likely low because waters identified as suitable for fertilization (i.e., high in chlorine but low in nutrients) are in the Southern Ocean and in the Indo-Pacific regions. Most of the experiments are also likely to occur on the high seas. Attempts to replicate these conditions in U.S. waters for such a test might trigger Clean Water Act obligations. In addition to the CWA, the Rivers & Harbors Act of 1899 ("Refuse Act") imposes strict liability for discharges of "refuse" into waters of the United States. 33 U.S.C. § 407 (2006). Although this statute has historically applied to the discharge of refuse or solid waste that poses a threat to navigability of U.S. waterways, federal courts have interpreted the Refuse Act to prohibit the unpermitted discharges of pollutants into U.S. waters. *See* New York v. New Jersey, 256 U.S. 296 (1921). If a climate engineering demonstration arguably requires discharge into U.S. waters of a material that might constitute a "refuse" (e.g., the large-scale deposition of iron or other nutrients into U.S. waters), that project may therefore require authorization from the U.S. Army Corps of Engineers.

[139] 33 U.S.C. § 1342; 40 C.F.R. § 122.2 (2011) (defining "pollutant" to include heat).

[140] *See supra* notes 80–82 and accompanying text (discussing climate engineering proposals to use marine heat pumps to reduce the surface temperature of ocean waters and thereby arguably reduce the risk of the formation of extreme storms or hurricanes).

[141] *See* discussion *supra* at note 64 (use of reflective ground cover for albedo enhancement).

[142] R. Lal, *Sequestering Atmospheric Carbon Dioxide*, 28:3 CRIT. REV. PLANT SCI. 90, 96 (2009) (suggesting that carbon sequestration in terrestrial ecosystems would provide potential total CO_2 drawdown of 50 ppm of atmospheric concentration over a period of five decades, and proposing that large scale terrestrial sequestration could provide a valuable complement to other geoengineering schemes).

gaseous or liquid form. Although some of that CO_2 will likely be used as a product or in other industrial processes, direct capture strategies may have to address the management or disposal of large volumes of captured CO_2. If direct capture systems use geologic sequestration to manage that CO_2, those sequestration wells will likely trigger EPA regulatory requirements under its underground injection well program.[143] State regulatory programs will also affect geologic sequestration aspects of any significant direct CO_2 capture systems.[144]

Climate engineering permitting under the Clean Water Act may pose some of the same conceptual challenges raised by the Clean Air Act. For example, the intentional release of materials into U.S. waters for an express remedial purpose may not constitute a discharge of a "pollutant" because the materials are not being discarded.[145] In addition, materials released into the ambient air for a climate engineering project may ultimately precipitate into U.S. waters, but that type of generalized deposition may not constitute a discharge from a "point source" that would trigger NPDES permitting requirements.[146]

3.1.3 Endangered Species Act

The federal Endangered Species Act[147] imposes stringent limits on the actions of governments and individuals that might result in the taking of an endangered or threatened species by directly harming individuals of that species or by damaging the species' critical habitat.[148] If a climate engineering project could potentially

[143] On December 10, 2010, EPA promulgated final rules under the Safe Drinking Water Act to govern the injection of CO_2 for geologic sequestration. 75 Fed. Reg. 77,229 (Dec. 10, 2010) (to be codified at 17 C.F.R. pt. 275). Storage of CO_2 in these wells may also trigger EPA regulatory requirements for greenhouse gas reporting. Mandatory Reporting of Greenhouse Gases: Injection and Geologic Sequestration of Carbon Dioxide, 75 Fed. Reg. 75,059 (Dec. 1, 2010) (to be codified at 29 C.F.R. pt. 403).

[144] *See, e.g.*, 16 TEX. ADMIN. CODE ANN.§ 5.301 (West 2011). These rules implemented Senate Bill 1387, 81st Legislature (Regular Session 2009) to "provide for the implementation of projects involving the capture, injection, sequestration, or geologic storage of carbon dioxide." 36 Tex. Reg. 4397 (July 8, 2011).

[145] *See* discussion *supra* at note 118.

[146] The Clean Water Act sets out much less onerous requirements for discharges from nonpoint sources into U.S. waters. 33 U.S.C. § 1329 (2011) (establishing nonpoint source management programs, but not requiring nonpoint sources to obtain discharge permits). In an analogous situation, however, at least one federal court has ruled that the generalized spraying of pesticides that precipitate into navigable waters constitutes a discharge of pollutants from a point source that triggers NPDES permitting requirements. Nat'l Cotton Council of Am. v. EPA, 553 F.3d 927 (6th Cir. 2009) (striking down EPA regulations attempting to exempt pesticide application from NPDES permit requirements).

[147] 16 U.S.C. § 1531 (2006).

[148] *Id.*§ 1538(a)(1)(B) (2006) (prohibiting the "taking" of any endangered species within the United States or the territorial sea of the United States); *id.* § 1532(19) (defining "taking" as "harass, harm, pursue,

affect a large region, that geographic area may include habitat for endangered or threatened species. In those circumstances, a claimant may seek to halt the project through a citizen suit and a request for injunctive relief if the proposed climate engineering project could injure any members of an endangered species or damage critical habitat.[149]

Obviously, such an action could face significant standing, causation, and evidentiary challenges. The U.S. Supreme Court has already held that plaintiffs cannot bring citizen suits under the Endangered Species Act (ESA) to challenge funding decisions for actions abroad that arguably threaten a listed species unless the plaintiffs show that they have suffered a concrete, specific, and actual or imminent injury arising from that action.[150] Persons challenging climate engineering projects under the ESA may face similarly challenging burdens of proof to demonstrate standing. Those burdens may be alleviated somewhat because climate engineering projects expressly seek to cause detectable changes in climate patterns. As a result, the defendant's own statements related to the project may remove the need to prove at least some causation issues – namely, whether the defendant's actions have resulted in altered climate effects. Plaintiffs would likely still have to show, however, that these climate effects resulted in some threat to the listed species at issue.

More important, the plaintiffs would also have to demonstrate that the climate engineering project's impact rises to the level of a "taking" through alteration of critical habitat or injury to individual members of the species. If the nexus between the climate engineering project and the injuries is too indirect, it may not demonstrate that the project proximately caused the injury within the statutory meaning of "take."[151]

Last, the U.S. Department of the Interior promulgated interpretative rules on May 14, 2008, for the proposed designation of polar bears as a threatened species.

hunt, shoot, wound, kill, trap, capture, or collect, or to attempt to engage in any such conduct"); *see also* Babbitt v. Sweet Home Chapter of Cmtys. for a Great Ore., 515 U.S. 687, 691 (1995) (upholding regulatory interpretation of "harm" to include "significant habitat modification or degradation where it actually kills or injures wildlife by significantly impairing essential behavioral patterns, including breeding, feeding or sheltering").

[149] This challenge could allege that the climate engineering project itself threatens to cause a taking of a protected species, which would violate the prohibition on any actions that "take" or "harass" protected species contained in Section 9 of the Endangered Species Act and its implementing regulations. In addition, if the project requires federal approval, funding, or permitting, the Endangered Species Act may require the federal agency performing the action to consult with the U.S. Fish & Wildlife Service or the National Marine Fisheries Service to assure that the federal action does not jeopardize the continued existence of a listed species or result in the destruction or adverse modification of critical habitat. 16 U.S.C. § 1536(a)(2) (2006); 50 C.F.R. § 402.01 (2011).

[150] Lujan v. Defenders of Wildlife, 504 U.S. 555 (1992).

[151] *Sweet Home Chapter*, 515 U.S. at 709 (O'Connor, J., concurring) (incorporating "ordinary requirements of proximate causation" and foreseeability into section 9 taking prohibition).

These rules sought to limit the scope of ESA listings and protections to exclude measures that addressed global climate change mitigation as a necessary step to protect critical habitat for endangered or threatened species.[152] The final polar bear listing rules expressly declined to use the Endangered Species Act to address broad climate change concerns.[153] A similar regulatory approach may drive the United States to use caution when adapting federal environmental statutes to oppose climate engineering projects.

3.1.4 National Environmental Policy Act

The National Environmental Policy Act (NEPA)[154] requires the federal government to undertake an environmental review of any major federal agency action that is likely to have a significant impact on the environment.[155] Although NEPA applies solely to governmental actions, it could play an important role if a climate engineering project required the federal government to undertake any significant discretionary permitting action or any other major actions related to the project.[156] The Council on Environmental Quality has expressly directed federal agencies to account for climate change implications in their review of governmental actions for

[152] Endangered and Threatened Wildlife and Plants; Determination of Threatened Status for the Polar Bear (*Ursus maritimus*) Throughout Its Range, 73 Fed. Reg. 28,211–28,303 (May 15, 2008) (to be codified at 50 C.F.R. pt. 17) [hereinafter Polar Bear Threatened Status]; Endangered and Threatened Wildlife and Plants; Special Rule for the Polar Bear, 73 Fed. Reg. 28,306–28,318 (May 15, 2008) (to be codified at 50 C.F.R. pt. 17) [hereinafter Polar Bear Special Rule]. The Interior Department has issued a new designation of critical habitat for polar bears. The habitat designation includes 187,157 square miles of protected habitat. Endangered and Threatened Wildlife and Plants, Designation of Critical Habitat for the Polar Bear (*Ursus maritimus*) in the United States, 75 Fed. Reg. 76,085, 76,137 (Dec. 7, 2010) (to be codified at 50 C.F.R. pt. 17); *see also In re* Polar Bear Endangered Species Act Listing & § 4(d) Rule Litig., 794 F. Supp. 2d. 65 (D.D.C. 2011) (upholding the final rule listing the polar bear as threatened under the Endangered Species Act).

[153] Polar Bear Threatened Status, *supra* note 152, at 28,247 ("Some commenters to the proposed rule suggested that the Service should require other agencies (e.g., the EPA) to regulate emissions from all sources, including automobiles and power plants. The science, law, and mission of the Service do not lead to such action. Climate change is a worldwide issue. A direct causal link between the effects of a specific action and 'take' of a listed species is well beyond the current level of scientific understanding . . ."), 28,300 ("Without sufficient data to establish the required causal connection – to the level of 'reasonable certainty' – between a new facility's GHG emissions and impacts to polar bears, section 7 consultation would not be required to address impacts to polar bears."); Polar Bear Special Rule, *supra* note 152, at 28,313.

[154] 42 U.S.C. § 4321–4370h (2006).

[155] *Id.* § 4332(2)(C) (2006); 40 C.F.R. §§ 1502.1–1502.25 (2011) (implementing environmental impact statement requirements).

[156] Notably, if the responsible federal agency had to conduct an environmental assessment or a full environmental impact statement, that action could also trigger a requirement for the agency to enter into the federal consultation process under section 7(a)(2) of the Endangered Species Act. *See* discussion *supra* note 149.

potential NEPA assessment.[157] If a federal agency must review a proposed climate engineering project for permitting, government financial assistance, or other support, it will likely conduct an environmental assessment of the project's purported impact on climate systems to determine whether it qualifies for a categorical exclusion – although an environmental assessment is not required prior to issuing a categorical exclusion[158] – or if it might have a significant impact that would require a fuller environmental impact statement.[159]

The federal government's environmental review may extend beyond an assessment of individual climate engineering projects. If a federal agency decides to craft a strategy for authorizing or supervising climate engineering projects, that policy decision may lead the agency to undertake a programmatic environmental impact statement (PEIS).[160] This PEIS could require a comprehensive assessment of the cumulative and global effects of a decision to allow or control climate engineering projects. That assessment would explicitly and expressly focus on the possible climate change effects that the projects might have on their targeted climate systems.

The applicability of NEPA requirements will turn heavily on the specific factual context for the climate engineering project as well as the nature of the federal government's action related to the project. For example, statements by proponents about a climate engineering experiment's intended regional or global effects might constitute a prima facie demonstration that the project will have a significant impact and thereby trigger the need for a full environmental impact statement.

[157] Memorandum from Nancy Sutley, Chair, Council on Environmental Quality, for Heads of Fed. Departments and Agencies, Draft NEPA Guidance on Consideration of the Effects of Climate Change and Greenhouse Gas Emissions (Feb. 18, 2010), *available at* http://pbadupws.nrc.gov/docs/ML1035/ML103510433.pdf.

[158] 40 C.F.R. § 1508.4 (2011).

[159] *See id.* § 1508.18 (2011) (defining "major Federal action"); *id.* § 1508.9 (outlining the purpose and contents of an environmental assessment); *id.* § 1508.4 (defining "categorical exclusion"). Importantly, the environmental assessment process must investigate whether the major federal action's consequences, when combined with other actions, might lead to a cumulative impact that could require a full environmental impact statement. *Id.* § 1508.7 (defining cumulative impact as "the impact on the environment which results from the incremental impact of the action when added to other past, present, and reasonably foreseeable future actions regardless of what agency (Federal or non-Federal) or person undertakes such other actions").

[160] A programmatic environmental impact statement assesses the environmental impact of broad governmental policies or initiatives. As a result, it can focus on agency program, area-wide actions in a region, or multiple actions that share a common geography or timing. The Commission on Environmental Quality's regulations do not define the term "programmatic environmental impact statement" specifically, but those rules provide enough flexibility for the definition of "environmental impact statements" to include programmatic impacts. 40 C.F.R. §§ 1502.4(b) (2011) (an "Environmental impact statement[] may be prepared … for broad federal actions such as the adoption of new agency programs or regulations"); *id.* § 1502.4(c) (allowing grouping of actions for EIS to include geography and generically similar actions or similar stages of technological development).

3.1.5 Marine Protection, Research and Sanctuaries Act

In contrast to other federal environmental statutes, the Marine Protection, Research and Sanctuaries Act (MPRSA, or "Ocean Dumping Act") has already been invoked to challenge climate engineering projects.[161] The MPRSA implements the United States' obligations under the London Convention to restrict the dumping of pollutants or refuse into the high seas, and it also sets out a comprehensive regulatory program to govern the placement of materials into the marine environment that might impair its health or ecological functions.[162] The MPRSA, as a result, applies to discharges into waters under U.S. jurisdiction as well as to acts on the high seas by ships under the U.S. flag.[163]

Because it applies to actions on the high seas, opponents invoked the MPRSA to fight Planktos' planned release of iron filings into the Pacific Ocean. Several environmental groups filed a petition with the EPA that contended the planned experiment would constitute the dumping of pollutants that violated the MPRSA, and they asked the EPA to intervene and halt the experiment. The EPA responded by notifying Planktos that the MPRSA could apply to the planned release, and it asked Planktos to confirm whether it would seek a permit or other authorization before proceeding with the project. Planktos responded that it would not trigger MPRSA obligations because it would use a vessel flying under a non-U.S. flag for the experiment. Planktos' response in part led the United States to alert the parties to the London Convention and seek consideration by the parties of a regime to govern ocean fertilization experiments.[164]

The MPRSA may offer a powerful initial platform to regulate climate engineering projects that involve actions in waters under U.S. jurisdiction or on vessels flying the U.S. flag. The MPRSA's express legislative purpose is to "regulate the dumping of all types of materials into ocean waters and to prevent or strictly limit the dumping into ocean waters of any material which would adversely affect human health, welfare, or amenities, or the marine environment, ecological systems, or economic potentialities."[165] Unless authorized by a permit, MPRSA generally prohibits (1) transportation of material from the United States for the purpose of ocean dumping; (2) transportation of material from anywhere for the purpose of ocean dumping by U.S. agencies or U.S.-flagged vessels; and (3) dumping of material transported from outside the United States into the U.S. territorial sea or into the contiguous zone to

[161] *See supra* at note 46 (EPA invoked MPRSA as potential regulatory basis to restrict Planktos' planned discharge of iron into ocean waters).

[162] 33 U.S.C. § 1401(a)–(c) (2006).

[163] *Id.* §§ 1441(a)(1)–(2).

[164] *See supra* notes 46–47 and accompanying text (describing EPA response to proposed Planktos project).

[165] 33 U.S.C. § 1401(b) (2006).

the extent that it may affect the territorial sea or the territory of the United States.[166] Given its broad scope and its express extraterritorial application to activities by U.S. vessels, the MPRSA may offer a strong and clear platform to challenge climate engineering projects that might otherwise lie outside the reach of other domestic federal environmental statutes.[167]

3.1.6 Other Statutes

This initial survey of federal environmental statutes has focused on major laws that offer the clearest opportunity to challenge climate engineering research projects. Several other federal statutes, however, could offer additional avenues for legal review if the specific climate engineering proposal fell within their coverage. For example, the Comprehensive Environmental Response, Compensation and Liability Act (CERCLA) might create liability for persons responsible for releases of hazardous substances as part of a climate engineering project.[168] Because liability would be strict as well as joint and several (if the release caused an indivisible harm), potentially responsible parties for response costs resulting from a climate engineering project might face the daunting task of proving which portion of those costs should be attributed to their activities. More important, the United States might

[166] *Id.* § 1401(c).

[167] Given MPRSA's origin as the United States' implementing legislation for the London Convention, the IMO's declaration that ocean iron fertilization projects constitute "dumping" (unless conducted for limited scientific purposes) would buttress an effort by the United States to regulate ocean iron fertilization under MPRSA. *See supra* notes 47–49 and accompanying text.

[168] 42 U.S.C. §§ 9601–9630 (2006). For example, the dispersal of large quantities of aerosols or engineered particulates into the stratosphere may constitute an arrangement for the disposal of those materials once they inevitably precipitate onto the ground. Similar arguments might be made for minerals or compounds dispersed onto the ocean surface for fertilization projects. If these materials fall within CERCLA's broad definition of "hazardous substance," researchers who arranged for the dispersal of those materials may face strict liability for costs incurred to respond to those releases. This risk could be especially problematic if the releases allegedly cause natural resource damages in addition to costs incurred to respond to the release. *See id.* § 9607(a); U.S. GOV'T ACCOUNTABILITY OFFICE, *supra* note 65, at 29 ("Although a stream of pure CO_2 is not a hazardous substance under CERCLA, an EPA official noted that injected CO_2 streams could contain hazardous substances, thus subjecting the parties injecting the CO_2 to liability for any release that did not qualify as federally-permitted release. In addition, if CO_2 enters groundwater, it might also cause hazardous substances, such as some metals, to be dissolved by the groundwater from enclosing strata. If that constitutes a release of hazardous substances from a 'facility,' such as the strata, then the owner of that facility could be liable for any cleanup costs caused by that release.").The targets of CERCLA actions, however, will likely be cases that exempt the dispersal of certain materials from CERCLA's definition of "release" if the intended use of those materials foresaw their dispersal and eventual placement onto land. In addition, if climate engineering projects obtain permits under the CAA or other federal statutes, releases pursuant to those permits may fall under relaxed requirements for federally permitted releases. *See id.* 42 U.S.C. § 9601(10).

have a broader scope to compel persons performing a climate engineering project to undertake emergency action to abate an imminent and substantial threat to human health and the environment. Although CERCLA fixes liability on potentially responsible parties for costs incurred to respond to a release of "hazardous substances,"[169] CERCLA authorizes the federal government to undertake any action needed to respond to a release of "pollutants" – a broader category than "hazardous substances."[170] Although this action may not result in liability for potentially responsible parties, it could nonetheless empower the government to impose substantial restrictions on an ongoing climate engineering project that arguably created a threat to human health or the environment.

The Migratory Bird Treaty Act imposes strict liability on persons whose activities cause the taking of a migratory bird, and that liability can be criminal.[171] A climate engineering project that unintentionally causes the deaths of migratory birds therefore might pose a risk of liability.[172] In addition, ocean fertilization projects in coastal

[169] 42 U.S.C. § 9607(a) (2006).

[170] *Id.* § 9604(a)(1). CERCLA defines "pollutant" much more broadly than the term "hazardous substance." *Id.* § 9601(33) (defining "pollutant" to "include, but not be limited to, any element, substance, compound, or mixture, including disease-causing agents, which after release into the environment and upon exposure, ingestion, inhalation, or assimilation into any organism, either directly from the environment or indirectly by ingestion through food chains, will or may reasonably be anticipated to cause death, disease, behavioral abnormalities, cancer, genetic mutation, physiological malfunctions (including malfunctions in reproduction) or physical deformations, in such organisms or their offspring"). CERCLA's definition of "pollutant," however, excludes petroleum. *Id.*

[171] The Migratory Bird Treaty Act (MBTA) implements the treaty obligations that the United States incurred in the Migratory Bird Treaty of 1918, Convention for the Protection of Migratory Birds, Aug. 16, 1916, United States–Great Britain, 39 Stat. 1702, T.S. 628, as well as subsequent treaties with Mexico, Japan, and Russia. Convention for the Protection of Migratory Birds and Game Mammals, Feb. 7, 1936, United States–Mexico, 50 Stat. 1311, T.S. 912; Convention for the Protection of Migratory Birds in Danger of Extinction and Their Environment, Mar. 4, 1972, United States–Japan, 25 U.S.T. 3329, T.I.A.S. 7990 (ratified 1973); Convention Concerning the Conservation of Migratory Birds and Their Environment, United States–Union of Soviet Socialist Republics, Nov. 19, 1976, 92 Stat. 3110, T.I.A.S. 9073. Section 703 of the MBTA prohibits the taking, killing, or unlawful possession of migratory birds without a permit. 16 U.S.C. § 703 (2006). A violation of this prohibition can constitute a misdemeanor punishable by a fine up to $15,000 and up to six months imprisonment. *Id.* § 707(a) (2011). Notably, the MBTA would treat such an offense as a strict liability misdemeanor that would allow the United States to prosecute responsible corporate officers without proof that they knew about the violation or acted negligently in supervising actions that led to injury to migratory birds. David M. Uhlmann, *After the Spill Is Gone: The Gulf of Mexico, Environmental Crime, and the Criminal Law*, 109 MICH. L. REV. 1413, 1444 n.204 (2011). Historically, the U.S. Department of Justice has not prosecuted environmental cases under strict liability theories unless the defendants have also acted at least negligently.

[172] *See* 16 U.S.C. § 703 (2006). For example, a demonstration project to adjust the acidity of marine waters may involve the addition of chemical buffering agents to affect the pH level of waters over a broad area. If these chemicals injured or killed migratory birds feeding in the area, the project's operator might arguably face civil and criminal liability under the MBTA.

waters that might affect marine sanctuaries could be subject to regulation under the National Marine Sanctuaries Act[173] or the Marine Mammal Protection Act.[174] The Offshore Continental Shelf and Lands Act may also provide a basis for citizen suits to challenge climate engineering projects that involve use of submerged lands in the U.S. territorial sea or exclusive economic zone.[175]

Notably, this initial overview of potential environmental challenges touches solely on federal statutory options. Claimants may find that state environmental laws offer richer opportunities to challenge climate engineering projects that might require an environmental impact statement or a state permit with more stringent emission or operating requirements.[176] In addition, states are not bound by federal determinations on standing that apply to limited jurisdiction federal courts, and as a result state courts can use thresholds for standing or justiciability that favor environmental claimants.[177]

3.2 *Potential Barriers to U.S. Judicial Review of Challenges to Climate Engineering Projects*

In addition to federal environmental statutory programs, federal or state common law nuisance claims may provide a viable avenue for judicial review of climate engineering projects. This field of law is in a state of high flux after the U.S. Supreme Court's recent decision in *American Electric Power Co. v. Connecticut*. The question presented was whether federal courts can hear federal common law claims alleging that climate change effects had created a public nuisance.[178] The Court concluded

[173] *Id.* §§ 1431–1445c1 (2006). The NMSA authorizes the Secretary of Commerce to designate and protect areas of the marine environment with special national significance, and it authorizes civil fines up to $100,000 per violation per day and damages against persons who injure marine sanctuary resources. *Id.* §§ 1436, 1437(d)(1), 1443(a)(1).

[174] *Id.* § 1361–1423H (2006). The MMPA generally prohibits the taking of marine mammals without a permit in waters of the United States or by U.S. citizens on the high seas. *Id.* § 1372(a).

[175] 43 U.S.C. § 1331–1356a (2006).

[176] Although federal environmental statutes can set minimum standards that states must meet, those statutes often also allow states to impose more stringent environmental standards or to require permits from sources exempt under federal law. As a result, states often have their own environmental statutes and regulatory systems that precede – and go beyond – federal environmental programs. For example, the California Environmental Quality Act has provided the basis for numerous citizen suits to challenge state actions where the government failed to properly account for climate change effects in state environmental impact statements. *See* Cal. Pub. Res. Code §§ 21000–21177; *see, e.g.*, Santa Clarita Org. for Planning the Env't v. City of Santa Clarita, 197 Cal. App. 4th 1042 (2011) (CEQA citizen suit objecting to adequacy of GHG mitigation measures for hospital expansion project).

[177] T. Hester, *A New Front Blowing In: State Law and the Future of Climate Change Public Nuisance Litigation*, 31 STAN. ENVTL. L. J. 101, 115–19 (2012).

[178] Am. Elec. Power Co. v. Connecticut, 131 S. Ct. 2527 (2011).

that passage of the Clean Air Act had displaced federal common law actions that might impose public nuisance liability for interstate emissions.[179]

The Court's ruling follows three federal appellate court decisions that had already undertaken searching scrutiny of climate change public nuisance claims – *Native Village of Kivalina v. ExxonMobil Corp.*,[180] *Comer v. Murphy Oil*,[181] and *Connecticut v. American Electric Power Co.*[182] – to address the role that federal courts can or should play in global climate change tort disputes. These decisions centered on the political question doctrine, standing, ability to prove causation, and displacement or preemption. The appellate courts rendered mixed decisions on climate change public nuisance claims, and the U.S. Supreme Court's decision in *American Electric Power* shed some much-needed light on this field of law.

Although the *American Electric Power* Court ultimately dismissed the plaintiffs' claims, the ruling does not automatically preclude the application of public nuisance tort principles to climate engineering projects. Most important, the decision presents climate engineering proponents with a Hobson's choice: either risk public nuisance liability by arguing that the Clean Air Act does not apply to climate engineering projects, or accept the prospect of Clean Air Act permitting obligations. In addition, the Court's rationale – that Congress displaced federal common law by giving the EPA the authority to regulate GHG emissions under the Clean Air Act – would not apply to climate engineering projects that create public nuisances without emitting significant GHGs – for example, albedo enhancement projects or solar radiation management without aerosols. The *American Electric Power* decision expressly did not overturn earlier precedents that plaintiffs claiming climate change damages could satisfy Article III threshold requirements for standing and avoid the political question doctrine.[183] In contrast to public nuisance damage lawsuits for climate alterations arising from past and global GHG emissions, climate engineering challenges may provide a clearer avenue to bring climate change tort actions into the federal courts. Although still raising important claims over climate change responsibilities and liability, these actions will neatly sidestep – or even reverse – the typical challenges raised against climate change nuisance suits under federal common law.

Before examining the application of federal tort liability theories to climate engineering projects, it is important to note that the primary challenges to climate engineering projects will likely rely on federal environmental statutes rather than

[179] *Id.* at 2537.
[180] Native Village of Kivalina v. ExxonMobil Corp., 663 F. Supp. 2d 863 (N.D. Cal. 2009).
[181] Comer v. Murphy Oil USA, 585 F.3d 855 (5th Cir. 2009), *vacated*, 607 F.3d 1049 (5th Cir. 2010).
[182] Connecticut v. Am. Elec. Power Co., 582 F.3d 309 (2d Cir. 2009).
[183] *American Electric Power* allowed the claims to proceed because the Court split 4–4 on whether the Second Circuit had correctly approved the plaintiffs' standing and justiciability claims. Notably, the

federal common law tort claims. Although statutory claims can still face standing and justiciability problems, those concerns are greatly reduced when Congress has established a statutory framework for judicial review. By doing so, Congress can exercise its power to define a property interest or procedural right that can become a legally protectable interest. An invasion of that statutory right thereby can support standing and justiciability. For example, the U.S. Supreme Court in *Massachusetts v. EPA* held that where Congress gives a procedural right to protect a plaintiff's concrete interests, the plaintiff "can assert that right without meeting the normal standing requirements of redressability and immediacy ... the litigant has standing if there is some possibility that the requested relief will prompt the injury-causing party to reconsider the decision that allegedly harmed the litigant."[184] Claimants attacking a climate engineering project could meet standing and justiciability requirements by showing a federal environmental statute provides them with a similar substantive or procedural right. As noted previously, U.S. environmental statutes could provide an array of possible options to contest climate engineering research or demonstration projects.[185]

Numerous other articles have surveyed the key challenges and procedural status of the three key climate change public nuisance lawsuits currently before the federal appellate courts,[186] and this chapter will only recount the key aspects of those cases as they might illuminate the role of public nuisance lawsuits in halting climate engineering efforts. It will also focus on the trial court decisions to some extent because their rationales offer the most insight into how federal trial courts will initially respond to climate engineering lawsuits. In each of the problematic areas for climate change public nuisance actions, a legal action seeking damages or injunctive relief against a climate engineering project would face significantly less difficulty in presenting a viable claim than a broader public nuisance action for damages caused by historic and ongoing anthropogenic greenhouse gas emissions.[187]

Court relied on its prior ruling in *Massachusetts v. EPA* that states deserved "special solicitude" on their standing claims because of their special role as sovereigns with an interest in their state's citizens and resources. Massachusetts v. EPA, 549 U.S. 497, 520 (2007). The availability of standing and justiciable claims for private plaintiffs alleging climate injuries remains controversial. *See infra* notes – and accompanying text.

[184] *Massachusetts*, 549 U.S. at 517–18 (2007).

[185] *See supra* Section 3.1.

[186] For some of the most recent analyses, see Robin Kundis Craig, *Adapting to Climate Change: The Potential Role of State Common-Law Public Trust Doctrines*, 34 VT. L. REV. 781 (2010); Randall S. Abate, *Public Nuisance Suits for the Climate Justice Movement: The Right Thing and the Right Time*, 85(2) WASH. L. REV. 197 (2010); Michael B. Gerrard, *What the Law and Lawyers Can and Cannot Do about Global Warming*, 16 SOUTHEASTERN ENVTL. L.J. 537 (2007); David Hunter & James Salzman, *Negligence in the Air: The Duty of Care in Climate Change Litigation*, 155 U. PA. L. REV. 1741 (2007); Shi-Ling Hsu, *A Realistic Evaluation of Climate Change Litigation through the Lens of a Hypothetical Lawsuit*, 79 U. COLO. L. REV. 701 (2008).

[187] This chapter does not explore whether prior tort litigation over damages from weather modification efforts might offer a precedent for liability for climate engineering projects. In particular, although

3.2.1 Political Question

The most threatening jurisprudential shoal for public nuisance climate change suits has been the political question doctrine. The political question doctrine, while much debated over its doctrinal justifications and exact formulation, holds generally that federal courts cannot entertain cases that present controversies or issues that either the U.S. Constitution has committed to the other two political branches or the judicial branch lacks the institutional capacity to resolve or enforce.[188] In particular, the political question doctrine can allow a federal court to dismiss requests for relief that would require the court to implement a long-term and complex remedial scheme in an area where the court lacks discernible legal standards to guide its supervision.[189] The political question doctrine has also been applied to cases that

efforts to modify precipitation patterns or affect local weather patterns have historically triggered lawsuits for trespass and negligence for damages allegedly caused by flooding or drought due to the projects, not one of those cases has yielded a final judgment for legal damages. Gregory N. Jones, Comment, *Weather Modification: The Continuing Search for Rights and Liabilities*, 1991 BYU. L. REV. 1163, 1167–70 (1991).

It may become relevant for climate engineering tort liability that some states have statutorily exempted weather modification activities from private or public nuisance liability. *See, e.g.*, COLO. REV. STAT. § 36–20–123 (2011); UTAH CODE ANN. § 73–15–7 (West 2011). Other states provide similar statutory exemptions from trespass liability for weather modification efforts. These statutory protections apply only if the defendant holds a permit or authorization from the state to perform weather modification. *See, e.g.*, COLO. REV. STAT. § 36–20–123 (2011) (permit required); *see also* N.D. CENT. CODE ANN. § 61–04.1–37(2) (West 2011) ("Dissemination of materials and substances into the atmosphere by a permittee acting within the conditions and limits of the permittee's permit shall not constitute trespass."); WIS. STAT. ANN. § 93.35(14)(b) (2011) ("Dissemination of materials and substances into the atmosphere by a permittee acting within the conditions and limits of his or her permit shall not give rise to the contention that the use of the atmosphere constitutes trespass."); COLO. REV. STAT. § 36–20–123(2)(a) (2011) ("Failure to obtain a permit before conducting [a weather modification] operation ... shall constitute negligence per se.").

[188] Baker v. Carr, 369 U.S. 186, 189 (1962). In this seminal opinion, the U.S. Supreme Court described the specific factors that identify a political question. The well-known *Baker* factors include "[i] a textually demonstrable constitutional commitment of the issue to coordinate political department; [ii] or a lack of judicially discoverable or manageable standards for resolving it; [iii] or the impossibility of deciding without an initial policy determination of a kind clearly for non-judicial discretion; [iv] or the impossibility of a court's undertaking independent resolution without expressing lack of the respect due coordinate branches of the government; [v] or an unusual need for unquestioning adherence to a political decision already made; [vi] or the potentiality of embarrassment from multifarious pronouncements by various departments on one question." *Id.* at 217.

For an analysis of the political question doctrine as it specifically relates to climate change cases, see James May, *Climate Change, Constitutional Consignment, and the Political Question Doctrine*, 85 DENV. U. L. REV. 919 (2008); shawn M. LaTourette, *Climate Change: A Political Question?*, 40 RUTGERS L.J. 219 (2008). The foreign policy aspect of the political question doctrine is likely to see fresh scrutiny by the courts as an increasing number of lawsuits swirl around the activities of American companies in theaters of war. *See* Note, *The Political Question Doctrine: Executive Deference, and Foreign Relations*, 122 HARV. L. REV. 1193 (2009).

[189] *See, e.g.*, California v. Gen. Motors Corp., No. C06–05755 MJJ, 2007 WL 2726871, at *14–16 (N.D. Cal. 2007) (dismissing federal common law tort action against automobile manufacturers for damages

turn on multifaceted nonlegal factors that ultimately rest on political judgments pertaining to the allocation of benefits or responsibilities.[190]

Climate change public nuisance suits are highly susceptible to political question challenges, and the three key cases have each spurred numerous motions to dismiss on political question grounds. Because each of the three suits raise different claims and seek varying types of relief, the trial courts have offered different rationales in their opinions granting each motion to dismiss. For example, in *American Electric Power*, the plaintiffs requested an injunction that would limit greenhouse gas emissions from coal-fired power plants in multiple northeastern states under a plan that would compel the plants to gradually reduce their emissions over decades of operation.[191] Not unexpectedly, the trial judge concluded that the plaintiff's request would force the trial court to make decisions that effectively allocated liabilities and influenced regional power generation over an extended period of time.[192] Judge Preska described this type of injunctive relief as squarely within the sphere of issues that the political question doctrine barred from federal court review:

> [A] non-justiciable political question exists when a court confronts "the impossibility of deciding without an initial policy determination of a kind clearly for non-judicial discretion." As the Supreme Court has recognized, to resolve typical air pollution cases, courts must strike a balance "between interests seeking strict schemes to reduce pollution rapidly to eliminate its social costs and interests advancing the economic concern that strict schemes [will] retard industrial development with attendant social costs." In this case, balancing those interests, together with the other interests involved, is impossible without an "initial policy determination" first having been made by the elected branches to which our system commits such policy decisions, viz., Congress and the President.[193]

arising from climate change due to GHG emissions by pointing in part to the lack of judicially manageable standards; "[t]he crux of this inquiry is not whether the case in unmanageable in the sense of being large, complicated or otherwise difficult to tackle from a logistical standpoint ... Rather, courts must ask whether they have the legal tools to reach a ruling that is 'principled, rational, and based upon reasoned distinctions'").

[190] For example, the federal courts have pointed to the lack of any judicially discernible and manageable standards to decline to review challenges to federal immigration programs or military policies. *See, e.g.*, Texas v. United States, 106 F.3d 661, 664–65 (5th Cir. 1997) (action to recover costs from federal government for expenditures related to undocumented aliens fell under political question doctrine because of lack of manageable standards); Carmichael v. Kellogg, Brown & Root Servs., 572 F.3d 1271, 1288–92 (11th Cir. 2009) (political question barred negligence action against military contractor because court could not identify readily ascertainable and judicially manageable standards). *Cf.* Crockett v. Reagan, 720 F.2d 1355, 1357 (D.C. Cir. 1983) (refusing to review decisions related to military aid because of equitable restraint doctrine).

[191] Connecticut v. Am. Elec. Power Co., 406 F. Supp. 2d 265, 267–70 (S.D.N.Y. 2005).

[192] *See id.* at 272–74.

[193] *Id.* at 272.

The Second Circuit reversed the trial court's dismissal order because the appellate panel concluded that the requested relief affected only a small number of power plants and that the courts could handle the admittedly complex allocation of liabilities and obligations required by such injunctive relief.[194] The federal courts, according to the Second Circuit, had long handled complex questions like these as part of their inherent award of equitable relief to multiple parties.[195]

Comer and *Kivalina* also yielded initial trial court rulings that dismissed the complaints because they posed political questions, but the courts diverged on their rationales. The *Comer* trial court stated that the plaintiffs' claim asked the court to "balance economic, environmental, foreign policy and national security interests and make an initial policy determination of a kind which is simply non-judicial."[196] The court further held that such "policy decisions are best left to the executive and legislative branches of the government, who are not only in the best position to make those decisions but are constitutionally empowered to do so."[197] Although the Fifth Circuit panel decision disagreed and concluded that the complaint raised no political question,[198] the full Fifth Circuit subsequently vacated that opinion without issuing a substantive analysis of its own to replace it.[199]

By contrast, the *Kivalina* trial court concluded that the limited relief sought by the plaintiffs nonetheless posed a political question because (1) the plaintiffs' claims rested on allegations of emissions and damages on a global scale that lacked any judicially discoverable or manageable standards, and (2) the issues raised by the plaintiffs' claims would require the trial court to make a fundamentally legislative policy judgment.[200] The district court gave no credit to the defendants' argument that the global warming issue may involve foreign policy and related economic issues, and therefore it failed to satisfy the first step in the *Baker* test. The court wrote that "the fact that this case 'touches foreign relations' does not ipso facto place it beyond the reach of the judiciary," and it noted that *Baker* itself cautions against sweeping generalities regarding foreign policy being textually delegated to the executive.[201]

Interestingly, the U.S. Supreme Court did not resolve this issue in *American Electric Power*. The Court noted briefly that four members found that at least some plaintiffs had standing under *Massachusetts v. EPA*,[202] and that "no other threshold

[194] Connecticut v. Am. Elec. Power Co., 582 F.3d 309, 326–30 (2d Cir. 2009).
[195] *Id.* at 326.
[196] Comer v. Murphy Oil USA, 585 F.3d 855, 860 n.2 (5th Cir. 2009).
[197] *Id.*
[198] *Id.* at 878–80.
[199] Comer v. Murphy Oil USA, 607 F.3d 1049 (5th Cir. 2010).
[200] Native Village of Kivalina v. ExxonMobil Corp., 663 F. Supp. 2d 863, 868, 873–77 (N.D. Cal. 2009).
[201] *Id.*
[202] Am. Elec. Power Co. v. Connecticut, 131 S. Ct. 2527, 2535 (2011).

obstacle bars review."[203] A footnote appended to this statement observed that the plaintiffs had renewed their political question objections made later in this chapter,[204] so presumably the four members believed that the case did not present a political question. The other four members would have found that the plaintiffs lacked standing, but the opinion does not disclose their opinion on political question and justiciability issues.[205] Because the Court split evenly (in light of Justice Sotomayor's recusal), the Court affirmed the Second Circuit's exercise of jurisdiction.[206]

A lawsuit seeking to halt a climate engineering project probably would not face the vulnerabilities to a political question attack described in the three public nuisance trial court opinions. Rather than seek a judicial determination on liabilities arising from global activities over decades arguably caused by thousands (if not millions) of other parties in both the United States and throughout the world, a judicial challenge to a climate engineering project could involve plaintiffs who challenge a discrete set of proposed actions by a limited and readily identifiable group of defendants that the court could easily address through injunctions or other equitable relief. Depending on the scope of the project, this relief would likely not require any continuing oversight by the court of complex technical activities with sweeping economic consequences, and the court's actions would not impinge on any overt textual commitment of the issue to either other governmental branch.[207]

3.2.2 STANDING

Standing has also posed a significant hurdle for federal public nuisance lawsuits where damages or injunctive relief have been sought for harms related to climate change.[208] Although standing pitfalls in climate change public nuisance litigation

[203] *Id.*

[204] *Id.* at 2535 n.6.

[205] *Id.* at 2535.

[206] More accurately, the Court lacked a majority that could reverse the Second Circuit's ruling. *Id.*

[207] Some aspects of climate engineering lawsuits may arguably ask the court to take actions that fall into the sphere of foreign affairs powers because they involve activities outside the United States. Unless those projects involve foreign governments or their instrumentalities, though, it is unlikely that these types of disputes will fall within the core activities that the U.S. Constitution textually commits to the legislative and executive branches.

If the United States itself chose to undertake a climate engineering project on any significant scale, however, the court could face many of the same issues raised in federal common law public nuisance tort actions against large GHG emitters. For example, if the U.S. government pursued a large-scale program to forestall an alleged climate emergency, the court hearing a challenge to that program could find itself wrestling with complex technical monitoring issues and foreign policy concerns.

[208] Because the trial court dismissed the plaintiffs' claims in *Connecticut v. American Electric Power* solely on political question grounds, it expressly declined to rule on whether the plaintiffs had standing to bring their claims. Connecticut v. Am. Elec. Power Co., 406 F. Supp. 2d 265, at 271 n.6 (S.D.N.Y. 2005). The vacated *Comer* appellate panel opinion concluded that the plaintiffs had standing because

have already spurred a large amount of scholastic analysis and commentary,[209] the basic principles of Article III standing illuminate why plaintiffs might face significant challenges in bringing claims for damages allegedly caused by generalized climate change attributable to specific defendants. As the U.S. Supreme Court has repeatedly noted, a plaintiff must meet three factors to demonstrate standing: an injury-in-fact (a specific and concrete invasion of a protectable interest held by the plaintiff), causation (a fairly traceable connection between the injury-in-fact and the defendant's conduct), and redressability (it is likely and not speculative that the plaintiff's injury will be remedied by the relief sought by the plaintiff).[210]

Given that GHG emissions worldwide contribute to global warming and that any effective relief arguably requires reductions in GHG emissions from a vast array of sources located throughout the world, these irreducible constitutional standing requirements pose an obvious hurdle that most federal climate change public nuisance claimants will have to overcome. The *Kivalina* trial court did not allow the case to go forward because of the political question doctrine discussed earlier in the chapter and because the plaintiffs could not show that any particular act by the defendants could be fairly traced to the plaintiffs' injuries. In particular, the judge noted that, "[e]ven accepting the allegations of the Complaint as true and construing them in the light most favorable to Plaintiffs, it is not plausible to state which emissions – emitted by whom and at what time in the last several centuries and in what place in the world – 'caused' Plaintiffs' alleged global warming injuries."[211] Allowing the suit to go forward, the court held, would make the dozen defendants responsible for the emissions released by "virtually everyone on

they needed to show only that the defendants' actions had contributed to (rather than solely or materially caused) global warming harms. Comer v. Murphy Oil USA, Inc., 585 F.3d 855, 864–65 (5th Cir. 2009). The *Comer* plaintiffs filed a petition for certiorari to the U.S. Supreme Court for review of the Fifth Circuit's decision to vacate the panel opinion; the Fifth Circuit subsequently lacked sufficient judges to conduct an en banc review. Comer v. Murphy Oil USA, 607 F.3d 1049 (5th Cir. 2010).

[209] Michael B. Gerrard, *Survey of Climate Change Litigation*, 238 N.Y.L.J. 63 (2007); Matthew E. Miller, Note, *The Right Issue, The Wrong Branch: Arguments against Adjudicating Climate Change Nuisance Claims*, 109 MICH. L. REV. 257 (2010); Richard O. Faulk & John S. Gray, *Defending against Climate Change Litigation: Threshold Issues*, 29 ANDREWS LIT. RPTR. 2–3 (2008), *available at* http://works.bepress.com/richard_faulk/16/; Holly Doremus, *The Persistent Problem of Standing in Environmental Law*, 40 ENVTL. L. REP. 10956 (2010).

[210] Sprint Commc'ns Co. v. APCC Servs., Inc., 554 U.S. 269, 273 (2008); Lujan v. Defenders of Wildlife, 504 U.S. 555, 560–61(1992).

[211] Native Village of Kivalina v. ExxonMobil Corp., 663 F. Supp. 2d 863, 880–81 (N.D. Cal. 2009). The *Comer* trial court reached a similar conclusion on standing by noting that "[t]hese are not injuries which are fairly attributable to these individual defendants." The court declined to impose liability for damages caused by "a larger group that [is] not before this Court, not only within this nation but outside of our jurisdictional boundaries as well." Transcript of Hearing on Defendants' Motions to dismiss at 36, Trial transcript at p. 36, Comer v. Murphy Oil, U.S.A., 2007 WL 6942285 (Aug. 30, 2007) (No. 1:05-CV-436-LG-RHW) (on file with author).

Earth."[212] The *American Electric Power* trial court added in a footnote that "because the issue of Plaintiffs' standing is so intertwined with the merits and because the federal courts lack jurisdiction over this patently political question, I do not address the question of Plaintiffs' standing."[213] The Second Circuit disagreed, finding that before delving into the merits of the case it had a duty to determine sua sponte whether the plaintiffs had Article III standing.[214] As noted above, the U.S. Supreme Court did not resolve the issue because it affirmed the Second Circuit's opinion in a 4–4 split that expressly relied on *Massachusetts v EPA* and its "special solicitude" for state plaintiffs bringing public nuisance claims.[215]

By contrast, plaintiffs seeking to challenge a proposed climate engineering project would have an easier burden of proof for standing. First, a climate engineering demonstration project will presumably involve an effort expressly designed to generate a measurable regional (or ultimately global) effect distinguishable from general climate change impacts. The plaintiffs in turn could attribute those effects and potential risks to the defendants' specific actions in the tests. As a result, plaintiffs could use the defendants' explanations to justify the basis for the experiment or project to build a prima facie case for both injury-in-fact and causation.[216] Proof of redressability also might not pose a major hurdle because the court presumably could address the alleged risks or injuries by enjoining the climate engineering project or awarding damages to compensate the specific injuries alleged by the plaintiffs.[217]

3.2.3 CAUSATION

Aside from the difficulties they have faced in showing that the defendants' actions could be "fairly traced" to alleged harms, climate change public nuisance plaintiffs

[212] *Village of Kivalina*, 663 F. Supp. 2d at 874.

[213] Connecticut v. Am. Elec. Power Co., 406 F. Supp. 2d 265, 271 n.6 (S.D.N.Y. 2005). The Second Circuit's panel opinion overruled this aspect of the trial court's opinion and found that the plaintiffs had standing to bring their claims because they need show only that the defendants "contributed to" the undifferentiated harms of global warming and that the court could grant some measure of relief, even if that relief could not result in measurable decreases in overall global warming effects. Connecticut v. Am. Elec. Power Co., 582 F.3d 309, 45–347 (2d Cir. 2009).

[214] *Am. Elec. Power Co.*, 582 F.3d at 333.

[215] Am. Elec. Power Co. v. Connecticut, 131 S. Ct. 2527, 2535 (2011).

[216] This outcome could vary based on the individual circumstances surrounding each project. For example, a small research project to demonstrate the technological feasibility of certain approaches may not be designed to cause any discernible effects on a regional or global level. Claimants challenging such a small project would have much greater difficulties establishing their standing.

[217] Although climate engineering opponents might face serious difficulties in quantifying the amount of harm or damages they might suffer from a research test or demonstration project, the federal courts have long issued injunctions to halt activities that might increase the risk of harm if that harm satisfied general or statutory tests for issuance of injunctions. *See* Winter v. Natural Res. Def. Council, 129 S. Ct. 365 (2008) (setting out standards for issuance of injunctions to halt alleged violations of NEPA requirements by the Navy's sonar tests).

will face even higher causation hurdles if their claims proceed to trial. The same features that make standing difficult to establish – the thorough mixing of CO_2 emissions on a global basis in a relatively short time period, the long residence time of CO_2 in the atmosphere, and the complex processes by which CO_2 and other GHGs can lead to multiple changes to climate and, in turn, to weather or marine conditions – will pose daunting challenges for public nuisance plaintiffs who wish to establish specific causation as well as causation-in-fact between the defendants' emissions and the alleged damages from climate change. By contrast, demonstration efforts and test projects for climate engineering research will have the express goal of altering climate globally or in a discrete region in measurable ways. The overt aims, design, and public statements for climate engineering projects may help reduce the evidentiary burdens to show that the projects caused, or might cause in the future, harms to individuals or the environment.[218]

3.2.4 PREEMPTION AND DISPLACEMENT

Although the trial courts in the *Kivalina*, *Comer*, and *American Electric Power* cases each dismissed the claims on political question or standing grounds, they also heard vigorous arguments that any federal common law public nuisance claims had been displaced by subsequent federal governmental actions that fully occupied the field. In particular, the defendants alleged that the executive branch's efforts to persuade other nations to reduce their GHG emissions through international negotiations demonstrated the executive's exercise of its constitutional authority over foreign affairs, and that any attempt by the federal courts to impose GHG emission limits through public nuisance verdicts could undermine the United States' negotiation position.[219] The defendants also argued that the failure of Congress to pass any GHG emission limits reflected a policy decision not to impose GHG emission limits that displaced any federal common law causes of action that might lead to conflicting results.[220] As EPA has promulgated an increasingly large array of regulatory

[218] Claimants would still have to link alleged climatic changes to actionable harms such as economic loss or aesthetic injuries before they could demonstrate individual or organizational standing. *See Am. Elec. Power Co.*, 582 F.3d at 346 (2d Cir. 2009). Again, the scale of the climate engineering project may directly affect the claimants' ability to show standing. *See* discussion *supra* at note 216.

[219] *Am. Elec. Power Co.*, 406 F. Supp. 2d at 273–74; Native Village of Kivalina v. ExxonMobil Corp., 663 F. Supp. 2d 863, 872–73 (N.D. Cal. 2009); Brief of Defendants-Appellees, Comer v. Murphy Oil USA, No. 07–60756 (5th Cir. Jan. 9, 2008), 2008 WL 8094253 at *48 n.19.

[220] *Am. Elec. Power Co.*, 582 F.3d at 378–88; Answering Brief for Defendants-Appellees, Kivalina v. ExxonMobil Corp., No. 09–17490 (9th Cir. June 30, 2010), 2010 WL 3299982 at *61–66; Comer v. Murphy Oil USA, 585 F.3d 855, 875–76 (reviewing displacement and preemption as component of political question analysis). Alternatively, those same defendants also contended that Congress had displaced federal common law for nuisance actions by promulgating the federal Clean Air Act (even if EPA chose not to exercise that authority). As discussed below, the U.S. Supreme Court ultimately

limits and permitting obligations for GHG emissions, the growing federal regulatory presence has strengthened the argument that federal common law in this arena is displaced.[221]

To the extent that federal environmental statutes might apply to climate engineering projects, federal common law tort plaintiffs may need to plead their cases carefully to sidestep displacement arguments. If they fail to persuade the court that federal environmental statutes can support challenges to climate engineering projects, the plaintiffs could argue in the alternative that the failure of environmental statutes and regulations to expressly address climate engineering concerns leaves undisturbed the federal courts' common law authority to hear tort claims. Given the lack of any express U.S. treaty, legislation, or regulation to address climate engineering, defendants may not be able to prove that current federal statutes and regulations have displaced the federal courts' authority to hear common law challenges to climate engineering projects that may affect specific plaintiffs.[222]

Last, federal common law may also provide a scaffold in U.S. courts for climate engineering legal attacks that rely on U.S. environmental treaties to which the United States is a party as well as other international obligations. As confirmed by long-standing U.S. Supreme Court precedent, federal common law incorporates customary international laws as the law of the United States for purposes of the Supremacy Clause.[223] In addition, treaties can become directly enforceable (if implemented by the Senate or if self-executing) as supreme federal law in U.S. courts. If climate engineering challenges assert that prior international conventions or treaties or international customary law prohibit those experiments, U.S. federal and state courts may provide a potentially viable forum to assert those claims.[224]

accepted this line of argument in *American Electric Power v. Connecticut*. Am. Elec. Power Co. v. Connecticut, 131 S. Ct. 2527, 2530–41 (2011).

[221] *Am. Elec. Power Co.*, 131 S. Ct. at 2530–41.

[222] A court deciding whether federal activities have preempted either conflicting state actions or the entire field under the Supremacy Clause will likely focus on these same analytical concerns. *See* U.S. CONST. art. VI § 2.

[223] The Paquete Habana, 175 U.S. 677 (1900) (in ruling that customary international law prohibited the capture and sale of a Spanish fishing vessel as a prize of war, the Court famously declared that "[i]nternational law is part of [U.S.] law, and must be ascertained and administered by the courts of justice of appropriate jurisdiction as often as questions of right depending upon it are duly presented for their determination. For this purpose, where there is no treaty and no controlling executive or legislative act or judicial decision, resort must be had to the customs and usages of civilized nations, and, as evidence of these, to the works of jurists and commentators who by years of labor, research, and experience have made themselves peculiarly well acquainted with the subjects of which they treat"); Sosa v. Alvarez-Machain, 542 U.S. 693, 729 (2004) ("[f]or two centuries we have affirmed that the domestic law of the United States recognizes the law of nations").

[224] The United States has already entered into one international convention that might limit climate engineering experiments if an experiment has military motives or implications. Under the Environmental Modification Treaty, the parties agree "not to engage in military or any other hostile

CONCLUSION

The challenge of climate engineering governance ultimately should require an international framework. Climate engineering projects will inherently affect multiple nations and the global commons, and they will cross jurisdictional lines in a way that will make it difficult for any regional or national regulatory scheme to effectively control risks posed by these projects.[225] Even viewed solely as a national regulatory initiative, the novel risks and aspects of climate engineering point out the need for an explicit federal legislative response that would give clear direction to both agencies and researchers on critical issues such as permitting, liability, and oversight.

In the absence of international action or federal legislative direction, however, U.S. environmental statutes and laws may provide a workable initial forum to lay the groundwork for risk management and governance of climate engineering projects that take place in the United States or which involve U.S. citizens or vessels. Researchers seeking to test or deploy climate engineering technologies will first have to determine whether federal and state environmental regulatory programs could apply to their projects. Although Congress clearly did not foresee these technologies when it passed the key federal environmental statutes, certain aspects of climate engineering projects may fall under current federal environmental regulatory authority. In particular, climate engineering projects that seek to reduce solar radiation influx through large-scale releases of sulfate aerosols from stationary sources may find themselves potentially subject to Clean Air Act regulation. To the extent federal environmental laws may oblige climate engineering researchers to seek

use of environmental modification techniques having widespread, long-lasting or severe effects as the means of destruction, damage or injury to any other State Party." Convention on the Prohibition of Military or Any Other Hostile Use of Environmental Modification Techniques, Art. I, § 1, *opened for signature* May 18, 1977, 1108 U.N.T.S. 151 (effective Jan. 17, 1980). If an individual sought to conduct a climate engineering demonstration or research project in a fashion that might constitute such a military or "hostile use," the United States may have a treaty obligation to take all constitutional steps to stop the project. *Id.* Art. IV ("Each State Party to this Convention undertakes to take any measures it considers necessary in accordance with its constitutional processes to prohibit and prevent any activity in violation of the provisions of the Convention anywhere under its jurisdiction or control."). The Convention does not provide for any private actions by citizens of member states to directly enforce its provisions.

[225] One concern not addressed here is whether principles of international and domestic law for transnational claims may raise additional opportunities for application of U.S. environmental laws and tort standards to climate engineering projects. For example, a foreign court may reach a judgment under its domestic law that would either seek to restrain or impose damages against operators of a climate engineering project. Attempts to enforce that judgment in the United States may raise complex issues involving the enforcement of foreign judgments. The prospect of multiple and overlapping domestic court judgments arising from a single climate engineering project raises the risk of a patchwork array of national laws that will yield conflicting direction and liability standards. It also may empower nations with the harshest liability standards to seek to constrain or entirely halt climate engineering projects sponsored in other nations.

authorizations or permits, the federal agencies in charge of those programs might need to begin drafting regulatory strategies and guidance that discuss the procedures and standards for their decisions to approve or reject these projects. Alternatively, federal agencies may also wish to explore their powers to halt objectionable climate engineering projects that pose unacceptable risks or spark strong public concern.

To the extent these federal and state environmental programs may not apply to specific climate engineering projects, challengers may instead turn to common law public nuisance causes of action to seek injunctions or damages. Although U.S. federal common law on climate change public nuisance is in a deep state of flux after the U.S. Supreme Court's ruling in *American Electric Power Co. v. Connecticut*, climate engineering tort challenges may sidestep the controversy. In contrast with public nuisance actions under federal common law over the effects of current and historical GHG emissions, climate engineering tort suits will present a better match with the U.S. courts' institutional constraints and constitutional competencies, although they will still test the U.S. courts' facility with highly complex and technical scientific issues. Absent earlier regulatory or legislative action to establish a framework for governing climate engineering efforts within U.S. jurisdiction, the federal and state courts should prepare for the bracing task of resolving domestic disputes over projects that are literally intended to reshape the global climate.

Aside from these immediate legal questions, advocates on both sides of the climate engineering debate will face deep and difficult questions of environmental policy and judicial review. Environmental petitioners, for example, might find themselves wrestling over whether to oppose projects that would counteract disruptive climatic change effects and reduce ongoing environmental damage. Alternatively, defendants may find themselves arguing that federal environmental laws do not apply to their actions because they have not altered the environment as much as they have attempted to preserve or restore it. They will likely contend that federal agencies and the courts should use a more generous or accommodating standard when reviewing climate engineering projects that serve, ultimately, a restorative goal.[226]

The federal judicial branch has been rightly categorized as the least dangerous branch because of the unique limits and fragility of judicial review and the judicial power to resolve cases and controversies.[227] Some climate engineering disputes may

[226] This issue evokes an even more challenging issue: Can environmental statutes drive the use of climate engineering techniques in certain circumstances? The ESA arguably mandates the use of habitat alteration or adaptation measures to save imperiled species. *See, e.g.*, P. Shirley & G. Lamberti, *Assisted Colonization under the Endangered Species Act*, 3 CONSERVATION LETTERS 45–52 (2010). That same legal rationale could extend to regional or global climate engineering technologies that would allow threatened or endangered species to avert certain extinction.

[227] *See* ALEXANDER BICKEL, THE LEAST DANGEROUS BRANCH: THE SUPREME COURT AT THE BAR OF POLITICS (1962).

squarely meet the definition of case or controversy under federal constitutional law, yet still raise questions over projects that literally and intentionally have global consequences. If so, the federal courts may find that even the most circumspect exercise of their judicial power to review climate engineering disputes could place the least dangerous branch squarely at the center of global efforts to address climate change. Climate engineering legal actions, as a result, could become an important crucible to test new legal theories for global environmental projects that invoke domestic or international mechanisms for liability and governance.

Index